CRITICAL ISSUES IN INSECT PEST MANAGEMENT

CRITICAL ISSUES IN INSECT PEST MANAGEMENT

Edited by

G.S. DHALIWAL
Department of Entomology
Punjab Agricultural University
Ludhiana, India

E.A. HEINRICHS
Department of Entomology
University of Nebraska
Lincoln, USA

COMMONWEALTH PUBLISHERS
New Delhi

Critical Issues in Insect Pest Management

Edition **2027**

ISBN : 81-7169-505-1

Published by :
COMMONWEALTH PUBLISHERS (P) LTD.
4831/24, Govind Lane,
Ansari Road, Darya Ganj,
New Delhi - 110002
Ph. : 23272541, 23242541
e-mail : commonwealthpublishers2016@gmail.com

Distributed by :
ARJUN PUBLISHING HOUSE
4831/24, Govind Lane,
Ansari Road, Darya Ganj,
New Delhi - 110002
Ph. : 23272541, 23242541
e-mail : arjunbooks2001@gmail.com

Laser Typesetting by :
Dimple Computers
Delhi

Printed at :
Milan Enterprises
Delhi

PREFACE

The concept of 'Integrated Pest Management' (IPM) is a means of controlling pests on the basis of sound biological knowledge and principles. IPM has been developed in response to a need, the need to cope with the diversity and complexity of biotic and abiotic constraints that limit production of food crops. The constantly changing pest populations due to the immigration of pests, the intensification of cropping and the development of biotypes attest to the need for dynamic insect management programmes. Pioneer entomologists, early in the 20th century, were the first proponents of the IPM approach to pest control but many times their programmes were not successful. However, with increased scientific knowledge, computer programmes, transportation and communication, in recent years, the effectiveness of IPM programmes has been enhanced. IPM is finally becoming an accepted practice in both developed and developing countries. In spite of the progress achieved, there still remain constraints to the development and implementation of IPM programmes in many regions of the world. This book addresses the issues limiting both the development and integration of IPM tactics and implementing their use at the farmers' level in combination with other crop management practices. The potential of modern approaches to IPM and their consequential environmental implications, and approaches for managing insect pests on a sustainable basis are discussed in the various chapters.

Chapter 1 sets the stage for the book by summarizing the historical developments of pest management as it has slowly evolved from a traditional to a more sustainable approach. It covers the era of traditional approaches from ancient times to 1938, the era of pesticides from 1939 to 1975, and the current era from 1976 onward. The importance of the 'farmer first' concept in IPM development and implementation is stressed. Development of the various control tactics; cultural control, biological control, host plant resistance, botanical insecticides, and pesticides is covered in chapters 2 to 7. These chapters discuss the progress achieved in developing the various tactics for use in different crops and the future research needs to improve their effectiveness in IPM systems. Chapter 8 covers the role that biotechnology approaches play in the development of control tactics for use in IPM systems; emphasis is on transgenic crops. Significant advances in understanding the roles of plant defense mechanisms in modifying insect pest behaviour have been made. Procedures for utilizing plant defense mechanisms in agricultural systems are discussed in Chapter 9. Chapter 10 'Integrated Pest Management: Springboard to Sustainable Agriculture,' stresses the point that putting all of the control tactics covered in chapters 2 to 9, together in a package does not constitute IPM. Successful IPM programmes are those developed through a 'farmer first' approach and become part of the farmers' crop production system. The 'Farmer Field School,' a learning by doing approach to IPM development and implementation is an outstanding example of sustainable IPM adoption by resource-poor farmers in the tropics. Chapter 11 highlights the challenges and opportunities of IPM for sustainable agricultural production systems in the 21st century.

The purpose of this book is to critically evaluate the existing technologies for insect pest management and indicate the future strategies for using them in IPM systems. As such, all of the disciplines involved in IPM development and implementation will find this book to be a valuable resource in their research programmes. In addition, this book will serve as a supplement to texts in university agricultural courses. We thank the authors for their contributions in meeting the purpose of this book.

G.S. Dhaliwal
Ludhiana, Punjab, India

E.A. Heinrichs
Lincoln, Nebraska, USA

June 5, 1998

CONTENTS

Preface (v)

The Contributors (xiii)

1. Insect Pest Management : From Traditional to Sustainable Approach 1

G.S. Dhaliwal, Ramesh Arora and E.A. Heinrichs

1.1. Historical Perspective 1

1.1.1. Era of Traditional Approaches (Ancient-1938) 2

1.1.2. Era of Pesticides (1939-1975) 2

1.1.3. Era of IPM (1976 onward) 5

1.2. Concept of Integrated Pest Management 5

1.3. Components of IPM 6

1.3.1. Forecasting System 6

1.3.2. Management Tactics 8

1.3.3. Decision Making Systems 9

1.3.4. Implementation 11

1.4. Achievements in IPM Implementation 15

1.4.1. Rice 16

1.4.2. Cotton 18

1.5. Conclusions 21

References 22

2. Cultural Control of Insect Pests in Tropical Crops 26

F.E. Nwilene and K.F. Nwanze

2.1. Introduction 26

2.2. Management Practices 27

2.2.1. Crop Residue Disposal 27

2.2.2. Destruction of Volunteer and Alternative Host Plants 35

2.2.3. Hand Picking of Deadhearts/Egg Masses/Insects 36

2.2.4. Time of Planting 37

2.2.5. Crop Rotation 39

2.2.6. Cropping Systems 40

2.2.7. Plant Spacing/Depth of Planting 43

2.2.8. Tillage and Mulching 44

2.2.9. Fertilizer Application 46

2.2.10. Water Management 48

2.2.11. Trap Crops 49

2.2.12. Other Methods 51

2.3. Future Research Needs 51

2.4. Conclusions 51

References 52

3. Biological Control in Agroecosystems through Management of Entomophagous Insects **67**
Miguel A. Altieri and Clara I. Nicholls

3.1. Introduction 67
3.2. The Role and Impact of Predators 68
3.3. Parasitoids : Biological Characteristics, Role and Impact 70
3.4. Strategies of Biological Control 72
3.4.1. Classical Biological Control 72
3.4.2. Augmentative Biological Control 74
3.4.3. Conservation and Habitat Management 78
3.5. Conclusions 83
References 83

4. Microbial Control of Insects : Use of Bacterial Insecticides **87**
Hyun-ku Lee, Hosun Cheong and Sarjeet S. Gill

4.1. Introduction 87
4.2. *Bacillus thuringiensis* Berliner 88
4.2.1. Classification 88
4.2.2. Toxin structure 89
4.2.3. Mechanism of Action 91
4.3. *Bacillus sphaericus* Neide 95
4.3.1. Classification and Biological Recycling 96
4.3.2. Toxin Structure 98
4.3.3. Mechanism of Action 100
4.4. Insect Resistance of Bacterial Insecticides 102
4.5. Commercialization 103
References 105

5. Host Plant Resistance to Insects **118**
Z.R. Khan and R.C. Saxena

5.1. Introduction 118
5.2. Historical Perspectives 119
5.3. Classification of Resistance 120
5.4. Factors Affecting Insect-Host Plant Interactions 122
5.5. Causes of Resistance 125
5.5.1. Biophysical Factors 125
5.5.2. Biochemical Factors 128
5.6. Abiotic Stresses and Expression of Resistance 131
5.6.1. Light 131
5.6.2. Temperature 131
5.6.3. Soil Fertility and Salinity 132
5.6.4. Relative Himidity 132

5.7. Genetics of Insect Resistance 133
5.8. Insect Biotypes 135
5.9. Future Considerations and Conclusions 136
References 140

6. Botanical Pest Control **155**
R.C. Saxena

6.1. Introduction 155
6.2. Plants with Insecticidal Properties 156
6.2.1. Pyrethrum 156
6.2.2. Rotenone and Rotenoids 157
6.2.3. Nicotine and Other Alkaloids 158
6.2.4. Ryania 159
6.2.5. Quassia 159
6.2.6. Sabadilla and Hellebore 159
6.2.7. Custard Apple 160
6.3. Plants with Photoxin Metabolites 160
6.4. Plants with Insect Hormonal Activity 160
6.5. Plants Affecting Insect Behaviour and Physiology 161
6.5.1. Rice 164
6.5.2. Maize, Sorghum and Millet 165
6.5.3. Grain Legumes and Vegetables 166
6.5.4. Flowers and Ornamentals 167
6.5.5. Fibre Crop 167
6.5.6. Forest Trees 168
6.5.7. Stored Products 168
6.5.8. Blood-sucking Pests 169
6.6. Botanical Pest Control and Insect Resistance 171
6.7. Conclusions 172
References 173

7. Pesticide Management : Issues and Strategies **180**
B.S. Parmar and P. Dureja

7.1. Introduction 180
7.2. Pesticides in Retrospect 180
7.3. Pesticide Management Issues 183
7.3.1. Inherent Toxicity 183
7.3.2. Pest Resistance 183
7.3.3. Pest Resurgence and Outbreaks of Secondary Pests 185
7.3.4. Destruction of Natural Enemies 186
7.3.5. Pesticide Residues 187
7.3.6. Effects on Non-target Organisms 188
7.4. Pesticide Management Strategies 189

7.4.1. Integration of Man-made and Natural Pest Control Agents 189
7.4.2. Use of Selective and High Potency Chemicals 190
7.4.3. Biocides to Bioregulators 192
7.4.4. Exploring Leads from Nature 194
7.4.5. Safer Formulations 198
7.4.6. Miscellaneous Strategies 200
7.5. Conclusion 202
References 202

8. Biotechnology and Insect Pest Management **209**
T.N. Ananthakrishnan

8.1. Introduction 209
8.2. Chemical Factors 210
8.3. Transgenic Crops 212
8.4. Baculoviruses in Insect Control 214
8.4.1. Genetic Implications 214
8.4.2. Vectors for Neurohormonal Genes 215
8.5. Proteinase Inhibitors 216
8.6. Conclusions 217
References 218

9. Behaviour and Development Modifying Approaches from Plant Defense Mechanisms **221**
Opender Koul

9.1. Introduction 221
9.2. Behaviour Modifications 222
9.2.1. Semiochemicals in Management Programmes 225
9.2.2. Phytochemicals and Behaviour Alterations 227
9.2.3. Antifeedant Approach and IPM 229
9.2.4. Critical Issues 232
9.3. Developmental Modifications 233
9.3.1. Growth Inhibitors 233
9.3.2. Insect Growth Regulators 235
9.3.3. Critical Issues 237
9.4. Conclusions 239
References 239

10. Integrated Pest Management : Springboard to Sustainable Agriculture **250**
Elske van de Fliert

10.1. Introduction 250
10.2. IPM: More Than Integration of Control Measures 251

10.2.1. The Old Paradigm : Observation Techniques and ETLs 251

10.2.2. The New Paradigm : Ecological Principles and Decision Making Skills 253

10.3. IPM Implementation 255

10.3.1. Characteristics of IPM 255

10.3.2. Characteristics of an IPM Farmer 256

10.3.3. IPM Farmer Field School : Learning by Doing 257

10.4. Beyond IPM 259

10.4.1. More Than Bugs That Bother Farmers 259

10.4.2. From IPM to Sustainable Agriculture 261

10.5. Conclusions 264

References 265

11. IPM in the 21st Century: Challenges and Opportunities 267

E.A. Heinrichs

11.1. Need for an Integrated Pest Management (PIM) Approach 267

11.11. Immigration of New Pests 267

11.1.2. Introduction of New Crops 267

11.1.3. Crop Intensification 268

11.2. Role of IPM in Crop Production 268

11.3. Constraints to IPM Implementation 270

11.4. Global IPM Implementation Strategies 271

11.5. The Asian Prototype for IPM Implementation 273

11.6. Conclusions 274

References 275

Index 277

THE CONTRIBUTORS

Miguel A. Altieri, Division of Biological Control, University of California, Berkeley, CA 94720-3112, USA

T.N. Ananthakrishnan, Entomology Research Institute, Loyola College, Chennai-600 034, India

Ramesh Arora, Department of Entomology, Punjab Agricultural University, Ludhiana-141 004, India

Hosun Cheong, Department of Entomology, University of California, Riverside, CA 92521-0127, USA

G.S. Dhaliwal, Department of Entomology, Punjab Agricultural University, Ludhiana-141 004, India

P. Dureja, Division of Agricultural Chemicals, Indian Agricultural Research Institute, New Delhi-110 012, India

Sarjeet S. Gill, Department of Entomology, University of California, Riverside, CA 92521-0127, USA

E.A. Heinrichs, Department of Entomology, University of Nebraska, Lincoln, NE 68585-0816, USA

Z.R. Khan, International Centre of Insect Physiology and Ecology (ICIPE), P.O. Box 30772, Nairobi, Kenya

Opender Koul, Insect Biopesticide Research Centre, 30, Prakash Nagar, Jalandhar-144 003, India

Hyun-ku Lee, Department of Entomology, University of California, Riverside, CA 92521-0127, USA

Clara I. Nicholls, Department of Entomology, University of California, Davis, CA 95616-8504, USA

K.F. Nwanze, International Crops Research Institute for the Semi-Arid Tropics (ICRISAT), Patancheru-502 324, Andhra Pradesh, India

F.E. Nwilene, International Crops Research Institute for the Semi-Arid Tropics (ICRISAT), Patancheru-502 324, Andhra Pradesh, India

B.S. Parmar, Division of Agricultural Chemicals, Indian Agricultural Research Institute, New Delhi-110 012, India

R.C. Saxena, International Centre of Insect Physiology and Ecology (ICIPE), P.O. Box 30772, Nairobi, Kenya

Elske van de Fliert, International Potato Centre (CIP-ESEAP Region), Jl. Bali 11, Widoro Baru, Ngropoh CC, Yogyakarta 55283, Indonesia

1

INSECT PEST MANAGEMENT: FROM TRADITIONAL TO SUSTAINABLE APPROACH

G.S. DHALIWAL AND RAMESH ARORA
Department of Entomology
Punjab Agricultural University, Ludhiana - 141 004, India

E.A. HEINRICHS
Department of Entomology
University of Nebraska, Lincoln, NE 68585 - 0816, USA

1.1. HISTORICAL PERSPECTIVE

Insects and land plants have coexisted for nearly 400 million years. During this period, plants developed a variety of physical and chemical defenses to prevent colonization by insects and other herbivores. Simultaneously, insects have also developed the ability to overcome these defenses. Thus, in natural ecosystems, insects and plants existed in a dynamic equilibrium. As soon as the land was cleared of natural vegetation and replaced by a few species of food plants, humans came into conflict with phytophagous insects. These insects, and other organisms feeding on the valuable crops which humans planted, were called pests (Dhaliwal and Arora, 1998). Broadly speaking, a pest is any plant or animal species that is in competition with humans for some resource. In agriculture, a phytophagous insect is labelled as a pest when its population is high enough to cause significant damage in yield or quality to any of the economic plants grown by man.

During ancient times, humans had to live with and tolerate the ravages of insects and other pests, but gradually learned to improve their condition through trial and error experiences. Over the centuries, farmers developed a number of mechanical, cultural, physical and biological control measures to minimize the damage caused by phytophagous insects. Synthetic organic insecticides developed during the mid-twentieth century initially provided spectacular control of these insects and resulted in the abandonment of traditional pest control practices. This was followed by the development of high yielding varieties of important crop plants. The intensive cultivation of these varieties, together with the application of increasing amounts of fertilizers and pesticides, has resulted in a manyfold increase in productivity. However, this technology package has also resulted in aggravation of pest problems in agricultural crops (Pimentel, 1977; Jayaraj, 1988; Bramble, 1989;

Dhaliwal and Arora, 1993). The increasing problems encountered with insecticide use resulted in the origin of the integrated pest management (IPM) concept. The history of agricultural pest control, thus, has three distinct phases, viz., the era of traditional approaches, the era of pesticides and the era of IPM.

1.1.1. Era of Traditional Approaches (Ancient-1938)

Cultural and mechanical practices like crop rotation, field sanitation, deep ploughing, flooding, collection and destruction of damaging insects/ insect infested plants, etc. developed by farmers through experience were among the oldest methods developed by humans to minimize the damage caused by insect pests (Smith *et al.,* 1976). This was followed by the use of plant products from neem, chrysanthemum, rotenone, tobacco and several other lesser known plants in different parts of the world. The Chinese were probably the pioneers in the use of botanical pesticides as well as biological control methods for the management of insect pests of stored grains and field crops (Dhaliwal and Arora, 1994). However, systematized work on many important tactics of pest control including the use of resistant varieties, biological control agents and botanical and inorganic insecticides was done in the U.S.A. from the end of the 18th to the end of the 19th century (Table 1.1). Remarkable success was achieved in the management of grape phylloxera caused by *Viteus vitifoliae* (Fitch) by grafting of European grape vine scions to resistant North American rootstocks during the 1880s. At around the same time, cottony cushion scale, *Icerya purchsi* Maskell which was causing havoc to citrus industry in California, USA was successfully controlled by release of the Vedalia beetle, *Rodolia cardinalis* (Mulsant) imported from Australia (Dhaliwal and Arora, 1998).

A number of synthetic inorganic insecticides containing arsenic, mercury, tin and copper were also developed towards the end of the 19th and the beginning of the twentieth century. With the development of these insecticides, the focus of research in entomology slowly shifted from ecological and cultural control to chemical control even before the development of synthetic organic insecticides (Perkins, 1980).

1.1.2. Era of Pesticides (1939-1975)

The inorganic synthetic insecticides were broad spectrum biocides and were highly toxic to all living organisms. These were followed in due course by the synthetic organic insecticides like alkyl thiocyanates, lethane; etc. The era of pesticides, however, began with the discovery of the insecticidal properties of DDT [2,2- (p-chlorophenyl) -1,1,1-trichlorethane] by Paul Muller in 1939. The impact of DDT on pest control is perhaps unmatched by any other synthetic substance and Muller was awarded Nobel Prize for this work in 1948.

DDT was soon followed by a number of other insecticides like HCH, chlordane, aldrin, dieldrin, heptachlor (organochlorine group); parathion, toxaphene, schradan, EPN (organophosphorus group) and allethrin (synthetic pyrethroid) during the 1950s and a large number of other popularly used organophosphates and carbamates in the ensuing decade.

Table 1.1. Landmarks in the history of agricultural insect pest management

Period	Landmark(s)
1	2
Ancient	The Chinese used chalk and woodash for the control of insect pests in enclosed spaces and botanical insecticides for seed treatment. They also used ants for biological control of stored grain as well as foliage feeding insects. In India, neem leaves were placed in grain bins to keep away troublesome pests. In Middle and Near East powder of chrysanthemum flowers was used as an insecticide.
1690	Tobacco extract was used as a plant spray in parts of Europe.
1782	"Underhill" variety of wheat reported resistant to Hessian fly in USA.
1858	Pyrethrum first used for insect control in the USA.
1888-89	Biological control of cottony cushion scale on citrus in the USA by use of Vedalia beetle imported from Australia.
1880-90	Control of grape phylloxera in Europe by grafting of European grapevine scions to resistant North American rootstocks.
1892	Lead arsenate used for the control of gypsy moth in the USA.
1911	Cotton Pest Act was enforced in Madras State. Under the act cotton stalks had to be removed by 1st August every year to minimise the incidence of pink bollworm.
1923	Multiple component suppression techniques involving the use of resistant varieties, sanitation practices and need based application of insecticides developed for the control of bollweevil in USA.
1939	● Insecticidal properties of DDT reported by Paul Muller in Switzerland. ● *Bacillus thuringiensis* Berliner first used as a microbial insecticide.
1941	Insecticidal activity of Hexachlorocyclohexane (HCH) discovered in France.
1946	Parathion, the first organo-phosphatic insecticide developed.
1948	"Doom" based on *Bacillus popilliae* Dutky and B. *lentimorbus* Dutky registered in the USA for the control of Japanese beetle larvae on turf.
1951	● R.H. Painter published his classic book *Insect Resistance in Crop Plants*. ● Introduction of first carbamate insecticide, Isolan.
1952-59	Concept of integrated control involving integration of chemical and biological control introduced.

Table 1.1 contd.

1	2
1959	Concept of economic injury level and economic threshold developed by V.M. Stern and coworkers.
1962	Rachel Carson published her book *Silent Spring* which dramatized the impact of misuse and overuse of pesticides on the environment.
1964	Paul DeBach published his book *Biological Control of Insect Pests and Weeds,* which established biological control as a separate discipline in Entomology.
1973	Development of first photo-stable pyrethroid, permethrin.
1975	● Elcar (*Helicoverpa* NPV) registered for the control of boll-worm and tobacco budworm on cotton. ● First insect growth regulator (Methoprene) registered for commercial use in the USA. ● R.L. Metcalf and W.H. Luckmann published the book *Introduction to Insect Pest Management* which was the first comprehensive book on IPM and established the concept on a firm footing.
1980	The interest in botanical pesticides revived and the First International Conference on Neem was held at Rottach-Egern, Germany.
1987	Development of first transgenic plant, reported by M. Vaeck and coworkers of Belgian Biotechnology Company, Plant Genctic Systems by transferring *B. thuringiensis* endotoxin gene to tobacco for the control of *Manduca sexta* (Johannsen).
1989	An IPM Task Force was established to garner international support for development and implementation of IPM programmes. A team of consultants appointed by the Task Force reviewed the status of IPM and made recommendations. The Task Force was later reconstituted as the Integrated Pest Management Working Group (IPMWG) in 1990.
1992	Concept of environmental economic injury levels proposed by L.P. Pedigo and L.G. Higley.
1994	A Task Force consisting of FAO, the World Bank, UNDP and UNEP co-sponsored the establishment of the Global IPM Facility with the Secretariat located at FAO, Rome.

Source: Modified after Dhaliwal and Arora (1994)

Due to their efficacy, convenience, flexibility and economy, these pesticides played a major role in increasing crop production. The success of high yielding varieties of wheat and rice that ushered in the 'green revolution' was partially due to the protection umbrella of pesticides (Pradhan, 1983). The spectacular success of these pesticides masked their limitations. The intensive and extensive use, misuse and abuse of pesticides during the ensuing decades caused widespread damage to the environment. In addition, insect pest problems in some crops increased following the continuous application of pesticides. This, in turn, further increased the consumption of pesticides resulting in the phenomenon of the pesticide treadmill (Altieri, 1995). The combined impact of all these problems together with the rising cost of pesticides provided the necessary feedback for limiting the use of chemical control strategy and led to the development of the IPM concept.

1.1.3. Era of IPM (1976 onward)

To overcome increasing problems associated with the strategy of exclusive and indiscriminate use of pesticides, Stern *et al.* (1959) propounded the concept of integrated control. The use of term 'pest management' was advocated by Geier and Clarke (1961). But the concept of IPM really came of age at the XV International Congress of Entomology during 1976. IPM represents ecologically sound approaches to pest control and automatically rejects the usage of broad spectrum and persistent biocides (Metcalf, 1980).

1.2. CONCEPT OF INTEGRATED PEST MANAGEMENT

Many components of integrated pest management (IPM) were developed in the late 19th and early 20th century (Table 1.1). Stephen A. Forbes, State Entomologist of Illinois (USA) was probably the pioneer in the application of ecological studies for dealing with insect problems in agriculture. A number of others concerned with crop protection also advocated this fundamental approach. By the early 1920s, a highly complex and sophisticated system involving the use of multiple component suppression techniques, *viz.*, resistant varieties, sanitation practices and chemical treatments with calcium arsenate at fixed population levels, was already developed for cotton boll weevil control in the USA.

The idea of managing insect pest populations was proposed by Geier and Clark (1961) who called this concept, 'protective management of noxious species' or 'pest management'. IPM refers to an ecological approach in pest management in which all available techniques are consolidated in a unified programme, so that pest populations can be managed in such a manner that economic damage is avoided and adverse side effects are minimized (NAS, 1969).

In fact, IPM is a dynamic and constantly evolving approach to minimise crop losses in which all the suitable management tactics and available surveillance and forecasting information are utilized to develop a holistic management programme as part of a sustainable crop production technology (Dhaliwal and Arora, 1998). Here it needs to be emphasized that the aim of IPM programmes should not be restricted to a mere efficient use of pesticides and product substitution (biorationals and botanicals in place of conventional insecticides), within an agricultural system that essentially remains unchanged (Table 1.2).

Rather, these programmes should aim at fundamental structural changes through a better understanding of ecological processes and synergy between crops (van Veldhuizen and Hiemstra, 1993).

1.3. COMPONENTS OF IPM

The major components of IPM include pest forecasting, management tactics, decision making and implementation.

1.3.1. Forecasting System

The forecasting system is based on surveillance and monitoring combined with a computer data base regarding weather predictions, costs and efficiencies of various management tactics, prices of inputs and farm produce, etc.

Monitoring phytophagous insects and their natural enemies is a fundamental took in IPM for taking management decisions. Monitoring requires estimation of changes in insect distribution and abundance, information about insects' life history and the influence of important biotic and abiotic factors on pest populations. A number of sampling tools have been developed to make inferences regarding the total insect costs and efficiencies of various management tactics, prices of inputs and farm produce, etc.

Monitoring phytophagous insects and their natural enemies is a fundamental tool in IPM for taking management decisions. Monitoring requires estimation of changes in insect distribution and abundance, information about insects' life history and the influence of important biotic and abiotic factors on pest populations. A number of sampling tools have been developed to make inferences regarding the total insect population as based on smaller collections (samples). Various sampling tools have been discussed in depth by Southwood (1978), Atwal and Singh (1990) and Reusink and Kogan (1994). A number of specialized sampling techniques becoming increasingly popular in IPM programmes like sequential sampling, variable intensity sampling, double sampling and binomial sampling have been discussed by Binns and Nyrop (1992).

Sampling techniques need to be modified to overcome some of the limitations of sampling as a component of IPM programmes. Because IPM is a continuous process, temporal changes in insect abundance should be explicity considered in sampling programmes. Secondly, two or more pest species causing the same type of damage may be simultaneously present in the field and make sampling difficult. Thirdly, few IPM programmes at the farmers' level attempt to estimate the abundance and impact of natural enemies on insect pest populations. There is an urgent need to develop efficient and cost reliable estimation procedures and forecasting models which incorporate the role of natural enemies in the decision making process (Wilson, 1985).

Table 1.2. Approaches to insect pest management: Retrospect and Prospect

S. No.	Parameter	Pest management system: Traditional	Industrial	Present IPM	IPM in sustainable agriculture
1.	Goal	Reduce losses due to pests	Eliminate or reduce pest species	Reduce costs of production	Multiple-ecological, economic and social
2.	Diversity	High	Low	Low to medium	High
3.	Ecosystem stability	Uncertain	Highly unstable	Unstable	Striving towards stability and equilibrium
4.	Spatial scale	Single farm	Single farm	Single farm or small region defined by pests	Biogeographic regions
5.	Time scale	Long term	Immediate	Single season	Long term steady state oscillatory dynamics
6.	Target	Single pest or closely related group of pests	Single pest	Several pests around a crop and their natural enemies	Fauna and flora of a cultivated area and linkages with non-cultivated ecologies
7.	Criteria for intervention	Past experience	Calendar date or presence of pest	Economic threshold	Multiple criteria
8.	Principal method	Cultural and mechanical measures	Pesticides	Resistant varieties, cultural practices, monitoring, product substitution, insecticide resistance, management and multiple interventions	Agroecosystem design to minimize pest outbreaks and mixed strategies including group action on an area wide basis to complement pest controls aimed at individual fields
9.	Research goal	Nil due to absence of organized effort	Improved pesticides	More kinds of interventions	Minimize need for intervention
10.	Extension technique	Nil	Transfer of technology (TOT)	TOT	Complimentarity between TOT and Farmer First (FF) mode
11.	Effect on environment	Usually negligible	Highly detrimental	Moderately detrimental	Negligible

Source: Dhaliwal and Arora (1994)

1.3.2. Management Tactics

Pest management tactics are either preventive or therapeutic. Preventive practice utilizes tactics to lower environmental carrying capacity (reduce the general equilibrium position) or increase tolerance of the host to pest injury. Prevention relies on an intimate understanding of the pest's life cycle, behaviour and ecology. The preventive tactics involve natural enemies, host resistance and cultural practices. In addition, quarantines are also an important component of preventive tactics. Therapeutic tactics are applied as a correction to the system when necessary. The objective of therapy is to decrease pest populations below the economic injury level. The only widely used therapeutic tactic is the use of conventional insecticides but other approaches like microbial agents, augmentation of natural enemies, use of growth regulators, etc. may also play a vital role (Pedigo, 1991).

Host plant resistance and cultural practices combined with natural biological control should form the core around which IPM systems are developed. A number of varieties, each having different bases of resistance, should simultaneously be in the field to reduce the chances of development of new insect biotypes. Multiple pest resistance has the potential to play a much more important rolc in crop pest management systems (Heinrichs, 1994a). Many of the traditional cultural practices may not be useful under modern intensive agriculture. Cultural and mechanical control techniques need to be modified and refined to make them compatible with intensive agriculture. A thorough understanding of the crop stress-insect interactions may enable us to manipulate the crop environment to the detriment of the pest (Heinrichs, 1988). The conservation and augmentation of natural enemies must receive priority. Genotypes favouring colonization by the natural enemies due to the presence of allelochemicals should be preferably utilized in IPM programmes. Transgenic plants developed through biotechnological approaches are likely to occupy an increasingly important place in future pest management programmes.

Pesticides will continue to play an important role in IPM programmes. But their use must be made compatible with other components of pest management and with society's requirement for maintaining environmental quality. This requires the development of selective insecticides exploiting weak links in insect physiology and behaviour. Efficacy of insecticides can also be increased by improvements in formulation and application technology.

In addition to conventional insecticides, therapeutic control may be achieved by use of parasitoids, predators, pathogens, behaviour modifying chemicals and selective toxicants. Methods for mass multiplication of parasitoids, predators and pathogens need further refinement, so that periodic releases of natural enemies become competitive with the cost of insecticides. Another important consideration in such releases is the difficulty in maintaining the quality of mass produced natural enemies. The global IOBC working group on 'Quality Control of Mass Reared Arthropods' is making efforts to develop quality control criteria and tests. Genetic improvement may help to enhance the efficiency of natural enemies by increasing host range and stability (Battu and Arora, 1997). Artificial selection, hybridisation and recombinant DNA techniques may be used for this purpose (Dhaliwal and Arora, 1998). Contrary to popular perception, biocontrol agents may also damage the environment, result in pest resurgence and secondary pest outbreaks. Therefore, ecological studies and environmental impact assessment of biocontrol agents must be undertaken before they are released in the field (Howarth, 1991).

Chemicals disrupting normal patterns of insect development, growth and behaviour, offer immense opportunities for suppressing insect populations in a selective manner. Many of these chemicals are essentially non-toxic to humans and domestic animals. These compounds are non-persistent and usually active at very low concentrations. Further improvements are needed in controlled release formulations, application technology and quality control in order to increase their utilization in IPM programmes. Plant derived products which act as repellents, antifeedants, oviposition deterrents, development disruptors and toxicants against phytophagous insects, may also serve as substitutes for synthetic pesticides.

1.3.3. Decision Making Systems

Pest management is a combination of processes that include decision making, taking action against a pest and obtaining the information to be used in reaching these decisions (Reusink and Onstad, 1994). In assessing, evaluating and choosing a particular pest control option, farmers are likely to take three major factors into account:

(i) Farmers' perception of the problem and of potential solutions is the most important factor. Here, the farmer's ability to identify pests, his assessment of likely and potential pest losses, and his opinion regarding the efficacy of different control options will affect the decision process.

(ii) The way in which control options are assessed will depend on the farmer's objectives. Subsistence farmers may opt for a guaranteed food supply while commercial farmers are more concerned with profit.

(iii) The number of options that a farmer can feasibly use depends on the constraints set by the resources available.

Evaluation of alternative pest control options. Various alternative pest control options could be evaluated for their cost effectiveness (Reichelderfer *et al.*, 1984).

(i) Determine from experimental results both the per hectare cost and a measure of effect of each alternative practice. If effectiveness can be measured in terms of output (yield and/or crop quality), use partial budgeting or other analytical techniques to evaluate alternatives. If effectiveness cannot be measured in these terms, proceed with determination of cost-effectiveness.

(ii) Using the same units in which effectiveness is measured (*e.g.*, reduction in pest numbers or damage), specify an effectiveness target that is appropriate, given the experimental data and information at hand (*e.g.*, a 50% reduction in pest population).

(iii) Multiply the cost of each practice times the effectiveness target, then divide that product by the actual level of effectiveness achieved by the practice. This gives a set of relative cost-effectiveness figures.

(iv) Compare the cost-effectiveness of alternative practices. The practice that has associated with it the least cost to achieve the effectiveness target is the most cost-effective practice.

Use of environmental economic injury levels. An environmental EIL evaluates a management tactic based on not only its direct costs and benefits to the user but also its effect on the environment (Pedigo and Higley, 1992).

$$\text{En EIL} = \frac{Pc + Ec}{VDIK}$$

Pc = Cost of management activity per unit of production

Ec = Environmental cost of management activity

V = Market value per unit of product

D = Damage per unit of injury

I = Crop injury per pest density

K = Proportionate reduction in injury

A number of techniques have been proposed to calculate the environmental costs of pesticides. These include the use of pest management ratings (Metcalf, 1975), contingent valuation (Higley and Wintersteen, 1992), environmental impact quotient (Kovach *et al.*, 1992) and integrated tabular approach (Dushoff *et al.*, 1994). A comparative account of all the techniques has been presented by Dhaliwal and Arora (1998).

1.3.4. Implementation

Although IPM has been accepted in principle as the most attractive option for the protection of agricultural crops from the ravages of insect and non-insect pests yet implementation at the farmers' level has been rather limited. Some of the important constraints to wider adoption of IPM and suggested measures to overcome them are discussed in this section.

A. Constraints in IPM implementation

The Consultant Group of the IPM Task Force has conducted an indepth study of the constraints on the implementation of IPM in developing countries and this section is largely based on the Consultants' Report (NRI, 1992).

(a) Institutional constraints. IPM requires an interdisciplinary, multi-functional approach to solving pest problems. Fragmentation between disciplines, between research, extension and implementation, and between institutes, all lead to a lack of institutional integration. Secondly, both the national programmes of developing countries and the donor agencies have lacked a policy commitment to IPM in the context of national economic planning and agricultural development. This has resulted in a low priority for IPM from national programmes and donors alike. Thirdly, the traditional top-down research in many cases does not address the real needs of farmers, who eventually are the end-users, and who select to adopt or reject the technology based on its appropriateness. Institutional barriers to research scientists in national programmes conducting on-farm research in developing countries are real, and need to be addressed.

(i) Informational constraints. The lack of IPM information which could be used by the farmers and by extension workers is a major constraint in implementation. While the individual control techniques are well known, little knowledge is available on using these in an integrated fashion under farm conditions. The lack of training materials, curricula and experienced teachers on the principles and practice of IPM is another major constraint.

(ii) Sociological constraints. The conditioning of most farmers and farm level extension workers by the pesticide industry has created a situation where chemicals are presented as highly effective and simple to apply. This acts as a major constraint in IPM implementation. There is a direct conflict between industry's objective of more sales and the IPM message of rational pesticide use. There is a need for private industry and the public sector extension agencies to work in a more complementary manner.

(iii) Economic constraints. A major constraint, even if IPM is adopted in principle, is the funding for research, extension and farmer training needed for an accelerated programme. IPM must be viewed as an investment and as with other forms of investment, requires an outlay. In the long run, IPM programmes may become self-generating due to savings on resource inputs for production.

(b) Political constraints. The relatively low status of plant protection workers in the administrative hierarchy is a constraint to general improvement in plant protection. Associated with the above are the morale and financial standing of these workers.

The continuance of pesticide subsidies by the government for political reasons and its tie up with the government-provided credit for crop production act as major constraints to farmers' acceptance of IPM. Various vested interests associated with the pesticide trade also act as a political constraint on the implementation of IPM. Indeed, the illicit financial rewards for licensing the import, formulation or distribution of agricultural chemicals are one of the major obstacles limiting the expansion of IPM activities (Jiggins, 1996).

B. Measures for Improving IPM Implementation

Acceleration of IPM implementation in developing countries requires farmers' participation, increased government support, legislative measures, improved institutional infrastructure and a favourable environment.

(i) Farmers' participation. It is not an exaggeration to say that the dawn of civilization started with farmer innovation. Ever since that day farmers have improved ways of growing crops through successive innovations. Prior to the emergence of crop protection sciences and even before the broad outlines of the biology of pests were understood, farmers evolved many cultural, mechanical and physical control practices for the protection of their crops from insect and non-insect pests (Smith *et al.,* 1976). Farmers' innovations were the only source of improvements in crop production and protection technology until formal research by scientists started complementing it during the late eighteenth and nineteenth century (Harverskort *et al.,* 1991). Unfortunately, with the advent of modern high-tech agriculture comprised of HYVs, fertilizers and pesticides, farmers have been completely displaced from the research and development process. Instead this role has been usurped by the private industry and the government agencies. The technology generated by scientists is being transferred through the extension agencies to the farmer. The new technology package has created a number of ecological and environmental problems. The alternative path of sustainable agriculture requires farmers' participation at every step of the research and development process in order to draw on his understanding of the local conditions and constraints, his innovativeness and his skills at making the best possible use of limited resources.

Placing the farmer at the centre of development process is wholly consistent with the IPM goal of making the farmer a confident manager and decision maker, free from dependence on a constant stream of pest control instructions from outside. The role of researchers, extension and non-government organizations (NGOs) and field workers is to act as consultants, facilitators and collaborators, stimulating and empowering the farmers to analyze their own situation, to experiment and to make constructive choices. A number of terms have been proposed for the new approach. These include: 'Farmer-first-and-last', 'farmer participatory research', 'people-centered technology development (PCTD)' and 'participatory technology development (PTD)' (Chambers *et al.*, 1991; Haverskort *et al.*, 1991). PTD serves to improve the experimental capacity of farmers and helps in development of locally-adapted improved technologies.

The approach has been used in the development of Farmers' Field Schools (FFSs) in IPM in Indonesia and the Philippines. The Field Schools are designed to build up farmers' understanding of agro-ecological relationships and to improve their capacities to systematically observe, document and interpret these. Experimental learning is based on the principles of adult learning, and follows a cycle of focused experience, systematic observation and measurement, critical reflection followed by planning and taking action. The action, in turn, begins a new cycle of experience-based learning and critical reflection. Thousands of farmers have been trained utilizing this approach and it is being tried on a pilot scale.

The evident advantages of this approach are a marked reduction in the use of chemical pesticides, with measurable benefits to the environment. Households save on cash, while crop yields tend to increase owing to better crop management (Jiggins, 1996).

(ii) Government support. Both the national programmes of developing countries and the donor agencies must have a policy commitment to IPM in the context of national economic planning and agricultural development. The costs to developing countries of not bringing their policies in line with the objectives of IPM are relatively greater than the costs of developed countries. National policies to promote IPM require close regulation at all stages related to the import and/or manufacture, distribution, use and disposal of pesticides. In the case of pesticides that do not meet prescribed standards for safety, persistence, etc., import and manufacturing bans should be enacted. At a minimum, the conditions laid out by the FAO Code of Conduct on the regulation, distribution and use of pesticides should be adopted. Pesticide subsidies need to be eliminated in order to make IPM an attractive alternative.

The funds so saved may be utilized for the implementation of IPM. Funds may also be diverted from some of the current research programmes to IPM-oriented plant protection programmes. Additional monetary resources may be generated through cooperation with bilateral/multilateral agencies willing to support such programmes (NRI, 1992).

(iii) Legislative measures. IPM is an information system and its adoption reduces pest control costs. The alternative to IPM is the indiscriminate use of broad spectrum synthetic organic pesticides. Unfortunately, while pesticide manufacturers and users (farmers) derive the full benefits from the use of these chemicals, they pass on the environmental and ecological costs of their use to society as a whole. If they are made to bear the full cost of the use of these toxicants, they may find IPM a more economical and attractive alternative. This could be achieved by enforcing suitable legislative measures.

Secondly, the success of an IPM programme in any geographical region depends upon its implementation by all farmers in area. Ideally, farmers may voluntarily adopt an IPM programme but some farmers may hold out. Such farmers called 'spoiler holdouts' may impair the success of a programme by failing to adopt a necessary practice thus causing damage to adjacent areas. Besides these, some farmers may free-ride and thus shift the costs of implementing and managing a programme to a group of participating farmers. To overcome 'spoiler holdouts' and 'free riders', it may be necessary to impose a programme upon an unwilling minority through suitable legislative measures (Tarlock, 1980).

(iv) Improved institutional infrastructure. IPM cannot be implemented unless there is a basic infrastructure for plant protection in a country. There is a need to develop and support national programme capabilities for on-farm testing and technology extrapolation. At the international level establishment of the Global IPM Facility to coordinate and monitor funding of IPM projects was needed to provide impetus to the implementation of IPM.

IPM is predominantly knowledge technology, the use of which requires training of the many groups involved. There is currently little training material for most of these groups including farmers, extension personnel and researchers. If IPM is to become the major approach for pest management in the developing world, this deficiency must be remedied (NRI, 1992).

Another aspect requiring greater attention is coordination of efforts within and between national research, training and implementation institutes/programmes and amongst international development agencies.

Lack of a reliable database has also hampered progress of IPM programmes. A reliable source of accurate information on the status of crops and pests in farmers' fields is necessary for many IPM activities. Most of the successful IPM programmes, both in developed and developing countries, have a reasonably accurate system of monitoring and evaluating various biological and environmental parameters in the agroecosystem. A reliable data base on crop yield and pest losses is required for planning and resource allocation at the national and international level. Systems analysis has been used as a problem diagnosis tool for IPM in developed-country cropping systems and may be used in the developing countries as well.

(v) Improved awareness. Increased education and awareness regarding the objectives, techniques and impact of IPM programmes is required at all levels including policy makers, planners, farmers, consumers and the general public. The importance and benefits of pesticides are being overemphasized by a multibillion dollar industry utilizing the services of not only their salesmen but also agricultural scientists, administrators and planners. There is not yet a strong market for IPM information. Policy makers and planners need to be convinced that without IPM current agricultural production systems are not sustainable. Similarly, much important information that might induce a farmer to adopt IPM is not immediately observable and is, therefore, not sought by him. A manufacturer has no incentive to recommend a programme that uses less pesticides, or even selective pesticides that kill a limited range of pests (Tarlock, 1980).

Consumer groups and the general public may also be able to support the implementation of the IPM programmes by demanding residue-free commodities. There is now a distinct market for organically produced food and other products. Non-Government Organizations (NGOs) and consumer groups need to be strengthened especially in developing countries, so that there is a public-oriented movement for implementation of IPM.

1.4. ACHIEVEMENTS IN IPM IMPLEMENTATION

Numerous attempts have been made during the last two decades to implement IPM in different regions of the world. As is to be expected, maximum efforts have been directed towards cotton, a crop which receives a disproportionately high amount of pesticides around the globe. There are many successful examples of IPM implementation on cotton in South and North America and Australia. In the Asian region, excluding China, there has been little progress and cotton ecosystems are in a crisis phase due to excessive usage of pesticides. However, in the same Asian region, success of IPM in rice provides one of the best examples of its implementation in the tropical developing countries. Both these crops provide excellent examples for understanding the progress and problems in IPM implementation.

1.4.1. Rice

Rice is the staple food of at least half of the world's population and is grown in approximately 148 million ha of land globally. Nearly 90 per cent of this area falls in the Asian region. Major pests attacking the crop in Asia are the brown planthopper, green leafhopper, hispa, stem borer, gall midge, whitebacked planthooper, leaffolder and foliage feeding caterpillars (Heinrichs, 1994b).

Major efforts in implementing IPM in irrigated rice have been carried out in Asia by the United Nations' Food and Agriculture Organization (FAO) through the Inter-country Programme for the Development and Application of Integrated Pest Control in Rice in South and South-east Asia. This programme remains one of the best examples of IPM implementation in the tropical region. It involves purposeful, direct efforts to change farmers' practices in contrast to some more indirect routes of IPM technology diffusion in many industrialized, temperate environments. The programme itself has evolved into its present transnational form from a relatively small project supported by Australia in the late 1970s, following large-scale pest outbreaks in several South-east Asian countries (NRI, 1992).

The first phase of the FAO programme (1980-86) focused on developing and testing the technical aspects of the IPM concept in its seven participating countries-Bangladesh, India, Indonesia, Malaysia, Philippines, Sri Lanka and Thailand. More recently, the project has been directed towards enhancing farmers' adoption of IPM. The programme is supported by Australia, the Netherlands and the Arab Gulf Fund. One significant accomplishment of the programme has been to cause policy changes within governments, in the form of official support of IPM as the means for national plant protection in the Philippines, Indonesia, India, Sri Lanka and Malaysia (NRI, 1992).

A case study of the National IPM Programme in Indonesia as a part of the regional programme during 1989-1991 provides an interesting scenario. Following research findings showing the relation between brown planthopper (BPH) outbreaks and high pesticide use, the Indonesian government banned the use of 54 broad-spectrum pesticides and declared IPM as the national pest control strategy. These measures, together with the gradual abolition of high subsidies on pesticides, created a favourable climate for the large scale implementation of IPM. To assess the impact of IPM on pest control and yield, IPM farmers were compared with those of non-IPM farmers (Table 1.3). Consequently, IPM farmers' expenditures on chemical insect control decreased. The number of farmers not applying pesticides was also significantly higher among IPM farmers. In spite of this, IPM farmers obtained higher yields than non-IPM farmers, which in addition to lower pest control expenditure resulted in higher returns from the crop. Additionally, there were less variations in fields among IPM farmers than among non-IPM farmers, indicating less riskier farm management under the IPM regime. Timeliness and adequacy of various cultivation and pest control practices, resulting from better monitoring and decision making skills, seem to be most important in this respect.

Table 1.3. Insect pest control practices, expenditures and yields of IPM and non-IPM farmers in two villages in Central Java, before, during and after IPM training

Parameter	Before training		During training		After one season		After two seasons	
	IPMF*	NIPMF	IPMF	NIPMF	IPMF	NIPMF	IPMF	NIPMF
Frequency of spray applications (no./season)	1.4	1.5	0.7	1.5	0.8	1.3	0.3	0.8
Frequency of granular applications (no./season)	0.4	0.3	0.7	0.9	0.7	0.8	0.6	0.7
Farmers not using pesticides (%)	26	31	41	19	46	24	50	43
Average insect control cost (Rs 1000/ha)	32	22	18	31	18	21	9	19
Average yield (t/ha)	5.38	5.11	5.77	4.64	3.70	3.11	6.38	5.68

*IPMF = IPM farmers, NIPMF = Non-IPM farmers
Source: van de Fliert (1993)

China, which had been experimenting with IPM for the control of rice pests since the early 1980s also joined the FAO project in 1989. During 1989-90 alone, nearly 160,000 farmers from over 2000 villages received IPM training. Compared with untrained farmers, IPM-trained farmers saved roughly a third of the pesticides in rice cultivation and still obtained 7 per cent higher yield. It was estimated that the investment in IPM training generated a return of more than 400 per cent. Encouraged by these results, the Ministry of Agriculture has set up a National Steering Committee for the comprehensive prevention and control of diseases and insect pests to protect the nation's rice crop and increase profits. The Committee conducts IPM tests, gives demonstrations and makes appraisals (Qinghua, 1995).

In Bangladesh, a non-government organization (NGO) CARE-Bangladesh began introducing IPM activities into an ongoing rice irrigation project in the late 1980s. Most of the farmers that CARE worked with were very small, owning less than half hectare of land. The farmers were trained in rice-fish culture and IPM. A survey among the farmers trained in rice-fish culture and IPM revealed that during 1992-93 virtually all of them stopped using pesticides and satisfactory control of insect pests was being provided by fish and naturally occurring parasitoids and predators. Farmers cultivating fish in their rice fields totally eliminated the use of pesticides even without training in IPM. However, only 20 per cent of the farmers could cultivate fish, mainly due to lack of adequate water, poor water-holding capacity of the soil and risk of flooding. Those farmers who could not cultivate fish still reduced their pesticide use by 76 per cent with IPM training. By harvest time, some of the farmers were able to harvest quantities of fish worth as much as their rice. These results underline the need for undertaking IPM programmes as part of overall rural development programmes (Kamp *et al.*, 1993).

In India, the first IPM programme in rice was started at Cuttack in 1975 covering an area of about 1000 ha in 10 villages. As a result of implementation of a number of cultural practices and ETL based applications of insecticides, the number of applications was reduced from 3-4 to usually one (Brader, 1979). Subsequently, an operational research project (ORP) was instituted on integrated control of rice pests at six locations in five provinces. This resulted in a reduction in the number of sprays and increased yield of the crop (Krishnaiah *et al.*, 1983). Economic threshold levels have now been established for all the major pests at several locations in the country.

A system of crop surveillance and forecasting for major rice pests is in operation in several countries including China, India, Indonesia, Malaysia, Republic of Korea, Pakistan, Philippines and Thailand (APO, 1993). However, judging from the current low percentage adoption, it would appear that there is much scope for increased IPM in almost all the rice growing areas. This may result in substantial savings to national treasuries as a result of elimination of pesticide subsidies. The Indonesian government estimated savings of about $ 50-60 million per year at 1987 prices as a result of reduction in pesticide subsidies from 85 to 15 per cent.

1.4.2. Cotton

The production of cotton is ancient and cotton fabrics as old as 3000

BC have been found in the Indus Valley. Presently, cotton is widely grown throughout the tropics, subtropics and warm temperate zones in small to large scale farms under rainfed and irrigated conditions. The maximum area under the crop is in China (27.2%), followed by USA (17.3%), Commonwealth of Independent States (CIS-erstwhile USSR) (11.6%), India (10.9%) and Pakistan (10.9%). Australia, Turkey and China with lint yields of 1623, 958 and 820 kg/ha, respectively are the leading countries in terms of productivity, while the Indian yield at 277 kg/ha is less than half of the world average (590 kg/ha).

Cotton ecosystems throughout the world include a wide variety of arthropods. The number of species found in the crop may range from a few hundred to more than a thousand. The vast majority of these species are parasitoids and predators of phytophagous species. Estimates of the number of pest species range from 20 to 60 but significant crop damage is caused by 5-10 key pests in most production systems (Luttrell *et al.*, 1994). The major pests in the Asian region include the pink bollworm, spotted bollworm. American bollworm, aphid, jassid, whitefly, thrips, leaf roller and tobacco caterpillar. In Australia, the major pests include *Helicoverpa spp.*, mites, thrips, aphids and mirids (*Creontiades* spp.). In the humid environments of Brazil and USA, the boll weevil, *Anthonomus grandis* Boheman is the most destructive pest. *Helicoverpa zea* (Boddie), *Heliothis virescens* (Fabricius), plant bugs, *Lygus lineolaris* (Palisot de Beauvois) and *Psallus seriatus* (Reuter) also cause serious damage. Species of *Spodoptera, Earias, Alabama, Dysdercus,* jassids, aphids and whitefly are of varying and local importance in different regions.

The cotton production systems of the world illustrate well the ecological and environmental problems associated with intensive insecticide use. Cotton growing in the Canete Valley, Peru is a well documented case that has become a classical example of integrated control. Even prior to the advent of synthetic organic insecticides, the increasing use of inorganic insecticides resulted in serious outbreaks of *H. virescens* and *Aphis gossypii* (Glover). The advent of synthetic organic insecticides resulted in extensive use of these chemicals. Both *H. virescens* and *A. gossypii* soon became resistant to these insecticides. A number of other pests like leafrollers, mealybugs, leaf perforator, etc. also became serious. The average number of insecticide applications was increased to 16 but still lint yields touched the lowest levels of 3.32 kg/ha. Integrated control measures involving the use of natural enemies (*Trichogramma* spp., lady beetles and carabid beetles), cultural practices (early maturing varieties, crop rotation, fixed sowing dates and clean-up campaign) and selective use of inorganic insecticides were implemented from 1956 onwards and by 1959 productivity reached a level of 800 kg/ha. The number of insecticide applications declined to 3 per season.

A similar situation in other cotton ecosystems was recorded in several countries including El Salvador, Guatemala and Nicaragua (Brader, 1979). In the 1960s and 1970s, worldwide efforts began to change cotton insect control from a unilateral dependence on chemical insecticides to a systems approach emphasizing diverse-insect-control tactics and ecological relationships.

In the United States, sound IPM programmes have been developed for the different cotton production regions. The most widely adopted practices are avoidance of pest pressure through the use of short-season varieties and conservation of natural enemies. For most pest species, threshold levels based on pest density and/or plant injury have been established to determine the need for pest control. Greater emphasis is being placed on microbial insecticides and transgenic cotton that express *Bt* endotoxin proteins (Luttrell *et al.*, 1994). There have been several comprehensive reviews on the present status and future prospects of IPM in cotton in the USA (Frisbie *et al.*, 1994; Luttrell, 1994).

In Australia, the introduction of some IPM principles arose through the development of the computer-based pest management programme, SIRTAC. The system incorporates sampling procedures, a series of linked plant and insect models, and a set of decision rules that, together with fixed or dynamic thresholds, provide objective pest management decisions. Pesticides continue to be the major tactic used but the average number of applications has decreased from 15-18 in the late 1970s to 10-12 today. Biological insecticides and safer chemicals are used early in the season to maintain beneficial insects. As in the USA, transgenic cotton and genetically engineered microbial pesticides are likely to become increasingly important in the coming years (Fitt, 1994).

In South and Central America, there has also been a significant reduction in pesticide application as a result of the introduction of IPM tactics. In many of these countries organic cotton grown without the application of synthetic fertilizers, pesticides and defoliants is now becoming a welcome alternative. Insect pests cause only limited damage in cotton and it fetches a 50 to 100 per cent higher price in North America and Europe. The efforts are being supported and certified by the International Federation of Organic Agricultural Movements (IFOAM) (Joosten and Eweg, 1993).

Only sketchy information is available regarding the pest management practices in the main cotton producing Asian countries, viz., China, India and Pakistan. The recent comprehensive review of cotton IPM systems of the world does not contain any information regarding the region (Luttrell *et al.*, 1994). However, China seems to have made successful efforts towards the implementation of IPM. The two principal approaches followed by the Chinese are a highly effective pest forecasting system and an elaborate system of cultural practices. Effective control in seed cotton storage has considerably reduced the importance of the pink bollworm. The adoption of early spring planting has reduced the status of spotted bollworms. Biological control tactics are also extensively used. Releases of *Dibrachys cavus* (Walker) are made in store houses for controlling pink bollworm. *Trichogramma* spp. are released on more than 600,000 ha annually to control *Helicoverpa armigera* (Hubrei). Lately, transgenic cotton is also being grown on thousands of hectares.

In India, the Indian Council of Agricultural Research sponsored a village level IPM Project to test and demonstrate the efficacy, practicability and economics of IPM in cotton in the Punjab during 1975. Later on, operational research projects (ORPs) on integrated control of cotton pests were started in several states which were executed by State Agricultural Universities. An evaluation of these projects revealed that there was reduction in pesticide use in the project areas (Simwat, 1994).

From 1994-95, Farmers' Field Schools (FFSs) have been started in Andhra Pradesh, Karnataka, Maharashtra, Punjab, Uttar Pradesh and several other states on the pattern of such schools in Indonesia. A total of 4692 schools have been conducted in rice, cotton, vegetables, pulses and oilseeds, where 1,40,068 farmers have been trained upto 1996. The yield increase in rice and cotton in IPM-areas varied from 6 to 42 per cent and 21 to 27 per cent, respectively, as compared to non-IPM areas. There was a reduction in pesticide consumption to the tune of 50-100 per cent and 30-50 per cent in rice and cotton, respectively in IPM-areas (Atwal and Dhaliwal, 1997).

1.5. CONCLUSIONS

During the last two decades IPM has moved from a peripheral position to the centre stage of agricultural production programmes. The desirability of controlling pests by the use of integrated pest management is no longer questioned. A variety of techniques have been developed and refined for controlling different insect pests. Farmer-centered methods have resulted in successful implementation of IPM in rice in parts of South-east Asia. Earlier, South-American countries showed the way to cotton IPM. Greater challenges, however, lie ahead.

As agriculture is further intensified, insect and non-insect pest problems are expected to become more serious. Increased emphasis on IPM is the only sensible option under these conditions. IPM implementation faces a number of institutional, informational, socioeconomic and political constraints (NRI, 1992). These problems could be overcome by increased farmers' participation, government support, legislature measures, improved institutional infrastructure and creating greater awareness (Dhaliwal and Arora, 1998). Hopefully, entomologists will work hand-in-hand with other agricultural scientists, environmentalists and farmers to develop and implement innovative IPM strategies targeted towards a sustainable crop production technology in the coming years.

REFERENCES

Altieri, M. 1995. Escaping the treadmill. *Ceres* **27** (4): 15-23.

APO (Asian Productivity Organization). 1993. *Pest Control in Asia and the Pacific.* Report of a Seminar, September 24-October 4, 1991, Tokyo, Japan.

Atwal, A.S. and Dhaliwal, G.S. 1997. *Agricultural Pests of South Asia and Their Management.* Kalyani Publishers, New Delhi, India.

Atwal, A.S. and Singh, B. 1990. *Pest Population and Assessment of Crop Losses.* Indian Council of Agricultural Research, New Delhi, India.

Battu, G.S. and Arora, R. 1997. Insect pest management through microorganisms. *In:* K.R. Dadarwal (ed.). *Biotechnological Approaches in Soil Microorganisms for Sustainable Crop Production.* Scientific Publishers, Jodhpur, India, pp. 223-246.

Binns, M.R. and Nyrop, J.P. 1992. Sampling insect populations for the purpose of IPM decision making. *A. Rev. Ent.* **37**: 427-453.

Brader, L. 1979. Integrated pest control in the developing world. *A. Rev. Ent.* **24**: 225-254.

Bramble, B.J. 1989. An environmentalist's view of pest management and the green revolution. *Trop. Pest Mgt.* **35**(3): 228-230.

Chambers, R., Pacey, A. and Thrupp, L.A. (eds). 1991. *Farmer First: Farmer Innovation and Agricultural Research.* Intermediate Technology Publications, London, UK.

Dhaliwal, G.S. and Arora, R. 1993. Changing status of insect pests and their management strategies. *In:* K. S. Gill, G. S. Dhaliwal and B.S. Hansra (eds). *Changing Scenario cf Indian Agriculture.* Commonwealth Publishers, New Delhi, India, pp. 98-145.

Dhaliwal, G.S. and Arora, R. 1994. Components of insect pest management: A critique. *In:* G. S. Dhaliwal and Ramesh Arora (eds). *Trends in Agricultural Insect Pest Management.* Commonwealth Publishers, New Delhi, India, pp. 1-55.

Dhaliwal, G.S. and Arora, R. 1998. *Principles of Insect Pest Management.* Kalyani Publishers, New Delhi, India.

Dushoff, J., Caldwell, B. and Mohler, C.L. 1994. Evaluating the environmental effect of pesticides: A critique of the environmental impact quotient. *Am. Ent.* **40**(3): 180-184.

Fitt, G.P. 1994. Cotton pest management: Part 3. An Australian perspective. *A. Rev. Ent.* **39**: 543-562.

Frisbie, R.E., Reynolds, H.T., Adkisson, P.L. and Smith, R.F. 1994. Cotton insect pest management. *In*: R.L. Metcalf and W.H. Luckmann (eds). *Introduction to Insect Pest Management.*John Wiley & Sons, New York, USA, pp. 421-468.

Geier, P.W. and Clarke, L.R. 1961. An ecological approach to pest control. *In: Proc. Eighth Technical Meeting.* International Union for Conservation of Nature and Natural Resources, *1960,* Warsaw, Poland, pp. 10-18.

Harverskort, B., Kamp, Johan van der and Waters-Bayer, A. (eds). 1991. *Joining Farmers' Experiments-Experiences in Participatory Technology Development.* Intermediate Technology Publications, London, UK.

Heinrichs, E.A. (ed.). 1988. *Plant Stress-Insect Interactions.* John Wiley & Sons, New York, USA.

Heinrichs, E.A. 1994a. Development of multiple pest resistant crop cultivars. *J. agric. Ent.* **11**(3): 325-353.

Heinrichs, E.A. (ed.). 1994b. *Biology and Management of Rice Insects.* Wiley Eastern Limited, New Delhi, India.

Higley, L.G. and Wintersteen, W.K. 1992. A novel approach to environmental risk assessment of pesticides as a basis for incorporating environmental costs into economic injury level. *Am. Ent.* **38**(1): 34-39.

Howarth, F.G. 1991. Environmental impacts of classical biological control. *A. Rev. Ent.* **36**: 485-509.

Jayaraj, S. 1988. Accentuation of insect and mite problems in the changing agriculture scenario in India: An overview. *National Seminar on Changing Pest Situation in the Current Agriculture Scenario of India.* July 14-16, 1988, Indian Agricultural Research Institute, New Delhi, India.

Jiggins, J. 1996. Women and the re-making of civil society. *For. Trees People Newsl.* **30:**18-22.

Joosten, L. and Eweg, E. 1993. The other colour of cotton. *ILEIA Newsl.* **9**(2): 20-21.

Kamp, K., Gregory, R. and Chowhan, G. 1993. Fish cutting pesticide use. *ILEIA Newsl.* **9**(2): 22-23.

Kovach, J., Petzoldt, C., Degni, J. and Tette, J. 1992. *A Method to Measure the Environmental Effect of Pesticides.* N. Y. Food Life Sci. Bull. No.139, New York, USA.

Krishnaiah, K., Quayum, M.A., Rao, C.S., Reddy, P.C., Charyulu, A.M.R.K. and Kalode, M.B. 1983. Integrated pest control in rice in Warangal District, Andhra Pradesh. *In*: S. Chelliah and M. Balasubramanian (eds). *Pest Management in Rice.* Tamil Nadu Agricultural University, Coimbatore, India, pp.283-293.

Luttrell, R.G. 1994. Cotton pest management. Part 2. A US perspective. *A. Rev. Ent.* **39**: 527-542.

Luttrell, R.G., Fitt, G.P., Ramalho, F.S. and Sugonyaev, E.S. 1994. Cotton pest management: Part 1. A worldwide perspective. *A. Rev. Ent.* **39**: 517-526.

Metcalf, R.L. 1975. Insecticides in pest management. *In*: R. L. Metcalf and W.H. Luckmann (eds). *Introduction to Insect Pest Management.* John Wiley & Sons, New York, USA, pp. 235-274.

Metcalf. R.L. 1980. Changing role of insecticides in crop protection. *A. Rev. Ent.* **25**: 219-256.

NAS (National Academy of Sciences). 1969. *Principles of Plant and Animal Pest Control. Vol. 3, Insect Management and Control.* Washington, DC, USA.

NRI (Natural Resources Institute). 1992. *Integrated Pest Management in Developing Countries: Experience and Prospects.* Chatham, UK.

Pedigo, L.P. 1991. *Entomology and Pest Management.* MacMillan Publishing Co., New York, USA.

Pedigo, L.P. and Higley, L.G. 1992. The economic injury level concept and cnvironmental quality. *Am. Ent.* **38**(1): 12-21.

Perkins, J.H. 1980. The quest for innovation in agricultural entomology. *In*: D.Pimentel and J.H. Perkins (eds). *Pest Control: Cultural and Environmental Aspects.* AAAS Selected Symposium 43, Westview Press, Colorado, USA, pp. 23-80.

Pimentel, D. 1977. Ecological basis of insect pests and weed problems. *In:* J.M. Cherrett and G.R. Sagar (eds). *Origin of Pest, Parasite, Disease and Weed Problems.* Blackwell Scientific Publications, Oxford, UK, pp. 3-31.

Pradhan, S. 1983. *Agricultural Entomology and Pest Control.* Indian Council of Agricultural Research, New Delhi.

Qinghua, Z. 1995. Careful control: IPM in China. *Ceres* **27**(4): 12-13.

Reichelderfer, K.H., Carlson, G.A. and Norton, G.A. 1984. *Economic Guidelines for Crop Pest Control.* FAO Plant Production and Protection Paper 58, Food and Agriculture Organization of the United Nations, Rome, Italy.

Ruesink, W.G. and Kogan, M. 1994. The quantitative basis of pest management: sampling and measuring. In: R.L. Metcalf and W.H. Luckmann (eds). *Introduction to Insect Pest Management.*John Wiley and Sons, New York, USA, pp. 355-391.

Ruesink, W.G. and Onstad, D.W. 1994. Systems analysis and modeling in pest management. *In:* R.L. Metcalf and W.H. Luckmann (eds). *Introduction to Insect Pest Management.* John Wiley & Sons, New York, USA, pp. 393-420.

Simwat, G.S. 1994. Modern concepts in insect pest management in cotton. *In:* G.S. Dhaliwal and Ramesh Arora (eds). *Trends in Agricultural Insect Pest Management.* Commonwealth Publishers, New Delhi, India, pp. 186-237.

Smith, R.F., Apple, J.L. and Bottrell, D.G. 1976. The origin of integrated pest management for agricultural crops. *In:* J.L. Apple and R.F. Smith (eds). *Integrated Pest Management.* Plenum Press, New York, USA, pp. 1-16.

Southwood, T.R.E. 1978. *Ecological Methods with Particular Reference to the Study of Insect Populations.* Chapman & Hall, London, UK.

Stern, V.M., Smith, R.F., van den Bosch, R. and Hagen, K.S. 1959. The integration of chemical and biological control of the spotted alfalfa aphid, I: The integrated control concept. *Hilgardia* **29:** 81-101.

Tarlock, A.D. 1980. Legal aspects of integrated pest management. *In:* D. Pimentel and J.H. Perkins (eds). *Pest Control: Cultural and Environmental Aspects.* AAAS Symp. 43, Westview Press, Colorado, USA, pp. 217-236.

van de Fliert, E. 1993. *Integrated Pest Management: Farmer Field Schools Generate Sustainable Practices.* Wageningen Agricultural University, Wageningen, The Netherlands.

van Veldhuizen, L. and Hiemstra, W. 1993. Cutting back: cure or prevent. *ILEIA Newsl.* **9**(2): 3-4.

Wilson, L.T. 1985. Estimating the abundance and impact of arthropod natural enemies in IPM systems. *In:* M.A. Hoy and D.C. Herzog (eds). *Biological Control in Agricultural IPM Systems.* Academic Press, Orlando, USA, pp. 303-322.

2

CULTURAL CONTROL OF INSECT PESTS IN TROPICAL CROPS[1]

F.E. NWILENE AND K.F. NWANZE

International Crops Research Institute for the Semi-Arid Tropics Patancheru - 502 324, Andhra Pradesh, India

2.1. INTRODUCTION

Cereals, legumes, root crops and cotton serve as staple food, feed and forage for millions of people and livestock in Africa, Asia and parts of Latin America. They are grown by the resource-poor farmers. Insect pests are generally recognised as important production-limiting factors in these crops, reducing yields on farmers' fields in combination with low and erratic rainfall, and poor soil fertility. Cultural practices have been the most important method for controlling pests and preventing crop losses in the developing world (Brader, 1979), Cultural control measures, including sowing time, crop residue disposal after harvest, crop husbandry, crop rotation, cropping systems, and water management pre-date pesticidal control by many years and are suitable for use by resource-poor farmers. Cultural control was ignored subsequent to the appearance of synthetic organic pesticides in the mid-1940s. The need for ecologically sound, effective, and economic methods for pest control has prompted renewed interest in cultural and other methods. With the increased awareness of integrated pest and crop management systems, it is essential to incorporate these techniques in a pest management system that would keep the pests below economic thresholds. The improvement of these crops in the tropics will rely on management practices that increase yields in the presence of pests.

The cultural control of insect pests is the tactical use of regular farm practices to delay or reduce insect pest attack (Lawani, 1982; Seshu Reddy, 1985). Cultural measures involve the manipulation of the agroecosystem to make it unfavourable for insect pests and more suitable for crop growth and natural enemies of the pests. Cultural practices have potential as components of IPM for farmers in the tropics because they are simple, effective, convenient, environmentally saf and socially acceptable, economically feasible (labour costs are comparatively low) and do not require any limitation posed by pesticides to the environment, human health and development of pest resistance.

[1] Approved as Journal Article No. 1946 by the International Crops Research Institute for the Semi-Arid Tropics (ICRISAT), Patancheru-502 324, Andhra Pradesh, India

This chapter examines outputs of research during the last 30 years on cultural control of insect pests for crops that are important food sources for people in developing countries.

2.2. MANAGEMENT PRACTICES

Several insect species have been reported as pests of cereals, legumes, root crops and cotton (Nwanze *et al.*, 1995; Youm *et al.*, 1996; Wightman *et al.*, 1990; Yaninek *et al*; 1990; Sidhu and Dhawan, 1985) Species of major economic importance and their management practices are listed in Tables 2.1 - 2.4.

2.2.1. Crop Residue Disposal

The removal or destruction of stubble and crop residues by burning, flooding or ploughing after the crop has been harvested reduces insect pest attack in the next season. For this measure to be effective, it must be adopted by all farmers in the target area, otherwise, remaining insect populations will continue to seek available hosts.

Cereal-based ecosystems. A review of field sanitation as a means for controlling stem borers is provided by Nwanze and Mueller (1989) and practices such as burning of stems were reported by Ingram *et al.* (1973) and Bowden (1976). They advocated burning all sorghum and maize residues immediately after harvest in order to control *Chilo partellus* (Swinhoe), *Busseola fusca* (Fuller) and *Sesamia botanephaga* Tams & Bowden. Partial burning of sorghum leaves after grain harvest (where the stalks are used for fencing, roofing, and traditional granaries) have indicated that 95 per cent of the *B fusca* larvae are killed without any damage to the stalks (Adesiyum and Ajayi, 1980). Heat generated from burning leaves is sufficient to kill the larvae within the stalks. Partial burning of pearl millet stalks immediately after harvest destroys 61-84 per cent of the larvae and 98-100 per cent of the pupae of *Coniesta ignefusalis* (Hampson) (NDoye and Gahukar, 1987). It was a common practice in the rural Sahel of West Africa to light fires around millet fields at night to reduce blister beetle populations feeding on the millet heads (Magema, 1984). In Ghana, larvae of *Eldana saccharina* Walker remain in sugarcane stubble after harvest and reinfest new shoots if care is not taken (Kumar and Sampson, 1982). The stubble larvae destroy the shoots in the ratooned field thereby reducing the tillers of the ratoons. Complete eradication of *C. partellus* and *B. fusca* on sorghum and maize by burning of stubble and crop residues has been suggested by Duerden (1953) in Tanzania. Destruction of sorghum residues by burning the stubbles, was an effective cultural control method against diapusing *B. fusca* and *C. partellus* on Rusinga Island, Kenya (Unnithan and Seshu Reddy, 1989). Studies by Kfir (1990) in South Africa have indicated that 70 per cent of *C. partellus* and *B. fusca*

Table 2.1. Cultural control methods against major insect pests of cereals in Africa, Asia and parts of Latin America

Insect pest	Crop(s) attacked	Region	Cultural practice(s)	Reference(s)
1	2	3	4	5
Stem borers				
Chilo partellus (Swinhoe)	Sorghum, maize, rice, sugarcane, finger millet, wheat	Asia, East and South Africa	Early planting, crop residue disposal, removal of volunteer and alternative host plants	Unnithan and Seshu Reddy, 1989; Roome, 1976; Singh *et al.*, 1992
C. suppressalis (Walker)	Rice, maize, sorghum, wheat, sugarcane	Asia, Pacific	Application of fertilizer (silica)	Djamin and Pathak, 1967
C. infuscatellus Snellen	Sugarcane, sorghum, maize	America, Asia	Crop residue disposal, sowing date, mulching with cane trash	Teetes, 1995; Hardyal, 1963
Eldana saccharina Walker	Maize, sugarcane, sorghum, rice, finger millet	Africa, south of the Sahara	Burning of stubble, burying crop residues, removal of alternative host plants	Kumar and Sampson, 1982
Busseola fusca (Fuller)	Maize, sorghum, sugarcane	Africa, south of the Sahara	Burning of stalks after grain harvest, early sowing, ploughing	Adesiyun and Ajayi, 1980; Kaufmann, 1983
Sesamia inferens (Walker)	Rice, sugarcane, maize, sorghum, wheat, barley, finger millet	Asia, West Africa, Pacific	Good husbandry, ploughing, early planting	Teetes *et al.*, 1983
Sesamia botanephaga Tams & Bowden	Maize, sorghum, rice, sugarcane, finger millet	West and East Africa	Crop residue disposal, rotation of sorghum with groundnut or cotton	Bowden, 1976; Sagnia, 1983
Sesamia calamistis	Maize, sorghum, finger millet,	Africa, south of	Intercropping maize/	Oloo and Ogeda, 1990

Table 2.1 contd...

1	2	3	4	5
(Hampson)	rice, sugarcane	the Sahara	sorghum with cowpea, good husbandry	
Coniesta ignefusalis (Hampson)	Pearl millet, sorghum, maize, sugarcane	West Africa	Sowing date, burning of stubble/stalks	Ajayi and Labe, 1990
Diatraea saccharalis (Fabricius)	Sugarcane, maize, sorghum, wheat, rice	North and South America	Ploughing and discing stubble, early planting	ICRISAT, 1985
Elasmopalpus lignosellus (Zeller)	Sorghum, maize, rice, sugarcane	Central and South America	Removal of crop residues by ploughing, early planting, destruction of weeds, soil moisture	ICRISAT, 1985; All and Gallaher, 1977
Ostrinia furnacalis (Guenee)	Maize, sorghum	Asia	Intercropping maize with groundnuts	IRRI, 1973
Seedling pest				
Atherigona soccata Rondani	Sorghum	West and East Africa, Asia	Early planting at the onset of the rainy season, crop rotation, burning of stubble/stalks, high seeding rate, nitrogenous fertilizers	Jotwani, 1982; Gahukar and Jotwani, 1980; Balasubramanian *et al.*, 1986
Panicle pests				
Contarinia sorghicola (Coquillett)	Sorghum	Asia, Africa, South America	Early and uniform planting over large areas, intercropping sorghum with leguminous crops	Hardas *et al.*, 1980; Teetes, 1995
Calocoris augustatus Lethiery	Sorghum	Africa, Asia	Early planting	Ramnath *et al.*, 1974
Heliocheilus albipunctella (de Joannis)	Pearl millet	West Africa	Deep ploughing of soil at the end of the season	Nwanze *et al.*, 1995; Vercambre, 1978

Table 2.1 contd...

1	2	3	4	5
Eurystylus immaculatus Odhiambo	Sorghum	West Africa	Intercropping sorghum with soybean	Ajayi and Tabo, 1995
Foliage feeders				
Spodoptera frugiperda (J.E. Smith)	Sorghum, pearl millet rice, maize, wheat	South America	Early planting, plant density, maize as a trap crop	Teetes, 1995; Castro and Pitre, 1988
Helicoverpa armigera (Hubner)	Sorghum, pearl millet	Africa, Asia	Crop rotation, ploughing the field after harvest exposing pupae to heat and predators	Teetes *et al.*, 1983
Mythimna separata (Walker)	Sorghum, pearl millet	Asia, Pacific Islands	Deep ploughing after harvest, proper and timely sowing of crop	Sharma and Davies, 1982; ICRISAT, 1985
Soil pest				
Holotrichia serrata (Fabricius)	Sorghum	Asia, North America	Ploughing after crop harvest and before planting, crop rotation, sowing date	Gahukar and Jotwani, 1980; Teetes, 1995

populations is sorghum and maize fields were destroyed by slashing the stubble after harvest. Ploughing or disking crop residues after slashing destroyed a further 24 per cent of the population on grain sorghum and 19 per cent on maize.

Table 2.2. Cultural control methods against major insect pests of legumes in Africa, Asia and parts of Latin America

Insect pest	Crop(s) attacked	Region	Cultural practice(s)	Reference(s)
1	2	3	4	5
Pod borers				
Helicoverpa armigera (Hubner)	Pigeonpea, chickpea, beans	Asia and East	Early planting, weed destruction,	Prasad *et al.*, 1986; Ombir and

Table 2.2 contd...

1	2	3	4	5
		Africa	intercropping, plant spacing, trap crop	Chauhan 1989; Chauhan, 1990; Naresh *et al.*, 1986; Abate, 1988
Maruca testulalis (Geyer)	Cowpea, pigeonpea	Africa, Asia, Latin America	Early maturing cultivars	Yadava *et al.*, 1988
Etiella zinckenella (Treitschke)	Pigeonpea, cowpea, green gram, mung	Asia, Africa, Latin America	Medium duration cultivars	Prasad *et al.*, 1989
Pod sucking bugs				
Anoplocnemis curvipes (Fabricius)	Cowpea	West Africa	Intercropping, weed destruction	Singh *et al.*, 1978; IITA, 1973
Clavigralla tomentosicollis Stal	Cowpea	Africa	Removal of alternative host plants	IITA, 1973
Riptortus dentipes (Fabricius)	Cowpea	Africa	Weed destruction	IITA, 1973
Leaf miners				
Aproaerema modicella (Deventer)	Groundnut, soybean, cowpea, chickpea	Asia	Intercropping, mulching	Logiswaran and Mohana-sundaram, 1985
Jassid				
Empoasca kerri Pruthi	Cowpea, urd, groundnut	Asia	Mulching	Cardona *et al.*, 1981
Armyworm				
Spodoptera litura (Fabricius)	Soybean, groundnut	North Africa, Asia	Removal of egg masses and conserva-tion of natural enemies, trap crop	Kamal, 1951; Wightman *et al.*, 1990
Aphid				
Aphis craccivora Koch	Groundnut, cowpea, beans	West and East Africa, Asia, Latin America	Sowing time, intercropping, removal of alternative host plants, plant spacing,	Evans, 1954; Farrel, 1976b; IITA, 1973; Lal *et al.*, 1989; Ranga Rao

Table 2.2 contd...

1	2	3	4	5
			water management, trap crops	*et al.*, 1991; Shoyinka, 1975; Lima and Goncalves, 1985
Pod/Stem flies				
Melanagromyza obtusa (Malloch)	Pigeonpea, cowpea	Asia	Early sowing	Prasad *et al.*, 1986
Ophiomyia centrosematis (de Meijere)	Beans, mung	East Africa	Organic fertilizer	Autrique, 1988
White grubs				
Holotrichia spp.	Groundnut	Asia	Crop rotation, tillage and mulching	NAS, 1969; Labana, 1987
Thrips				
Frankliniella schultzei (Trybom)	Groundnut	Asia, East Africa	Intercropping	Rajapakse and Jayasena, 1992
Megalurothrips sjostedti (Trybom)	Cowpea	Africa, Latin America	Intercropping	Gahukar, 1989

Table 2.3. Cultural control methods against insect pests of root crops in Africa, Asia and parts of Latin America

Insect pest	Crop(s) attacked	Region	Cultural practice(s)	Reference(s)
1	2	3	4	5
Phthormaea operculella (Zeller)	Potato	Asia	Plant density, irrigation, weed species, e. g., *Latana* sp. has repellent properties, free water on the store walls	Akhade *et al.*, 1970; Temerak, 1983; Raman and Booth, 1984
Myzus persicae (Sulzer)	Potato	Asia, Pacific	Application of fertilizer	Jansson and Smilowitz, 1986
Cylas formicarius (Fabricius)	Potato	Asia, Africa	Removal of alternate *Ipomoea* sp. host plants,	Talekar, 1989; Lema and Hahn, 1987; Alghali and

Table 2.3 contd...

1	2	3	4	5
			deep-rooting, hand picking insects	Bockarie, 1994
Leptinotarsa decemlineata (Say)	Potato	USA	Crop rotation, mulching	Wright, 1984; Brust, 1994
Mononychellus tanajoa (Bondar)	Cassava	Latin America, Africa	Early planting, good husbandry	Yaninek *et al.*, 1990
Phenacoccus manihoti Matile-Ferrero	Cassava	Latin America, Africa	Early sowing, mulching	Singh, 1982
Aonidomytilus albus (Cockerell)	Cassava	Asia	Crop residue disposal (phytosanitation)	Anonymous, 1996
Aleurotrachellus socialis Bondar	Cassava	Latin America	Intercropping	Gold *el al.*, 1991a
Bemisia tabaci (Gennadius)	Cassava	Asia, East Africa	Intercropping	Ghosh *et al.*, 1986

Table 2.4. Cultural control methods against major insect pests of cotton in Africa, Asia and parts of Latin America

Insect pest	Crop attacked	Region	Cultural practice(s)	Reference(s)
1	2	3	4	5
Pectinophora gossypiella (Saunders)	Cotton	West and East Africa, Asia, USA	Burning of cotton sticks, timely sowing, proper stacking of cotton sticks, plant spacing, deep ploughing	Nandal *et al.*, 1984; Sidhu and Dhawan, 1985; Henneberry *et al.*, 1990
Helicoverpa armigera (Hubner)	Cotton	Asia, East Africa	Synchronous sowing, trap crops, croo residues, intercropping	Jayaraj *et al.*, 1990; Singh and Sidhu, 1990; Suharto, 1989
Anthonomus grandis Boheman	Cotton	USA	Crop residues	Norman and Cartwright, 1986
Amrasca biguttula (Ishida)	Cotton	Asia	Plant spacing	Sidhu and Dhawan, 1985
Protostrophus spp.	Cotton	South Africa	Removal of weeds	Ingram *et al.*, 1973

The cultural technique of conserving one third of sugarcane residue and respreading the same after burning the whole ratooned area resulted in lower sugarcane leafhopper, *Pyrilla perpusilla* Walker incidence in India (Joshi and Sharma, 1989). The traditional storage of sorghum stalks for fodder increases the carry over of *C. partellus* and Taley and Thakare (1980) and Gahukar and Jotwani (1980) recommended that the practice of burning stubbles and feeding stalks to cattle from the previous season before the onset of monsoon rains can reduce the carry over of stem borer. Singh *et al.* (1992) stressed the importance of chaffing of sorghum stalks before February and the timely removal of stubble and earheads and their proper disposal in order to reduce the carryover of *C. partellus* larval and pupal populations.

Resource-poor farmers of central America usually feed their cattle maize and sorghum stubble during the dry season, which helps to reduce the diapausing larval populations (Quezada, 1979; CATIE, 1986).

Legume-based ecosystems. Studies such as those by Wightman *et al.* (1990) have attempted to experimentally investigate the effects of groundnut, *Arachis hypogaea* L. residue disposal on termites in India. Many farmers according to them do not clear their fields of crop residues after harvest, but the farmers could reduce the amount of food available to sustain termites through the dry season by incorporating residues into the soil, composting them off the field or, as a last resort, burning them.

Rood crop ecosystems. Whitefly, *Aleurodichus dispersus* Russell on the foliage and cassava scale, *Aonidomytilus albus* (Cockerell) on the stem are two important sucking pests of cassava in India (Anonymous, 1996). Phytosanitation has been recommended as an approach to control heavily infested plants by roguing and burning.

Cotton-based ecosystems. In South Africa, Niles and Greeff (1989) considered the practice of ratooning or leaving cotton to regrow for many years without replanting to be responsible for an increase in pests which eventually become uncontrollable. Ratoon cotton bridges the gap between the old and the new plantings at a time when stainers in the genus *Dysdercus* are able to multiply in the presence of adequate food and rising temperatures. This extra generation at the start of the season allows stainer populations to attain huge numbers to the detriment of the new crop.

In Sudan, the destruction of cotton residues and weeds has been considered essential in preventing infestation by pink bollworms (Ripper and George, 1965). This practice of phytosanitation is enforced by legislation. Similar legislation has been enacted in Kenya and Nigeria. In India, Jayaraj *et al.* (1990) recommended the removal of cotton residues to avoid carryover of pests to the next season. Cotton sticks stored by farmers are the main source of pink bollworm carryover. It is reported that 82 per cent of the population which damages the main crop comes from stacked cotton, followed by cotton seed (12%), seed material (5%) and gin trash (1%). Nandal *et al.* (1984) observed high mortality of larvae in June, if the boll bearing portion was oriented eastwards while in May and June they should be oriented towards the southwest. Burning of cotton sticks and gin trash is an important practice to prevent pest carryover.

Boll weevil, *Anthonomus grandis* Boheman, is the most destructive pest of cotton in the Lower Rio Grande Valley (LRGV) of Texas. Recommended procedures for cotton residue disposal in the LRGV (Norman and Cartwright, 1986) emphasize mechanical stalk destruction (i.e., initial shredding followed by immediate ploughing with conventional implements (moldboard, lister, disk). The authors found that stalk-shredding machines kill a high percentage of the weevils in the field after harvest.

2.2.2. Destruction of Volunteer and Alternative Host Plants

Volunteer cereal host plants, wild sugarcane, wild sorghum and other wild host plants should be removed and destroyed, together with their stubble. They can act as hosts for oviposition and places for insects to hide, thus shielding them from natural enemies and serve as sources for carryover populations into the next growing season. Poor weed control drastically reduces crop yields as a result of competition for light, nutrients, space and moisture. Weed management practices need to be developed to suit the changing agricultural and economic situation of the farmer and the region. Studies of borer species and associated natural enemies on their original/surrounding grass hosts have been initiated at ICRISAT Asia Centre, India. These studies could provide the basis for further advances in borer management, including plant breeding.

Cereal-based ecosystems. Evidence from studies in Iran (Rezwany and Schahosseini, 1977) on the ecology of the rice stem borer, *Chilo suppressalis* (Walker) showed that in the absence of rice, oviposition sometimes occurred on the weed, *Echinochloa crus-galli* (L.) P.B. After rice harvest, some of the larvae remained in the stubble whereas others migrated to weeds at the edge of the rice fields.

The removal of wild grasses by hoeing in pearl millet fields reduces the larvae of certain defoliators such as *Spodoptera* spp. and grasshoppers which are likely to develop on wild grasses in the Sahelian zone of West Africa (NDoye and Gahukar, 1987). Similarly, in Botswana, Sudan grass was often infested with *C. partellus* and can be a significant carryover source of the pest from one season to the next (Roome, 1976). In Sudan, crop residues and weeds, whether a volunteer, standover or abandoned crop must be destroyed after harvest. This practice of phytosanitation is enforced by legislation. In sugarcane fields in West Africa, the elimination of weeds may help to reduce moth borer infestations if grasses are important alternative hosts (Kumar, 1984). Cereal crops that are free from weeds suffer lower armyworm damage than weed-infested crops in India (Sharma *et al.*, 1982).

Legume-based ecosystems. *Vicia sativa* is common weed associated with chickpea in northern India. Destruction of this weed at a time when maximum eggs were laid substantially reduced the incidence of *Helicoverpa* (Ombir and Chauhan, 1989). Studies in Nigeria have shown (IITA, 1973) that weeds in cowpea and soybean fields are positively correlated with a higher percentage of seed damaged by hemipterous bugs and the moth, *Laspeyresia*. In northern Nigeria, *Aphis craccivora* Koch - the vector of groundnut rosette virus, occurs on the weed *Euphorbia hirta* through the dry season and destruction of the weed has been recommended in order to reduce the aphid population transmitting the disease in the following season. Most of Asia's better farmers now sow in rows (row cropping) to make weeding and cultivation easier.

Root crop ecosystems. The introduction of IPM techniques such as the removal of alternate *Ipomoea* sp. host plants from sweet potato surroundings and hilling up the crop to close soil cracks to control the sweet potato weevil, *Cylas formicarius* (Fabricius) in Penghu Island, Taiwan, has been recommended by Talekar (1989), because it is difficult to develop a sweet potato cultivar resistant to this pest. According to him, these two PM techniques can reduce plant and root damage from 100 and 50 per cent to 75 and 30 per cent, respectively.

Cotton-based ecosystems. The destruction of weeds in cotton fields has been shown to significantly reduce damage caused by the leaf-eating weevils, *Protostrophus* spp. to cotton in Botswana (Ingram *et al.*, 1973).

2.2.3. Hand Picking of Deadhearts/Egg Masses/Insects

Physical measures such as hand picking of deadhearts, destruction of egg masses, and jarring insects from plants play an important role in the control of some pests but are hardly practical on crops grown on a large scale. However, their use may be particularly attractive in the tropics where labour costs are comparatively low.

Cereal-based ecosystems. Central shoots of young cereal seedlings, e.g., sorghum show early damage referred to as "deadheart" and these contain a large number of young stem borer larvae, ready for dispersal to other plants. Removal and destruction of deadhearts in sorghum, if carried out by farmers over large areas, has been found to be a successful practice in reducing infestation of *C. partellus* (Seshu Reddy, 1985). In Buldana district, Maharashtra, India, farmers minimized shoot fly and stem borer damage by hand picking deadheart and throwing the central shoot outside the field or by hand picking the central shoot and dipping into an insecticide (endosulfan 5 ml/litre of water) (Duale *et al.*, 1995).

Legume-based ecosystems. Hand picking of *Spodoptera litura* (Fabricius) egg masses on soybean daily was recommended to farmers as a cultural control method against this pest in Egypt. Careful studies on the effect of egg removal on natural enemies, however, indicated that picking egg masses once every three days was more favourable than removing the masses every day because it allowed the beneficial fauna to survive (Kamal, 1951).

Root crop ecosystems. Hand picking of insect pests (e.g., *Cylas* spp.) on sweet potato has been widely employed by farmers in three villages in south Sierra Leone (Alghali and Bockarie, 1994).

Cotton-based ecosystems. Egyptian cotton leafworm is one of the main pests on the cotton crop in the Arab Republic of Egypt. Hand picking of egg masses is the main control method against this insect (Clapham, 1980). Teams of small children go through the cotton crop once every three days to collect the egg masses which are counted and then destroyed. An economic appraisal of the mass trapping method by Gubbins and Campion (1982) suggested that, provided it was technically feasible, it would be considerably cheaper than the system of egg-mass collection.

2.2.4. Time of Planting

This practice follows the principle of sowing the crop to avoid the egg-laying period of a particular pest, to enable seedlings to become established before the onset of attack, to reduce the period of susceptibility during which insects will attack, and/or to harvest before a certain pest becomes abundant. It is, however, not always true that early planting reduces pest abundance, for this depends on the life history of the insect in question, the crop and locality. In general, adjustment of sowing dates when well applied can contribute to reduced pest incidence.

Cereal-based ecosystems. In the sorghum tracts of the semi-arid tropics, planting after the first rains is the first step the farmer takes to ensure a good crop (Nwanze and Mueller, 1989). In the Sudan savanna of northern Nigeria, sorghum and millet are sown by farmers as early as possible in the rainy season. Yields are greater when an early onset of season is accompanied by the early sowing. The grains start developing at the end of the normal rainy season - with the result that they are not attacked by molds and insects (Kassam *et al.*, 1976). Ajayi and Labe (1990) found that pearl millet sown between 10 July and 2 August at Samaru, Nigeria suffered low damage by the pyralid, *C. ignefusalis,* and appeared to be optimum for grain yields. Ajayi and Tabo (1995) reported that sowing date of sorghum affects the damage by head bug, *Eurystylus oldi* Odhiambo at Bagauda, Nigeria. Youm and Gilstrap (1993) found that population densities of *Heliocheilus albipunctella* (de Joannis) were reduced by delaying millet planting dates. However, population densities of *C. ignefusalis* were not reduced by adjusting millet planting dates, but were affected by millet growth stage. Sithole (1987) found that infestations of sorghum by the stem borer complex and shoot flies were highest (32 and 31%, respectively) for the last sowing date and lowest for the first sowing date in Zimbabwe. Nwanze (1985) reported that early planting of millet with the first rains results in lower stem borer, *C. ignefusalis* damage than the late crop in Niamey, Niger. Sorghum planted during the first half of May matured early enough to avoid panicle-inhabiting insects as compared to a late sown crop in June in Texas (Archer *et al.*, 1990). In China, farmers have used a careful selection of sowing dates to prevent damage by the rice borer, *Schoenobius incertellus* Walker; the wheat stem maggot, *Meromyza americana* Fitch and *Chilo infuscatellus* Snellen (NAS, 1969).

Evidence from studies in India (Nwanze *et al.*, 1997b) has shown that under no protection, early sowing of sorghum genotypes during the rainy and post-rainy seasons produced the same control of shoot fly infestation or higher grain yield compared to those in insecticide protected treatments. Synchronous sowing of sorghum cultivars over a large area helps reduce yield losses caused by shoot fly (Ramnath *et al.*, 1974), sorghum midge (Hardas *et al.*, 1972) and head bug (Thimmaiah *et al.*, 1972). *C. partellus* causes extensive damage to sorghum sown in June compared with that sown in July or August (Panwar and Sarup, 1979). The highest incidence of leaf injury, deadheart formation, and stem tunnelling by *C. partellus* has been observed in sorghum sown in February and March and the lowest damage in the crop sown in June and October at Coimbatore, India (Mahadevan and Chelliah, 1986). Chaudhary and Sharma (1990) reported that early planting of maize should be done before April to prevent infestation by stem borer. Sorghum and millet sown at the onset of rainy season are least prone to attack by *Atherigona soccata* Rondani in India (Shah and Garg, 1990; Dhaliwal *et al.*, 1992).

Legume-based ecosystems. Rosette virus disease transmitted by *A. craccivora* is a serious pest of groundnut in sub-Saharan Africa and in the Indian subcontinent (Booker, 1963). However, it has been reported that groundnuts sown early in the wet season suffer less attack than those sown later in Tanzania (Evans, 1954). At Morogoro, Tanzania, unsprayed cowpea produced higher yields when planted in January than at later dates, because the crop escaped attack by two major pests, *Ootheca bennigseni* Weise and *Acanthomia* sp. (Enyi, 1974). Nangju *et al.* (1979) observed that early planting of cowpea (April-July) gives a much higher yield than late planting (September-November) at Ibadan, Nigeria. They noted that the populations of major cowpea pests were low at the start of the rainy season and reached their peaks in September or October.

Pigeonpea planted before the end of July contained negligible numbers of blister beetle, *Mylabris phalerata* Pallas and consequently gave a higher yield than a crop sown later (Dutta and Singh, 1989). Prasad *et al.* (1986) reported that pod damage by *Helicoverpa armigera* (Hubner) and *Melanagromyza obtusa* (Malloch) was lowest (2.6%) on early-sown pigeonpea (10 July) than on the late-sown crop (25 September).

Root crop ecosystems. Low thrips population on early planted potatoes in the Philippines produced higher marketable yield than those that were planted late (Barroga, 1992). These findings suggest that a reduction in thrips populations did not necessarily reduce damage and an increase in the number of thrips did not necessarily increase damage. Therefore, thrips damage appeared to be a more significant factor in yield reduction than the thrips population. Early planting of cassava at the beginning of the cropping season can reduce the damage caused by cassava mealybug, *Phenacoccus manihoti* Matile-Ferrero and cassava green mite, *Mononychellus tanajoa* (Bondar) in Africa (Singh, 1982; Yaninek *et al.*, 1990). Cassava sustains the least amount of green mite damage when planted early in the growing season and harvested following the dry season.

Cotton-based ecosystem. Synchronous sowing (mid-April to end of May) of cotton over a large area has been recommended by Jayaraj *et al.* (1990) to reduce the incidence of *H. armigera* and pink bollworm on cotton in India. Early sowing, as soon as the rains or temperatures allow, will often enable the bolls to mature before significant stainer invasions occur. Natural hosts offer sufficient food, especially in the monsoon Savannah regions, to sustain stainer populations early in the crop season and migration to cotton may occur too late to cause damage to developing bolls.

2.2.5. Crop Rotation

Crop rotation is an effective practice to avoid build-up of insect pests in the tropics. The most common rotations involve cereals, legumes and root crops. This approach seems to be most appropriate for soil-borne pests such as white grubs and is unlikely to be an effective control method for stem borer, *C. ignefusalis* (Youm *et al.*, 1996). Crop rotation is a useful cultural practice with a pest that has an overwintering stage in the soil under affected crops.

Cereal-based ecosystems. Evidence from studies in the Gambia (Sagnia, 1983) has shown that population of *Sesamia nonagrioides* (Lefebvre) and *B. fusca* can cause serious injuries to sorghum plants, and the planting of a pulse crop such as groundnut in rotation with sorghum has been recommended. In Texas, rotation of cereals with legumes has been recommended for the *Diatraea* complex (Anonymous, 1979). Literature on crop rotation of reducing stem borer incidence and damage is limited.

Legume-based ecosystems. The use of legumes in rotation with cereals has been found to greatly reduce white grub injuries (NAS, 1969). It is observed that some leguminous crops are unfavourable to the development of white grubs that feed on the root crops.

Root crop ecosystems. Crop rotation is one cultural practice currently in use on Long Island, New York that contributes to Colorado potato beetle (CPB) control. Rotation of potato for one year to a non-host grain crop (rye or wheat) significantly reduced the pest status of *Leptinotarsa decemlineata* (Say) in the following year's potato crop (Wright, 1984). The greatest effect of crop rotation on CPB is against the overwintered adults and their progeny in May and June.

Cotton-based ecosystems. Cotton-groundnut is a good rotation for the management of pest nematodes in the soil and to increase groundnut yields without the use of nematicides (Rodriguez-Kabana *et al.*, 1987).

2.2.6. Cropping Systems

Cropping systems such as intercropping or mixed cropping, alley cropping, strip cropping, etc. when carefully selected can reduce pest incidence by: (i) acting as a barrier, a hazard or a camouflage; (ii) acting as alternative hosts, diverting the pest away from the crop at risk; (iii) benefiting natural enemies of the pest or minimizing the risks involved in monocultures. The effects of intercropping on pest populations and crop damage are well documented and are believed, among other factors, to be due to increased diversity in the agroecosystem and non-host effects. However, farmers have rarely adopted any of our recommendations on improved intercropping configurations. As scientists, we should begin by asking ourselves the question "why" ? Much information is available on intercropping and perhaps a shift in our approach involving increased farmer participatory inputs should be examined. Cropping systems and crops that encourage natural enemies should be identified and incorporated into pest management systems.

Cereal-based ecosystems. Intercropping sorghum with soybean decreased head bug, *E. oldi* incidence on sorghum by up to 22 per cent in 1989 and 8-49 per cent in 1991, at Bagauda, Nigeria (Ajayi and Tabo, 1995). The incidence of *C. partellus, B. fusca, E. sacharina* and *Sesamia calamistis* Hampson on maize and sorghum monocrops and maize/sorghum dicrop occurred earlier and increased over time, whereas intercropping these two cereals with a legume such as cowpea in Ghana caused a significant delay in borer colonization and establishment (Amoaka-Atta *et al.*, 1983). Similarly, there was a reduction in populations of *C. partellus, B. fusca, E. saccharina* and *S. calamistis* by intercropping sorghum/maize with cowpea in Kenya (Oloo and Ogeda, 1990; Dissemond and Hindorf, 1990). Ampong-Nyarko *et al.* (1994) in Kenya studied the ovipositional behaviour of *C. partellus* in maize-sorghum-cowpea, sorghum-cowpea and maize-cassava intercropping systems. They found that oviposition on nonhost crops was dependent on intercropping with host crops and concluded that *C. partellus* oviposition on nonhost crop was one mechanism by which borer abundance can be reduced.

Sorghum is generally intercropped with pigeonpea, cotton, cowpea, safflower or other pulses in India to reduce the damage caused by *A. soccata* and *Contarinia sorghicola* (Coquillett) (Hardas *et al.*, 1980). Intercropping sorghum with other crops has been shown to reduce *C. partellus* damage on sorghum, mungbean, urdbean, pigeonpea (Singh and Singh, 1974); cowpea (Omolo and Seshu Reddy, 1985; Mahadevan and Chelliah, 1986) and lablab bean (Mahadevan and Chelliah, 1986). Chand and Sharma (1977) reported that growing maize with legumes reduced *C. partellus* damage on maize.

In the Philippines, intercropping of groundnuts with maize markedly decreased corn borer, *Ostrinia furnacalis* (Guenee) damage in maize. The decrease was attributed to predation by the spider, *Lycosa* spp. in the presence of groundnut (IRRI, 1973).

Legume-based ecosystems. Intercropping chickpea with barley or wheat controlled *H. armigera* to a level below its economic threshold (Ghosh *et al.*, 1989; Chauhan, 1990). The lowest pod damage by *H. armigera* occurred when pigeonpea was intercropped with black gram (Dash *et al.*, 1987). The incidence of groundnut leaf miner, *Aproaerema modicella* (Deventer) was highly reduced when groundnut was intercropped with cowpea or black gram at the ratio of 3:1 (Logiswaran and Mohanasundaram, 1985) and with pearl millet at a 4:1 ratio (Baskaran and Thangavelu, 1990; Sathiyanandam *et al.*, 1992). The latter case increased natural enemy activity and reduced the requirement for one round of insecticide spraying and increased yield. Evidence from studies (Dhuri *et al.*, 1986) showed that intercropping black gram with sorghum/ pigeonpea caused a reduction in pest numbers and increase in the number of predators. Chickpea sown with linseed followed by chickpea+wheat+linseed was the best combination to reduce pod borer damage (Yadava, 1987). Prasad and Chand (1989) found that intercropping chickpea with barley and wheat suppressed numbers of *H. armigera* by 59 and 47 per cent, respectively and gave net financial returns of Rs 3260 and 3144 ha^{-1}, respectively. Singh *et al.* (1978) recommended the intercropping of cowpea with maize to reduce coreid bug, *Anoplocnemis curvipes* (Fabricius) damage to cowpea pods and blister beetle, *Mylabris farquharsoni* Blair damage to cowpea flowers. They observed that this bug did not lay its eggs on cowpea plants but laid eggs on maize planted with cowpea. The presence of maize tassels at the time of cowpea flowering attracted the beetles to the tassels.

Intercropping peas with other legumes (*Lathyrus sativus,* chickpea, faba beans and lentil) in Chile did not affect the adult population or oviposition of the pea weevil, *Bruchus pisorum* (Linnaeus), but resulted in a 10 per cent reduction in grain damage when peas were associated with *L. sativus* and lentil (Araya *et al.*, 1992). Also, intercropping groundnut with other cereals resulted in a low incidence of thrips, *Frankliniella schultzei* (Trybom) in Sri Lanka (Rajapakse and Jayasena, 1992).

Groundnut intersown with field beans. *Phaseolus vulgaris,* in Malawi suffered less rosette virus infection than did groundnut in monoculture (Farrell, 1976b). Cowpea intercropped with pearl millet under different configurations, affected the activity of the thrips, *Megalurothrips sjostedti* (Trybom) and *F. schultzei,* on cowpea in Senegal (Gahukar, 1989).

Root crop ecosystems. Experiments in Colombia have shown that populations of the whitefly, *Aleurotrachelus socialis* Bondar were lower on cassava intercropped with cowpea than on cassava in monoculture (Gold *et al.*, 1991a). They also found that cassava intercropped with sunnhemp suffered less root damage from the burrowing bug, *Cyrtomenus bergi* Froeschner than casava monoculture. Potato-native tomato and potato-onion associations consistently registered the lowest potato tuber moth, leaf and tuber infestations compared to sole potato plots in Peru (Raymundo and Alcazar, 1984).

In India, Ghosh *et al.* (1986) reported that minimum populations of the aleyrodid, *Bemisia tabaci* (Gennadius), occurred on cassava grown with *Eucalyptus* and maximum on cassava grown with coconut. They also found that the overall population of spider mite, *Tetranychus* spp., was lowest in cassava-banana polyculture as compared to cassava monoculture. Intercropping potato and maize in the ratio of 9:1 has been shown to reduce thrips population on potato in the Philippines (Barroga, 1992).

Cotton-based ecosystems. Intercropping cotton with rice in Orissa, India produced the lowest infestation to cotton bolls by *Earias vittella* (Fabricius) (Dash *et al.*, 1987). Evidence from studies by Suharto (1989) in the Philippiness showed a significantly lower number of *H. armigera* (eggs and larvae) on cotton intercropped with maize or sorghum than on cotton monoculture or those with soybean or peanut intercrop at 77 days after planting. Jayaraj *et al.* (1990) recommended intercropping of cotton with less-preferred crops such as mungbean, urdbean, soybean, castor, sorghum and growing of cowpea, onion, maize and coriander for conservation of natural enemies.

Intercopping cotton serially with trap crop (rape-spring corn-summer corn) in North China facilitates the immigration and development of the insect predators and parasites since the wheat fields adjacent to cotton fields serve as sources of natural enemies of the pests (Yang, 1987).

2.2.7. Plant Spacing/Depth of Planting

Spacing helps to maintain optimum plant stands and may increase the effectiveness of the natural enemies in reducing the pest populations. The role of spacing or plant density in pest control depends on the crop species, insect species and other factors.

Cereal-based ecosystems. Infestation of wheat by the wheat-stem sawfly, *Cephus cinctus* Norton, decreased as row spacing was reduced (Luginbill and McNeal, 1958). Using a high seeding rate helps to maintain sorghum stands and reduce *A. soccata* damage (Gahukar and Jotwani, 1980). However, pearl millet planted in Senegal in narrow (40×10 cm) or wide (80×80 cm) spacing with 1 or 3 plants per hill did not influence the incidence of *C. ignefusalis* and *H. albipunctella* (Gahukar, 1989). But increasing seedling density and closer spacing led to a reduction in the proportions of empty earheads of rice plants caused by the stem borer, *S. incertellus* (Ahmed and Rao, 1965). In contrast, Chiang and Hudson (1972) noted that an increase in maize plant density led to increase in populations of *Ostrinia nubilalis* (Hubner). A positive relationship between the density of maize plants and southwestern corn borer, *Diatraea grandiosella* Dyar, has been reported by Zepp and Keaster (1977) and between sorghum plants and incidence of *C. partellus* by Singh (1986).

Legume-based ecosystems. Field studies conducted in India to determine the effect of planting density and chickpea genotype on infestations of *A. craccivora* show fewer aphids on plants sown 30×10 cm than on plants sown 60×20 cm apart, irrespective of the genotype sown (Lal *et al.*, 1989). It has been shown that groundnut planted more closely (20×10 cm) caused a significant reduction in the incidence of groundnut leaf miner and resulted in an increase in percentage parasitism compared to groundnut planted at a normal spacing of 30×10 cm (Logiswaran and Mohanasundaram, 1985). At a plant density of 10 pants/m^2, the larval population of *H. armigera* was drastically reduced on chickpea compared to a higher density of 30 plants/m^2 (Naresh *et al.*, 1986). Experiments in Malawi (Farrell, 1976a) showed that aphids reproduced less rapidly on closely spaced groundnuts, suggesting a form of pseudoresistance induced by the physiologic state of the host. Farmers used wide spacing - in part to fit with the row spacing adopted for maize and tobacco.

Root crop ecosystems. Studies in the Deccan plateau of India by Akhade *et al.* (1970) revealed that infestation due to potato tuber moth was reduced from 18.1 to 9.1 per cent as a result of planting seed potatoes at 10 cm below soil level instead of 6 cm.

Cotton-based ecosystems. Studies conducted in India indicate that two row spacings (60 and 45 cm) in combination with three plant (F414 variety of hirsutum cotton) spacings (45, 30 and 22.5 cm) affected the incidence of cotton jassid, *Amrasca biguttula* (Ishida) and pink bollworm, *Pectinophora gossypiella* (Saunders). The incidence of jassid and pink bollworm was significantly reduced and there was an increase in yield of seed cotton in 60×45 and 60×30 cm compared to 45×45, 45×30 and 45×22.5 cm spacings (Sidhu and Dhawan, 1985).

2.2.8. Tillage and Mulching

Soil tillage after harvest and before planting reduces the potential for carryover and build up of pest populations from one season to the next through mechanical damage by exposing larvae to both desiceation and predation by insect predators and parasitoids, birds and rodents. It will also provide a rugged surface and reduce the effect of wind erosion on the soil. Tillage destroys stubble, weeds and volunteer hosts that may provide food and breeding sites for insect pests, such as cereal stem borers during the off-season to re-infest newly planted crops. The effects of tillage on insect pests depend on the method and frequency of tillage and vary with insect species.

Mulching may be looked upon as a way to increase the diversity of the ecosystem, as it involves the addition of a new element, the mulch material into the system. It suppresses weeds, improves the soil structure and leads to improved water economy through reduced evaporation from the soil surface. Disadvantages of mulches are their cost and decreased efficiency as the plant canopy develops and covers them.

Cereal-based ecosystems. Evidence from field studies on pearl millet in Niger (Nwanze *et al.*, 1995) has shown that deep ploughing (30 cm) at the end of the cropping season reduced the number of surviving *H. albipunctella* pupae in the soil (>50%), with highest pupal mortalities in the top 10 cm of the profile. *H. albipunctella* survives the dry season as diapausing larvae in the soil with the majority at depths of 10-20 cm (Vercambre, 1978). A reduction in *C. partellus* population and its carryover to the next crop as a result of off-season tillage was observed in India (Gahukar and Jotwani, 1980). They suggested that this practice exposed the stem borers to their natural enemies and harsh climatic conditions.

In contrast to other workers, Cheshire and Griffin (1985) indicated that predators of lesser corn stalk borer larvae were more abundant in no-tillage than in conventional tillage cropping systems. Infestations of lesser corn stalk borer were lower in no-tillage systems than in insecticide-treated conventional tillage systems (All and Gallaher, 1977). They noted that increased soil moisture was responsible for prohibiting infestation of this pest. In Nigeria, Kaufmann (1983) studied the effects of zero-tillage and conventional tillage on the infestation of maize by *E. saccharina, S. calamistis, B. fusca* and *Mythimna unipuncta* (Haworth). She found that in the first season of May-August, there were five times as many borers in the zero-tillage maize plot compared with the conventional tillage plot and attributed it to the maize debris (stem, cobs, roots) left in the zero-tillage plot which constituted suitable breeding sites for all four borer species.

Mohyuddin and Greathead (1974) observed that untreated crop residues were often used to mulch the next sorghum crop in Uganda. However, studies by Gill (1963) indicated that mulching increased the incidence of *C. infuscatellus* in the ratoon crop of sugarcane. Cheshire and All (1979) have shown that mulching of wheat and rye residue provides food to lesser corn stalk larvae, which diverts them from attacking the maize crop. There are few studies on the use of organic mulches to suppress insect pests. Some of the these studies do, however, show that mulching may result in increased densities of generalist predators (Ramert, 1996) and reduced herbivore infestation (Brust, 1994).

Legume-based ecosystems. Studies have shown that mulching groundnut with rice straw reduced the infestation of groundnut leaf miner on groundnut and resulted in an increase in percentage parasitism (Logiswaran and Mohanasundaram. 1985). Labana (1987) reported that the soil should be ploughed twice during May and June at or before sowing to expose white grubs, *Holotrichia* sp., in the soil to harsh weather. An experiment carried out at ICRISAT Asia farm showed that mulches (neem cake, sunnhemp litter, the weed, *Celosia argentia* and the chopped foliage and branches of another weed, *Ipomaea fistulosa*) had a profound influence on the level of damage caused to dry groundnut pods by termites (Gold *et al.,* 1991b). The data showed clearly that the *Ipomaea* mulch and neem cake gave drying pods a high level of protection from termites, *Microtermes obesi* Holmgren and *Odontotermes* spp. *I. fistulosa* (morning glory bush) is a common roadside plant in India and the leaves are not eaten by insects or goats. A neem cake mulch applied to row crops was shown to have no beneficial effect on yield or foliar insect management and was very costly (Ranga Rao *et al.,* 1991). A mulch made of *C. argentia,* a common weed that survives in the vicinity of areas of high termite activity gave some protection, but sunnhemp litter apparently increased the level of termite activity, compared with bare ground. There is no evidence to link such mulches with benefits to the growing crop.

Mulching with aluminium foil and rice straw significantly reduces adult leafhopper, *Empoasca* spp., colonization of dry beans, possibly as a result of increased light reflection (Cardona *et al.,* 1981).

The larvae of soybean stem borer, *Dectes texanus* LeConte spend nearly ten months inside the soybean plant and the girdled stubble. Cultural control by burial of stems with deep ploughing or row bedding methods proved more effective in reducing larval survival and adult emergence than insecticide granules and sprays (Campbell and Van Duyn, 1977).

Root crop ecosystems. The use of deep-rooting sweet potato lines or earthing up the base of the plants of shallow-rooting lines (re-ridging) reduces weevil, *Cylas* spp., damage to sweet potato tubers in many parts of Africa (Lema and Hahn, 1987). Sweet potato lines resistant to the weevils are also available in many parts of Africa. However, each control method used individually does not provide a satisfactory control. A low input control package being developed at the IITA, Nigeria shows that the integration of the above-mentioned control techniques produces the best overall results (Lema and Hahn, 1987). Studies by Brust (1994) indicated that mulching potato with wheat straw significantly reduced Colorado potato beetle eggs and larvae in mulched plots due to attack by coccinelids and chrysopids on eggs and first and second instar larvae of the potato beetle.

Cotton-based ecosystems. Deep ploughing has been shown to significantly reduce overwinter survival of boll weevil and pink bollworm infesting cotton bolls (Henneberry *et al.,* 1990). The practice has been mandated by law in southern Texas as a means of destroying root systems of cotton stalks that would otherwise produce lush regrowth (which constitutes a near-optimal overwintering habitat for pests such as boll weevil) during much or all of the fall and winter period (Summy *et al.,* 1988). Moreover, deep ploughing may significantly reduce the incidence of certain cotton pathogens, e.g., bacterial blight, *Xanthomonas malvacearum* and root rot, *Phymatotricum omnivorum* (Presley and Bird, 1968).

2.2.9. Fertilizer Application

The use of fertilizers in subsistence farming is known to increase plant nutrition and susceptibility of crops to insects. The depletion of soil minerals by erosion and over-cropping encourages the use of fertilizers to enhance plant nutrition. Nitrogen content is regarded as a critical aspect of plant quality because it is often a limiting nutrient for both plants and herbivorous insects. However, the exact role of fertilizers in insect pest control depends on species of insect, host plant, soil type, etc.

Cereal-based ecosystems. Evidence from field trials carried out in Bolivia indicated that nitrogenous fertilizers increased the borers, *Diatraea rufescens* Box, *Diatraea dyari* Box and *D. saccharalis* incidence in sugarcane (Teran, 1979). Application of nitrogen fertilizer encourages high infestation of sorghum by head bug (Ajayi and Tabo, 1995) which is usually more pronounced on susceptible genotypes than on resistant ones. Fertilizer dosage at 50-300 kg ha^{-1} or urea at 50-200 kg ha^{-1} applied to plots sown with two pearl millet genotypes (Souna and IBV 800) resulted in significant increase in *C. ignefusalis* infestation and larval abundance in Senegal (Gahukar, 1992). Starks *et al.* (1971) observed more *C. partellus* incidence on sorghum when both nitrogenous and phosphorous fertilizers were used. Both fertilizers contributed to increased yield of grain, but the stem borers prevented a maximum response to soil fertility. However, increased infestation following nitrogen application does not, apparently, always lead to yield losses.

In contrast, high larval mortality of *C. suppressalis* has been reported on rice varieties with high silica content at the International Rice Research Institute in the Philippines (Djamin and Pathak, 1967). Silica plays an important role in stengthening the cell walls of the plants. High silica content in the plant seemed to interfere with feeding and boring by the larvae, and caused damage to the mandibles. Studies in India on shoot fly incidence with fertilizer application have indicated that infestations by *A. soccata* were lowered in plots treated with potash (K) alone and nitrogen+potash (Mote and Ramshe, 1987), as was the incidence of deadhearts and tunnelling caused by *C. partellus* (Balasubramanian *et al.,* 1986). In Sudan, the use of triple superphosphate fertilizer increases millet yield and reduces the incidence of *H. albipunctella* larvae by reducing the time interval from planting to heading (Hughes and Rhind, 1988). It is suggested that more attention be given to fighting insect pests through soil improvement than through the use of insecticides.

Legume-based ecosystems. Three species of bean fly, *Ophiomyia spencerella* Greathead, *O. phaseoli* (Tryon) and *O. centrosematis* (de Meijere), attack beans in Burundi. Coordination of appropriate organic fertilizer (manure) reduces bean fly incidence (Autrique, 1988) and inoculation of soybean seed and fertilizers insignificantly affected the population of insects such as bean fly, leafhopper, leaffolders, and cutworm (Aganon, 1989).

Root crop ecosystems. Experiments carried out at Kafr El-Sheikh region in Egypt revealed that potash fertilization applied at 50 or 75 kg ha^{-1} significantly decreased the infestation of sugarbeet weevil, *Lixus juncii* Boheman (Aly, 1988). Nitrogen fertilizer applied to potatoes (56-140 kg ha^{-1}) increased peach-potato aphid, *Myzus persicae* (Sulzer) population growth whereas excess nitrogen (224 kg ha^{-1}) reduced population growth (Jansson and Smilowitz, 1986). In southeastern Nigcria, NFKMg (12:12:17:2) suppressed termites in yam (*Dioscorea rotundata* cv. Nwopoko) and improved yield significantly (Atu, 1993).

Cotton-based ecosystems. Jayaraj *et al.* (1990) recommended the use of nitrogenous fertilizers as a cultural management against *Helicoverpa* in India. The use of potash has been reported to reduce the incidence of bacterial blight on cotton hybrids (Mudholkar *et al.,* 1984).

2.2.10. Water Management

Soil moisture is an important factor in insect development because of its effect on plant vigour and growth. Increasing this moisture through irrigation may affect insect populations by causing mechanical damage to their development. Water management might not provide the most practical method of control because fertilizer may be lost, weed control may be reduced, drying time may be too short to control larval populations effectively, rainfall may prevent field drying, costs associated with water management may be prohibitive and yield loss due to plant water stress may occur.

Cereal-based ecosystems. Water management as a cultural control tactic in rice has been shown to have potential as a substitute for the use of insecticides to control the rice water weevil, *Lissorhoptrus oryzophilus* Kuschel in Louisiana, USA (Quisenberry *et al.,* 1992). This study involved the removal of water to alter the habitat necessary for larval survival because the larvae feed on rice roots. Similarly in Colombia, water drainage in rice fields (weekly flush as against permanent flooding) resulted in fewer eggs, mines and pupae of the rice leafminer, *Hydrellia wirthi* Korytkowsky (Pantoja *et al.,* 1993). Rivnay (1967) reported that in the absence of irrigation in Israel only two or three generations of the oriental corn borer, *Chilo agamemnon* Bleszynski, developed. But with irrigation, the borer was able to pass through six generations.

All and Gallaher (1977) reported that increased soil moisture deterred infestations of *E. lignosellus* in sorghum. Deol (1964) observed that frequent irrigation reduced the percentage of sugarcane plants attacked by *C. infuscatellus*. Flooding of rice paddies for at least five days after harvest also kills many larvae and pupae of rice stem borers (NAS, 1969). Chowdry and Sharma (1960) found that the frequency of irrigation had an effect on the incidence of *Sesamia inferens* (Walker) on wheat as compared to methods of irrigation which showed no impact. Nwanze *et al.* (1996) reported that manipulation of soil water content during the vulnerable early stages of crop growth by reduced irrigation controlled shoot fly attack in irrigated postrainy season sorghum. Less irrigation water during early plant growth gave the same or better control of shoot fly damage (9.84% deadheart compared to 84.32% in the control) and the same or higher grain yield (75% increase in biomass and 80% increase in grain yield) than in insecticide-protected plots with full irrigation.

Legume-based ecosystems. Overhead irrigation led to an 80 per cent decrease in populations of *Tetranychus* sp. within 5 days in three groundnut fields at Patancheru, India without recourse to pesticide application (Ranga Rao *et al.*, 1990). This reduction was sufficient to allow the labourers to harvest without suffering from skin irritation caused by *Tetranychus* sp. Though overhead irrigation is not within the reach of most of groundnut farmers in developing countries, it is envisaged that this type of technology would help in developing future pest management systems. The only outbreak of *Tetranychus* sp. on the ICRISAT farm in recent years was associated with furrow irrigation combined with insecticide application. Recent studies on groundnut aphids indicated that aphid populations after rainfall declined by 90 per cent (Ranga Rao *et al.*, 1991). Irrigation also significantly reduced the incidence of insect pests and delayed the appearance of the major pests in chickpea in India (Mehto *et al.*, 1987).

Root crop ecosystems. Irrigation can not only reduce tuber infestation by the potato tuber moth, *P. operculella* (Temerak, 1983), but can also increase the number of predatory natural enemies (Sorokin, 1982). The presence of free water on the store walls or on potato tubers themselves reduces tuber damage caused by tuber moth in Peru (Raman and Booth, 1984). However, the presence of a free water film on tuber surfaces for prolonged periods greatly enhances the risk of bacterial soft rotting, particularly at elevated storage temperatures.

Cotton-based ecosystems. The use of judicious water management to prevent excessive vegetation growth of cotton against *Helicoverpa* has been recommended by Jayaraj *et al.* (1990) in India. *Helicoverpa* was not regarded earlier as a major pest of cotton but significant changes in cropping pattern, cropping intensity, excessive use of insecticides, nitrogenous fertilizers and irrigation rendered cotton agroecosystems more conducive for *Helicoverpa* multiplication (Singh and Sidhu, 1990).

2.2.11. Trap Crops

Interculture in cropping systems could reduce, to varying extents, pest incidence and act as a buffer to losses caused by pests. The effects may depend on the interaction of host, non-host and insect and the spatial arrangement of the intercrops.

Cereal-based ecosystems. Sarup *et al.* (1986) studied the possibility of utilizing the preferred host, sorghum as a trap to control the muscids, *A. soccata* and *Atherigona naqvii* Steyskal in spring sown maize. They found that interposing sorghum plants in the maize crop as a trap did not produce differences in percentages of deadhearts in maize between treatments with varying sorghum densities. Evidence from studies in Honduras (Castro and Pitre, 1988) has shown that maize could be used as a trap crop in reducing larval infestations of *Spodoptera frugiperda* (J.E. Smith) on sorghum in areas of low rainfall where sorghum is grown as a main crop. Wild hosts, especially grasses, act as trap plants in minimizing maize streak virus transmitted by leafhoppers in the genus *Cicadulina* (Homoptera: Cicadellidae) on maize in southeastern Nigeria (Nwilene, 1988; Dabrowski *et al.*, 1991).

Legume-based ecosystems. Trap crops (maize, pigeonpea, hyacinth bean, sunflower and lupin) have been used to control *H. armigera* on haricot bean (*P. vulgaris*) in Ethiopia. The percentage of haricot pods damaged by *H. armigera* was reduced for all the trap crop combinations compared with the control, but this was only significant for maize (Abate, 1988). Lima and Goncalves (1985) in Brazil and Shoyinka (1975) in Nigeria demonstrated that maize planted as a barrier around cowpea reduced the incidence of cowpea aphid-borne mosaic virus, a non-persistent aphid-transmitted virus.

Farmers are aware of trap crops and often sow castor or sunflower at the borders of groundnut fields in Andhra Pradesh, India to attract *Spodoptera* moths because they lay eggs preferentially on the leaves of the trap crop where egg masses can be easily seen and destroyed before hatching. Farmers in coastal Andhra Pradesh dig holes around their fields to trap migrating larvae (Wightman *et al.*, 1990).

Rust (1977) reported that border rows of snap beans planted prior to soybeans functioned to attract and hold emerging overwintering adult Mexican bean beetle, *Epilachna varivestis* Mulsant. Destruction of the trap crop generation of Mexican bean beetles resulted in protection of the adjacent soybean fields. Trap crops planted prior to lima beans failed to attract and hold the beetle.

Root crop ecosystems. Studies conducted in Georgia (Seal *et al.*, 1992) have shown that baited traps of a maize-wheat seed mixture significantly reduced populations of elaterids from the sweet potato hills. They found that elaterids were more abundant in raised beds than between raised beds and repeated ploughing of the soil reduced insect abundance in a field.

Cotton-based ecosystems. The use of trap crops such as *Nicotiana rustica* or okra has been recommended by Jayaraj *et al.* (1990) to reduce *Helicoverpa* damage to cotton in India. They further recommended the planting of two rows of okra (one at the time of sowing cotton and the other after 3 weeks) to trap eggs of *Helicoverpa* and *Earias* sp. However, in Punjab, Singh and Sidhu (1990) recommended against planting okra and pigeonpea around cotton to avoid dispersal of *Helicoverpa* from okra and pigeonpea to cotton.

2.2.12. Other Methods

Some of the approaches used by Nigerian farmers to prevent termites from coming into the fields are: planting termite grass (*Vetiveria nigrotana*) around ornamentals and *Digitaria* sp., *Cymbopogon shoenanthus* and *Pennisetum purpureum* around farms; burying cassava meal (fufu); hiring of drummers to drive them away; and burying dead goats or fish viscera in the fields (Malaka, 1972).

2.3. FUTURE RESEARCH NEEDS

Because cultural methods are free from the drawbacks experienced with the use of pesticides, there is a need to re-examine every technique for the control of pests in the light of modern technology. Nowhere is there more need for work than on the development of novel cultural practices for managing insect pests in the developing world. The emphasis must be on continually altering the environment to suit the needs of man. Methods of cultural control without doubt provide very useful tools in integrated pest management strategies and with an increasing awareness of the biology of the pest as well as the effect of crop phenology on the pest and its natural enemies it may be possible to develop unique cultural management strategies of pest control.

We believe that much remains to be learnt about the range of interactions between pests and crops operating in the cropping systems practised in Asia, Africa and parts of Latin America. There is an urgent need to explore the potential of plants and/or natural products for stem borer and head bug control. Extensive studies in farmers' fields are required. Research efforts should be directed to identify suitable improved cultural control practices for easy adoption by resource-poor farmers. The transfer of research technology that is readily accessible and affordable to the farmer should be intensified. Research on the methodology of producing pest-free seed for farmers and enlightening them on the importance of using such seed should be vigorously pursued. Cropping systems and cultivars that encourage natural enemies should be identified and fitted into pest control programmes. Information on proper timing of sowing and pest severity relationship be provided to the resource-poor community.

2.4. CONCLUSIONS

Important cultural operations, such as the disposal of crop residues, sowing date, good husbandry, crop rotation, fertilizer, cropping systems and water management are discussed in relation to pest control of cereals, root crops and cotton in the tropics of Africa, Asia and parts of Latin America. Crops may never be totally free from insect attack, but judicious use of crop protection practices, along with biological agents and insect resistant crop varieties will limit damage from insects and stabilize yields where rainfall and soil nutrients are not limiting factors for production. The efficacies of some of these measures need to be ascertained, taking effectiveness, cost and labour constraints into account. The concept of IPM emphasizes increased application of pest management strategies that are compatible with subsistence agriculture (optimization of farmer-oriented control options) and a sustainable environment. However, research should be intensified and take into account variation in local environmental conditions, and insect pests as well as the nature of the crop.

Although cultural management practices have been identified for various pests, utilization by farmers is limited, due perhaps to lack of awareness and lack of technology transfer. More awareness about the array of available cultural control measures with their advantages and limitations should be highlighted by extension officers as they have tremendous potential for reducing pest populations in traditional agriculture and integrate very well with other control methods such as host plant resistance and biological control. These three components provide a strong base for ecologically oriented pest-management systems. A concentrated effort should be made at developing well-focused IPM strategies for farmers. Collaborative research between agricultural scientists (entomologists, pathologists, agronomists, breeders, geneticists, physiologists, etc. and their extension partners) is required, using existing cultural measures as model, to develop suitable improved cultural packages. It is also necessary to promote increased awareness of and to stimulate efforts for application-oriented cultural control research, that is compatible with subsistence agriculture throughout the tropics.

REFERENCES

Abate, T. 1988. Experiments with trap crops against African bollworm, *Heliothis armigera*, in Ethiopia. *Ent. exp. Appl.* **48**(2): 135-140.

Adesiyun, A.A. and Ajayi, O. 1980. Control of the sorghum stem borer, *Busseola fusca*, by partial burning of the stalks. *Trop. Pest Mgt.* **26**:113-117.

Aganon, T.M. 1989. Cultural control against soybean insect pests. Pest Control Council of the Philippines, Baguio City, Philippines.

Ahmed, M.K. and Rao, R.S. 1965. The influence of cultural practices on the stem borer (*Schoenobius incertellus* Walker) attack in rice. *Madras agric. J.* **52**: 311-313.

Ajayi, O. and Labe, D.A. 1990. The effect of sowing date and planting method on stem borer damage in dauro millet. *Ann. appl. Biol.* **117**(3): 487-494.

Ajayi, O. and Tabo, R. 1995. Effect of crop management practices on *Eurystylus immaculatus* on sorghum. *In*: K.F. Nwanze and O. Youm (eds). *Panicle Insect Pests of Sorghum and Pearl Millet: Proceedings of an International Consultative Workshop.* 4-7 October, 1993, ICRISAT Sahelian Center, Niamey, Niger, pp. 233-240.

Akhade, N.N., Tidke, P.M. and Patkar, M.B. 1970. Control of potato tuber moth, *Gnorimoschema operculella* Zell. in Deccan plateau through insecticides and depth of planting. *Indian J. agric. Sci.* **40**(12): 1071-1076.

Alghali, A.M. and Bockarie, A. 1994. Cultural practices and perceptions in sweet potato farming and their implications for pest management in southern Sierra Leone. *Int. J. Pest Mgt.* **40**(1): 27-30.

All, J.N. and Gallaher, R.N. 1977. Detrimental impact of no-tillage corn cropping systems, involving insecticides, hybrids and irrigation on lesser corn stalk borer infestations. *J. econ. Ent.* **70**: 361-365.

Aly, F.A. 1988. Effect of nitrogenous and potash fertilization rates on sugar beet infestation with main insects [Egypt]. *J. agric. Sci. Mansoura Univ. Egypt* **13**(1): 357-361.

Amoaka-Atta, B., Omolo, E.O. and Kidega, E.K. 1983. Influence of maize, cowpea and sorghum intercropping systems on stem/pod-borer infestations. *Insect Sci. Applic.* **4**(1-2): 47-57.

Ampong-Nyarko, K., Seshu Reddy, K.V. and Saxena, K.N. 1994. *Chilo partellus* Swinhoe (Lepidoptera: Pyralidae) oviposition on non-hosts: a mechanism for reduced pest incidence in intercropping. *Acta Oecol.* **15**(4): 469-475.

Anonymous 1979. Insect and mite pests of grain sorghum-management approaches. Texas Agricultural Extension Service, College Station, Texas A&M University, Texas, USA, 24 pp.

Anonymous 1996. Science and Technology. *The Hindu,* May 23, 1996, p. 28.

Araya, S.F., Gerding, P.M. and Medina, S.V. 1992. Effect of associated crops and plant extracts on the behaviour of the pea weevil, *Bruchus pisorum* L. *Agro-Ciencia* **8**(2): 79-85.

Archer, T.L., Losade, J.C. and Bynum, E.D., Jr. 1990. Influence of planting date on abundance of panicle-feeding insects associated with sorghum. *J. agric. Ent.* **7**(3): 233-239.

Atu, U.G. 1993. Cultural practices for the control of termite (Isoptera) damage to yams and cassava in southeastern Nigeria. *Int. J. Pest Mgt.* **39**(4): 462-466.

Autrique, A. 1988. Synthesis of research works on bean fly conducted in Burundi. Seminaire Regional sur I' Amelioration du Haricot dans la Region des Grands Lacs. Bukavu (Zaire).

Balasubramanian, G., Balasubramanian, M., Sankaran, K. and Manickam, T.S. 1986. Effect of fertilization on the incidence of shoot fly and stem borer on rainfed sorghum. *Madras agric. J.* **73**(8): 471-473.

Barroga, S.F. 1992. Effects of planting dates, corn intercropping and chemical spraying on thrips population and yield of potato. *In*: E.T. Rasco Jr., F.B. Aromin, V.D.R. Amante and P.J.S. Lopez (eds). *Southeast Asian Program for Potato Research and Development.* Compilation of Annual Reports, July 1991-June 1992, Potato, Manila, Philippines, pp. 172-187.

Baskaran, R.K.M. and Thangavelu, S. 1990. Non-conventional method to minimize the incidence of groundnut leaf miner. *Groundnut News* **2**(2): 6.

Booker, R.H. 1963. The effect of sowing date and spacing on rosette disease of groundnut in northern Nigeria with observations on the vector, *Aphis craccivora. Ann. appl. Biol.* **52**: 125-131.

Bowden, J. 1976. Stem borer ecology and strategy for control. *Ann. appl. Biol.* **84**: 107-111.

Brader, L. 1979. Integrated pest control in the developing world. *A. Rev. Ent.* **24**: 225-254.

Brust, G.E. 1994. Natural enemies in straw-mulch reduce Colorado potato beetle populations and damage in potato. *Biological Control: Theory and Applications in Pest Management* **4**(2): 163-169.

Campbell, W.V. and Van Duyn, J.W. 1977. Cultural and chemical control of *Dectes texanus texanus* on soybeans. *J. econ. Ent.* **70**(2): 256-258.

Cardona, C., van Schoonhoven, A., Gomez, L., Garcia, J. and Garzon, F. 1981. Effect of artificial mulches on *Empoasca kraemeri* Ross and Moore populations and dry bean yields. *Envir. Ent.* **10**: 705-707.

Castro, M. and Pitre, H. 1988. Potential for using maize as a trap crop for the fall armyworm, *Spodoptera frugiperda* (Lepidoptera: Noctuidae), where sorghum and maize are intercropped on subsistence farms. *Florida Ent.* **71**(93): 273-278.

CATIE (Centro Agronomico Tropical de Investigacion Y Ensenanza). 1986. Alternative de Manejo para el sistema maiz+maicillo, Tejutla, El Salvador: descripcion y validacion en fincas pequenas, Turrialba, Costa Rica. (In Es.) Informe Tecnico no. 76. Turrialba, Costa Rica.

Chand, P. and Sharma, N.N. 1977. Influence of crop association on insect pest incidence. *Proc. Indian natn. Sci. Acad.* **43**: 108-114.

Chaudhary, R.N. and Sharma, V.K. 1990. Incidence and abundance of maize stalk borer, *Chilo partellus* Swinhoe in spring season. *Indian J. Ent.* **52**: 165-169.

Chauhan, R. 1990. Present status of *Heliothis* on pulses and strategies for its management in Haryana. *In*: *Proc. First National Workshop on Heliothis Management: Current Status and Future Strategies*.30-31 August, 1990. Directorate of Pulses Research, Kanpur, U.P., India, pp. 52-56.

Cheshire, J.M. and All, J.N. 1979. Feeding behavior of lesser corn stalk borer larvae in simulations of no-tillage, mulched conventional tillage and conventional tillage corn cropping systems. *Envir. Ent.* **8**: 261-264.

Cheshire, J.M. and Griffin, J. 1985. Evidence that mulch density influences lesser corn stalk borer damage to no-tillage grain sorghum. *Sorghum Newsl.* **27**: 61.

Chiang, H.C. and Hudson, A.C. 1972. Population fluctuations of the European corn borer, *Ostrinia nubilalis,* at Waseca, Minnesota, 1940-70. *Envir. Ent.* **1**: 7-16.

Chowdry, S. and Sharma, R.G. 1960. Effect of methods and frequency of irrigation on the incidence of stem borer of wheat (*Sesamia inferens* Wlk.) at irrigation research farm, Madhipura. *Indian J. Agron.* **4**: 264-268.

Clapham, W.B. 1980. Egyptian cotton leafworm: integrated control and the agricultural production system. *Agric. Envir.* **5**(3): 201-211.

Dabrowski, Z.T., Nwilene, F.E. and Kumar, R. 1991. First regular observations on leafhoppers, *Cicadulina* spp., vectors of maize streak virus (MSV) in southeastern Nigeria. *Insect Sci. Applic.* **12**(1/2/3): 249-261.

Dash, A.N., Mahapatra, H., Pradhan, A.C. and Patnaik, N.C. 1987. Effect of mixed and intercropping on occurrence of some pests in Orissa. *Envir. Ecol.* **5**(3): 526-530.

Deol, D.S. 1964. Effect of nitrogen and irrigation on the incidence of some sugarcane pests. *In*: *Proc. 5th All India Conference of Sugarcane Research and Development Workers*. Coimbatore, India, pp. 302-305.

Dhaliwal, J.S., Mehndiratta, P.D. and Brar, G.S. 1992. Overcoming the shoot fly menace in fodder sorghum. *Indian Fmg.* **41** (10): 5.

Dhuri, A.V., Singh, K.M. and Singh, R.N. 1986. Effect of intercropping on population dynamics of insect pests of black gram, *Vigna mungo* (L.) Hepper. *Indian J. Ent.* **48**(3): 329-338.

Dissmond, A. and Hindorf, H. 1990. Influence of sorghum/maize/cowpea intercropping on the pest, disease and weed situation at Mbita/ Kenya. *In*: *Proc. Integrated Pest Management in Tropical and Subtropical Cropping Systems*. 3 Frankfurt am Main, Germany, pp. 843-851.

Djamin, A. and Pathak, M.D. 1967. Role of silica in resistance to Asiatic rice borer, *Chilo suppressalis* Walker in rice varieties. *J. econ. Ent.* **60**: 347-351.

Duale, A.H., Rao, R. and Rao, G.D.N. 1995. Incidence of shoot pests in Maharashtra. Report of a Diagnostic Survey, Crop Protection Division (Cereals Entomology), ICRISAT, Patancheru, Andhra Pradesh, India. 9 pp.

Duerden, J.C. 1953. Stem borers of cereal crops at Kongwa, Tanganyika, 1950-52. *East African Agric. J.* **19**: 105-119.

Dutta, M. and Singh, B.V. 1989. Blister beetle (*Mylabris phalerata*), a serious pest of pigeonpea in the lower hills of Uttar Pradesh. *Int. Pigeonpea Newsl.* **10**: 29.

Enyi, B.A.C. 1974. Effects of time of sowing and phosphamidon (Dimecron) on cowpea (*Vigna unguiculata*). *Exptl. Agric.* **10**: 87-95.

Evans, A.C. 1954. Groundnut rosette disease in Tanganyika. I. Field studies. *Ann. appl. Biol.* **41**: 189-206.

Farrell, J.A.K. 1976a. Effects of groudnut crop density on the population dynamics of *Aphis craccivora* Koch (Hemiptera: Aphididae) in Malawi. *Bull. ent. Res.* **66**: 317-329.

Farrell, J.A.K. 1976b. Effects of intersowing with bean on the spread of groundnut rosette virus by *Aphis craccivora* Koch (Hemiptera: Aphididae) in Malawi. *Bull. ent. Res.* **66**: 331-333.

Gahukar, R.T. 1989. Pest and disease incidence in pearl millet under different plant density and intercropping patterns. *Agric. Ecosyst. Envir.* **26**(1): 69-74.

Gahukar, R.T. 1992. Effect of various fertilizers and rates on insect pest/pearl millet relationship in Senegal. *Trop. Agric.* (Trinidad) **69** (2): 149-152.

Gahukar, R.T. and Jotwani, M.G. 1980. Present status of field pests of sorghum and millet in India. *Trop. Pest Mgt.* **26**: 128-151.

Ghosh, P.K., Devendra, P., Prasad, R. and Shaw, S.P. 1989. Management of pod borer, *Heliothis armigera* (Hub.) in chickpea. *In*: *Maximizing Crop Production in Rainfed and Problem Areas*. Indian Agricultural Research Institute, New Delhi, India, pp. 122-126.

Ghosh, S.P., Pillai, K.S. and Thankappan, M. 1986. Cassava based multiple cropping system II. Incidence of pests and diseases. *J. Root Crops* **12**(2): 83-89.

Gill, M.S. 1963. Effect of trash mulch on the growth and yield of ratoon crop of sugarcane. *West Pakistan J. agric. Res.* **1**(2): 149-156.

Gold, C.S., Vargas, O. and Wightman, J.A. 1991a. Intercropping effects on insect pests of cassava in Colombia and of groundnut in India. *In*: G.K. Veeresh, D. Rajagopal and C.A. Viraktamath (eds). *Advances in Management and Conservation of Soil Fauna: Proc. 10th International Colloquim of Soil Zoology*. 7-13 August, 1988, Bangalore, Karnataka, India, pp. 329-332.

Gold, C.S., Wightman, J.A. and Pimbert, M.P. 1991b. Effects of mulches on foraging behaviour of *Microtermes obesi* and *Odontotermes* spp. in India. *Insect Sci. Applic.* **12**(1-3): 297-303.

Gubbins, K.E. and Campion, D.G. 1982. Economic aspects of pheromone trapping techniques for the control of Egyptian cotton leafworm. *Outlook Agric.* **11**: 62-66.

Hardas, M.G., Bhagwat, V.R. and Shamagat, B.K. 1980. Influence of associate cropping systems on the incidence of sorghum pest complex. *College Agric. Nagapur Mag.* **51-52**: 19-27.

Hardas, M.G., Supare, N.K. and Chopde, P.R. 1972. Studies on the seasonal incidence of sorghum midge (*Contarinia sorghicola*) infesting sorghum hybrids. *Sorghum Newsl.* **15**: 88-89.

Hardyal, S.G. 1963. Influence of hot weather irrigation and trash mulch on the incidence of certain sugarcane borers and weeds. *Indian J. Sugarcane Res. Dev.* **7**(3): 161-163.

Henneberry, T.J., Meng, T., Jr., and Bariola, L.A. 1990. Overwintering, survival and emergence of boll weevil (Coleoptera: Curculionidae) in cotton bolls in Arizona. *J. econ. Ent.* **83**: 1879-1882.

Hughes, D. and Rhind, D. 1988. Control of millet head worm in the Western Sudan. *Trop. Pest Mgt.* **34**(3): 346-348.

ICRISAT (International Crops Research Institute for the Semi-Arid Tropics). 1985. *Proc. International Sorghum Entomology Workshop*. 15-21 July, 1984, Texas A&M University, College Station, TX, USA, Patancheru, A.P., India.

IITA (International Institute of Tropical Agriculture). 1973. *Report-Grain Legume Improvement Program*, 1972. Ibadan, Nigeria.

Ingram, W.R., Irving, N.S. and Roome, R.E. 1973. *A Handbook on the Control of Agricultural Pests in Botswana*. Government Printer, Gaborone, Botswana, 129 pp.

IRRI (International Rice Research Institute). 1973. *Annual Report*, 1972. Los Banos, Laguna, Philippines.

Jansson, R.K. and Smilowitz, Z. 1986. Influence of nitrogen on population parameters of potato insects: abundance, population growth and within-plant distribution of the green peach aphid, *Myzus persicae* (Homoptera: Aphididae). *Envir. Ent.* **15**: 49-55.

Jayaraj, S., Uthamasamy, S., Sundarababu, P.C. and Dhandarpani, N. 1990. Status of *Heliothis* on cotton and strategies for its management. *In: Proc. First National Workshop on Heliothis Management.* 30-31 August, 1990, Directorate of Pulses Research, Kanpur, U.P., India, pp. 141-160.

Joshi, P.K. and Sharma, S.K. 1989. Augmentation and conservation of *Epiricania melanoleuca* Fletcher, for the population management of sugarcane leafhopper, *Pyrilla perpusilla* Walker, under arid conditions of Rajasthan. *Indian Sugarcane* **39**(8): 625-628.

Jotwani, M.G. 1982. Factors reducing sorghum yields-insect pests. *In: Sorghum in the Eighties: Proc. International Symposium on Sorghum.* 2-7 November, 1981, ICRISAT Asia Center, Patancheru, A.P., India, pp. 251-255.

Kamal, M. 1951. The biological control of the cotton leaf worm, *Spodoptera litura* Fab. in Egypt. *Bull. Soc. Fouad I d'Ent.* **35**: 221-270.

Kassam, A.H., Dagg, M., Kowal, J.M. and Khadr, F.H. 1976. Improving food crop production in the Sudan savanna zone of northern Nigeria. *SPAN* **8**(6): 341-347.

Kaufmann, T. 1983. Behavioral biology, feeding habits and ecology of three species of maize stem borers: *Eldana saccharina* (Lepidoptera: Pyralidae), *Sesamia calamistis* and *Busseola fusca* (Noctuidae) in Ibadan, Nigeria. *J. Georgia ent. Soc.* **18**(2): 259-272.

Kfir, R. 1990. Prospects for cultural control of the stalk borers, *Chilo partellus* Swinhoe and *Busseola fusca* Fuller in summer grain crops in South Africa. *J. Ent. Soc. S. Africa* **53**(1): 41-47.

Kumar, R. 1984. *Insect Pest Control with Special Reference to African Agriculture.* Edward Arnold Publishers Ltd., London, Great Britain.

Kumar, R. and Sampson, M. 1982. Review of stem borer research in Ghana. *Insect Sci. Applic.* **3**(2/3): 85-88.

Labana, K.S. 1987. It is time to control white grub. *Indian Fmg.* **37**(4): 31.

Lal, S.S., Yadava, C.P. and Dias, C.A.R. 1989. Effect of planting density and chickpea cultivars on the infestations of the black aphid, *Aphis craccivora* Koch. *Madras agric. J.* **76**(8): 461-462.

Lawani, S.M. 1982. A review of the effects of various agronomic practices on cereal stem borer populations. *Trop. Pest Mgt.* **28**: 266-276.

Lema. K.M. and Hahn, S.K. 1987. Integrated management of the sweet potato weevils in Africa: an example of low input package. *Proc.*

International Congress of Plant Protection, 5-9 October, 1987, Manila, Philippines.

Lima, J.A.A. and Goncalves, M.F.B. 1985. O uso de barreira viva visando o controle do "Cowpea aphid-borne mosaic virus" em culture de feijao-de-corda (*Vigna unguiculata*). *Fitopatologia Brasileira* **10**: 321.

Logiswaran, G. and Mohanasundaram, M. 1985. Effect of intercropping, spacing and mulching in the control of groundnut leaf miner, *Aproaerema modicella* Deventer (Gelechiidae: Lepidoptera). *Madras agric. J.* **72**(12): 695-700.

Luginbill. P., Jr. and McNeal, F.H. 1958. Influence of seeding density and row spacings on the resistance of spring wheats to the wheat stem sawfly. *J. econ. Ent.* **51**(6): 804-808.

Mahadevan, N.R. and Chelliah, S. 1986. Influence of intercropping legumes with sorghum on the infestation of the stem borer, *Chilo partellus* Swinhoe in Tamil Nadu, India. *Trop. Pest Mgt.* **32**(2): 162-163.

Magema, N.S. 1984. Synthese d' observations. Ennemis des cultures vivrieres (hivernage 1983). (In Fr.) Project CILSS. Nouakchott, Mauritania: Comite Permanent Interetats do Lutte la Secheresse dans le sahel. 55 pp.

Malaka, S.L.O. 1972. Some measures applied in the control of termites in parts of Nigeria. *Nigerian Ent. Mag.* **2**: 137-141.

Mehto, D.N., Singh, K.M. and Singh, R.N. 1987. Influence of irrigation on succession and population build up of insect pests in chickpea (*Cicer arietinum* Linn.). *Indian J. Ent.* **49**(3): 297-329.

Mohyuddin, A.I. and Greathead, D.J. 1974. An annotated list of the parasites of graminaceous stem borers in East Africa with a discussion of their potential in biological control. *Entomophaga* **15**: 241-274.

Mote, U.N. and Ramshe, D.G. 1987. Effect of fertilizers and plant population on the incidence of shootfly (*Atherigona soccata* Rondani) on some sorghum cultivars. *Indian J. Ent.* **49**(4): 490-495.

Mudholkar, N.J., Deshmukh, R.K. and Meshram, M.K. 1984. Effect of potassium levels on yields, quality characters and disease incidence on cotton hybrids under rainfed conditions. *Golden Jubilee and 49th Annual Convention of Indian Society of Soil Science.* 7-10 December, 1984. Indian Agricultural Research Institute, New Delhi, India.

Nandal, A.S., Singh, J.P., Singh, Z. and Lather, B.P.S. 1984. Effect of direction of stacking cotton sticks on the mortality of the diapausing larvae of pink bollworm. *Indian J. agric. Sci.* **54**: 671-673.

Nangju, D., Nwanze, K.F. and Singh, S.R. 1979. Planting date, insect damage and yield of cowpea in western Nigeria. *Exptl. Agric.* **15**: 1-10.

Naresh, J.S., Malik, V.S. and Kaushik, S.K. 1986. Larval population and damage of *Heliothis armigera* Hubner estimated on different plant densities of chickpea during five weeks. *Bull. Ent.* **27**(1): 70-73.

NAS (National Academy of Science). 1969. *Insect Pest Management and Control.* NAS Publication. 1695(3): 208-242.

NDoye, M. and Gahukar, R. 1987. Insect pests of pearl millet in West Africa and their control. *In: Proc. International Pearl Millet Workshop.* 7-11 April, 1986, ICRISAT Asia Center, Patancheru, A.P., India, pp. 195-205.

Niles, G.A. and Greeff, M.S. 1989. *Ratoon Cotton in Perspective.* Technical Communication No. 217, Department of Agriculture and Water Supply, Republic of South Africa.

Norman, J.W. and Cartwright, B. 1986. *Management of Cotton Insects in the Lower Rio Grande Valley.* Texas Agricultural Extension and Services, B-1210.

Nwanze, K.F. 1985. Some aspects of pest management and host plant resistance in pearl millet in the Sahel. *Insect Sci. Applic.* **6**(3): 461-465.

Nwanze, K.F. and Mueller, R.A.E. 1989. Management options for sorghum stem borers for farmers in the semi-arid tropics. *In: Proc. Int. Workshop on Sorghum Stem Borers.* 17-20 November, 1987, ICRISAT Center, Patancheru, India, pp. 105-116.

Nwanze, K.F., Klaij, M.C. and Markham, R.H. 1995. Possibilities for integrated management of millet earhead caterpillar, *Heliocheilus albipunctella. In*: K.F. Nwanze and O. Youm (eds). *Panicle Insect Pests of Sorghum and Pearl Millet: Proc. International Consultative Workshop.* 4-7 October, 1993, ICRISAT Sahelian Center, Niamey, Niger, pp. 263-271.

Nwanze, K.F., Nwilene, F.E. and Reddy, Y.V.R. 1997. Interaction of genotype and sowing date on shoot fly incidence and grain yield of sorghum. *African Ent.* (In press).

Nwanze, K.F., Reddy, Y.V.R., Nwilene, F.E., Soman, P., Laryea, K.B. and Jayachandran, R. 1996. Reduction of shoot fly damage in irrigated post-rainy season sorghum by manipulating irrigation. *Ann. appl. Biol.* **129**: 391-403.

Nwilene, F.E. 1988. Distribution pattern, population dynamics of leafhoppers, *Cicadulina* spp. and their role in maize streak virus transmission in southeastern Nigeria. M. Phil. Thesis, Rivers State University of Science and Technology, Port Harcourt, 205 pp.

Oloo, G.W. and Ogeda, K. 1990. The incidence of *Chilo partellus* Swinh. (Pyralidae) and the contribution of natural enemies to its mortality under intercropping system in Kenya. *Trop. Pest Mgt.* **36**(3): 244-248.

Ombir and Chauhan, R. 1989. Management of *Heliothis armigera* in chickpea through removal of weed *Vicia sativa*. In: *National Symposium on New Frontiers in Pulses Production and Development.* 10-12 November, 1989, Directorate of Pulses Research, Kanpur, U.P., India, p. 19.

Omolo, E.O. and Seshu Reddy, K.V. 1985. Effects of different sorghum based cropping systems on insect pests in Kenya. *In: Proc. International Sorghum Entomology Workshop.* 15-21 July, 1984, Texas A&M University, College Station, TX, USA, ICRISAT, Patancheru, A.P., India.

Pantoja, A., Salaza, A., Mejia, O.I., Velazquez, J.G. and Duque, M.C. 1993. Cultural practices to manage the rice leafminer *Hydrellia wirthi* (Diptera: Ephydridae) in Colombia. *J. econ. Ent.* **6**: 1820-1823.

Panwar, V.P.S. and Sarup, P. 1979. Relationship between successive dates of sowing of maize and damage caused by *Chilo partellus* Swinhoe affecting grain yield. *J. ent. Res.* **3**: 9-24.

Prasad, D. and Chand, P. 1989. Effect of intercropping on the incidence of *Heliothis armigera* Hubner and grain yields of chickpea. *J. Res. Birsa agric. Univ.* **1**: 15-18.

Prasad, D., Chand, P. and Srivastava, G.P. 1986. Effect of sowing dates on the infestation of insect pests and grain yield of pigeonpea. *Indian J. Ent.* **48**: 230-231.

Prasad, D., Chand, P. and Haque, M.F. 1989. Incidence of pod boring insects in different cultivars of pigeonpea. *J. Res. Birsa agric. Univ.* **1**: 79-80.

Presley, J.T. and Bird, L.S. 1968. Diseases and their control. *In*: F.C. Elliott, M. Hoover and W.K. Porter (eds). *Advances in Production of Quality Cotton: Principles and Practices.* Iowa State University Press, Ames, Iowa, pp. 347-366.

Quezada, J.R. 1979. Poblaciones remanentes de barrendadores en canas de maiz. (In Es.) San Salvador. El Salvador: Universidad de El Salvador, Facultad de Ciencias y Humanidades. 22 pp.

Quisenberry. S.S., Trahan, G.B., Heagler, A.M. and McManus, B. 1992. Effect of water management as a control strategy for rice water weevil (Coleoptera: Curculionidae). *J. econ. Ent.* **85**: 1007-1014.

Rajapakse, R. and Jayasena, K.W. 1992. Incidence and management of groundnut thrips causing bud necrosis disease in the southern region of Sri Lanka. *In*: S.N. Nigam (ed.). *Groundnut: A Global Perspective*. Proc. International Workshop, ICRISAT Center, Patancheru, Andhra Pradesh, India, p. 494.

Raman, K.V. and Booth, R.H. 1984. Integrated control of potato moth in rustic potato stores. *In*: *Proc. Symposium of the International Society for Tropical Root Crops*. 6th Lima, 1983, International Potato Center (CIP), Lima, pp. 509-515.

Ramert, B. 1996. The influence of intercropping and mulches on the occurrence of polyphagous predators in carrot fields in relation to carrot fly (*Psila rosae* (F.)) (Dipt., Psilidae) damage. *J. appl. Ent.* **120:** 39-46.

Ramnath, B., Verma, B., Rao, S.B.P. and Mittal, S.P. 1974. Effects of dates of planting on sorghum varieties. *Sorghum Newsl.* **17**: 65.

Ranga Rao, G.V., Ranga Rao, D.V., Reddy, M.P. and Wightman, J.A. 1990. Overhead irrigation and the integrated management of spider mites in a groundnut crop. *Int. Arachis Newsl.* **8:** 23-24.

Ranga Rao, G.V., Wightman, J.A., Wadia, K.D.R., Ranga Rao, D.V. and Nageswara Rao, R.C. 1991. Influence of water stress on groundnut aphids. *Int. Arachis Newsl.* **9:** 14.

Raymundo, S.A. and Alcazar, J. 1984. Control of potato tuber moth (*Phthorimaea operculella*) through crop association. *In: Proc. Symposium of the International Society for Tropical Root Crops*. 6th Lima, 1983. International Potato Center (CIP), Lima, pp. 517-521.

Rezwany, N. and Schahosseini, D. 1977. Biology and ecology of the rice stem borer (*Chilo suppressalis* Walker). *Ent. Phytopathol. Appl.* **43:** 1-5.

Ripper, W.E. and George, M. 1965. *Cotton Pests of the Sudan*. Blackwell Scientific Publications, Oxford, UK.

Rivnay, E. 1967. A contribution to the biology of the maize borer *Chilo agamemnon* Blesz. *Israel J. Ent.* **2:** 15-27.

Rodriguez-Kabana, R., Ivey, H. and Backman, P.A. 1987. Peanut-cotton rotations for the management of *Meloidogyne arenaria*. *J. Nematol.* **19**(4): 484-486.

Roome, R.E. 1976. Resistance in sorghum varieties to attack by larvae of *Chilo partellus* Swinhoe (Sorghum stalk borer). *In: Crop Protection in Botswana: Biennial Report 1971-73*. Ministry of Agriculture, Division of Agricultural Research, Gaborone, Botswana, pp. 111-115.

Rust, R.W. 1977. Evaluation of trap crop procedures for control of Mexican bean beetle (*Epilachna varivestis*) in soybeans and lima beans. *J. econ. Ent.* **70**(5): 630-632.

Sagnia, S.B. 1983. Possible integrated pest management tools for the effective control of cereal stem borers in the Gambia. *Insect Sci. Applic.* **4**(1-2): 217-219.

Sarup, P., Siddiqui, K.H. and Marwaha, K.K. 1986. Possible limitations in exploring the trap of preferred host sorghum for the control of the shoot fly complex (*Atherigona soccata* Rondani and *A. naqvii* Steyskal) in spring sown maize. *J. ent. Res.* **10**(2): 119-124.

Sathiyanandam, V.K.R., Baskaran, R.K.M. and Thangavelu, S. 1992. Effect of nonconventional methods for managing groundnut leaf miner. *In:* S.N. Nigam (ed.). *Groundnut: A Global Perspective.* Proc. International Workship, ICRISAT Center, Patancheru, Andhra Pradesh, India, p. 491.

Seal, D.R., Chalfant, R.B. and Hall, M.R. 1992. Effects of cultural practices and rotational crops on abundance of wireworms (Coleoptera: Elateridae) affecting sweet potato in Georgia. *Envir. Ent.* **21**(5): 969-974.

Seshu Reddy, K.V. 1985. Integrated approach to the control of sorghum stem borers. *In: Proc. International Sorghum Entomology Workshop,* 15-21 July, 1984, Texas A&M University, College Station, TX, USA, ICRISAT, Patancheru, A.P., India, pp. 205-215.

Shah, N.K. and Garg, D.K. 1990. Management of shoot fly, *Atherigona falcata* Thomson, by adjusting of sowing time in barnyard millet. *Indian J. Ecol.* **17**(2): 197-198.

Sharma, H.C. and Davies, J.C. 1982. *Studies on Pearl Millet Insects.* Sorghum Entomology Progress Report-7. International Crops Research Institute for the Semi-Arid Tropics, Patancheru, Andhra Pradesh, India.

Sharma, H.C., Bhatnagar, V.S. and Davies, J.C. 1982. *Studies on Mythimna separata at ICRISAT.* Sorghum Entomology Progress Report-6. International Crops Research Institute for the Semi-Arid Tropics, Patancheru, India.

Shoyinka, S.A. 1975. Attempted control of virus incidence in cowpeas by the use of barrier crops. Nigerian Society of Plant Protection, NSPP Abstracts, Occasional Publication **1**:27.

Sidhu, A.S. and Dhawan, A.K. 1985. Effect of different spacings on incidence of cotton jassid and pink bollworm on F 414 variety of Hirsutum cotton. *J. Res. Punjab agric. Univ.* **22**(3): 483-487.

Singh, J. and Sidhu, A.S. 1990. Present status of *Heliothis* on cotton and strategies for its management in Punjab, Haryana and

Rajasthan. *In*: *Proc. First National Workshop on Heliothis Management*. 30-31 August, 1990, Directorate of Pulses Research, Kanpur, U.P., India, pp. 170-179.

Singh, K.M. and Singh, R.N. 1974. The population build up of *Pyrilla perpusilla* Walker on sorghum and pearl millet under dryland conditions at Delhi. *Indian J. Ecol.* **1**: 12-16.

Singh, S.P. 1986. Screening of forage sorghum genotypes for resistance to shoot fly, *Atherigona soccata* Rondani and stem borer, *Chilo partellus* Swinhoe and estimation of avoidable losses. Ph.D. Thesis, Haryana Agricultural University, Hisar, Haryana, India, 135 pp.

Singh, S.P., Verma,A.N. and Lodhi,G.P. 1992. Larval and pupal population of *Chilo partellus* Swinhoe in different sorghum plant parts at crop harvest and moth emergence during off-season. *Crop Res.* **5**(2): 359-362.

Singh, S.R., van Emden, H.F. and Taylor, T.A. 1978. *Pests of Grain Legumes: Ecology and Control.* Academic Press, London.

Singh, T.P. 1982. The mealybug problem and its control. *In*: *Workshop on Root Crops in Eastern Africa.* IDRC, Ottawa, Canada, pp. 70-72.

Sithole, S.Z. 1987. The effect of date of planting on shoot fly and stem borer infestations on sorghum. *In*: *Proc. Third Regional Workshop on Sorghum and Millet for Southern Africa, Lusaka, Zambia.* 6-10 October, 1986, SADC/ICRISAT Sorghum and Millet Improvement Program, Bulawayo, Zimbabwe, pp. 174-183.

Sorokin, N.S. 1982. Irrigation and natural enemies. *Zashchita Rastenij* **2**: 30.

Starks, K.J., Schumaker, G. and Eberhart, S.A. 1971. Soil fertility and damage to *Chilo zonellus* on grain sorghum. *J. econ. Ent.* **64**: 740-743.

Suharto 1989. Abundance of *Heliothis armigera* and its natural enemies in cotton intercropped with cereals or legume. M.S. Entomology Thesis, University of the Philippines, College, Laguna, Philippines.

Summy, K.R., Cate, J.R. and Hart, W.G. 1988. Overwintering strategies of boll weevil in southern Texas: reproduction on cultivated cotton. *Southwest Ent.* **13**: 159-164.

Talekar, N.S. 1989. Integrated control of sweet potato weevil. *Int. Sweet Potato Newsl.* **2**(1): 3.

Taley, Y.M. and Thakare, K.R. 1980. Note on the population dynamics in carryover of *Chilo partellus* Swinhoe. *Indian J. agric. Sci.* **58**: 635-637.

Teetes, G.L. 1995. Sorghum management options to reduce insect-induced strees. *In*: K.F. Nwanze and O. Youm (eds). *Panicle Insect*

Pests of Sorghum and Pearl Millet: Proc. of an International Cosultative Workshop. 4-7 October, 1993, ICRISAT Sahelian Center, Niamey, Niger, pp. 241-246.

Teetes, G.L., Seshu Reddy, K.V., Leuschner, K. and House, L.R. 1983. *Sorghum Insect Identification Handbook.* Information Bulletin No. 12, ICRISAT, Patancheru, A.P., India.

Temerak, S.A. 1983. A preliminary survey on the soil inhabiting predacious arthropods associated with pupae of the large sugarcane borer, *Sesamia cretica* Led. in sorghum field. *Bull. Soc. ent. Egypt* **62**: 153.

Teran, F.O. 1979. Sugarcane nutrition modifies infestation by *Diatraea* spp. *Ent. Newsl. Int. Soc. Sugarcane Technol.* **6**: 20-23.

Thimmaiah, G., Desai, K.S.M., Panchabhavi, K.S.M. and Malipatil, M.B. 1972. A note on the effect of the late sowing on the incidence of sorghum earhead-bug (*Calocoris angustatus* Lethiery) in North Mysore region. *Sorghum Newsl.* **15**: 34.

Unnithan, G.C. and Seshu Reddy, K.V. 1989. Incidence, diapause and carryover of the cereal stem borers on Rusinga Island, Kenya. *Trop. Pest Mgt.* **35**(4): 414-419.

Vercambre, B. 1978. *Raghuva* spp., *Masalia* sp., chenilles des chandelles du mil en Zone Sahelienne. *Agron. Trop.* **33**: 62-67.

Wightman, J.A., Dick, K.M., Ranga Rao, G.V., Shanower, T.G. and Gold, C.G. 1990. Pests of groundnut in the semi-arid tropics. *In*: S.R. Singh (ed.). *Insect Pests of Tropical Food Legumes.* John Wiley & Sons, Chichester, England, pp. 243-322.

Wright, R.J. 1984. Evaluation of crop rotation for control of Colorado potato beetles (Coleoptera: Chrysomelidae) in commercial potato fields on Long Island. *J. econ. Ent.* **77**(5): 1254-1259.

Yadava, C.P. 1987. Ecological studies on gram pod borer, *Heliothis armigera* Hubner in chickpea in relation to its biotic and abiotic factors. Ph.D. Thesis, Kanpur University, Kanpur, U.P., India.

Yadava, C.P., Lal, S.S. and Sachan, J.N. 1988. Assessment of incidence and crop loss due to pod borers of pigeonpea (*Cajanus cajan*) of different maturity groups. *Indian J. agric. Sci.* **58**(3): 216-218.

Yaninek, J.S., Gutierrez, A.P. and Herren, H.R. 1990. Dynamics of *Mononychellus tanajoa* (Acari: Tetranychidae) in Africa: Effect on dry matter production and allocation in cassava. *Envir. Ent.* **19**(6): 1767-1772.

Yang, Q.H. 1987. Studies on the integrated control of cotton insects in North China. *Proc. Int. Congress of Plant Protection.* 5-9 October, 1987, Manila, Philippines.

Youm, O. and Gilstrap, F.E. 1993. Population dynamics and parasitism of *Coniesta* (=*Haimachia*) *ignefusalis, Sesamia calamistis and Heliocheilus albipunctella* in millet monoculture. *Insect Sci. Applic.* **14**(4): 419-426.

Youm, O., Harris, K.M. and Nwanze, K.F. 1996. *Coniesta ignefusalis* (*Hampson*), *the Millet Stem Borer. A Handbook of Information.* Information Bulletin No. 46. ICRISAT, Patancheru, A.P., India.

Zepp, D.B. and Keaster, A.J. 1977. Effects of plant densities on the girdling behaviour of the southwestern corn borer. *J. econ. Ent.* **70**(6): 678-680.

3

BIOLOGICAL CONTROL IN AGROECOSYSTEMS THROUGH MANAGEMENT OF ENTOMOPHAGOUS INSECTS

MIGUEL A. ALTIERI
Centre for Biological Control
University of California, Berkeley, CA 94720-3112, USA

CLARA I. NICHOLLS
Department of Entomology
University of California, Davis, CA 95616-8504, USA

3.1. INTRODUCTION

By the close of the 20th century, agriculturalists should have learned an important ecological lesson, i.e., plant communities that are modified to meet the special food and fibre needs of humans become subject to heavy pest damage and generally the more intensely such communities are modified, the more abundant and serious the pests (Altieri, 1994). Large monocultures composed of genetically similar or identical plants that have been selected for increased palatability are highly vulnerable to adapted herbivores (Price, 1981). Moreover, agricultural practices commonly used in the management of monocultures (pesticides, chemical fertilizers, etc.) tend to disrupt the natural enemies of herbivores and often exacerbate pest problems (Papavizas, 1981).

The inherent ecological stability and self-regulating characteristics of natural ecosystems are lost when humans simplify natural communities through the shattering of the fragile thread of community interactions. This breakdown, however, can be repaired by restoring the shattered elements of community homeostasis through the addition or enhancement of functional biodiversity in agroecosystems. One of the most important reasons for restoring and/or maintaining biodiversity in agriculture is that it performs a variety of ecological services. One of such services is the regulation of abundance of undesirable organisms through predation, parasitization and competition (van Driesche and Bellows, 1996). Probably, every insect population in nature is attacked to some degree by one or more natural enemies, thus predators and parasites act as natural control agents resulting in the regulation of herbivore numbers in a particular ecosystem. This regulation has been termed biological control and has been defined by DeBach (1964) as "the action of parasites, predators, or pathogens in maintaining another organism's population density at a lower average than would occur in their absence". As practiced, *biological control* can be self-sustaining and distinguishes itself from all other forms of pest control by acting in a density-dependent manner, i.e., natural enemies increase in intensity and destroy a larger population of the population as the density of that population increases and vice-versa (DeBach and Rosen, 1991).

In a strict ecological sense, *applied biological control* can be considered a strategy to restore functional biodiversity in agroecosystems by adding, through classical and augmentative biocontrol techniques, "missing" entomophagous insects or by enhancing naturally occurring predators and parasitoids through conservation and habitat management. In this chapter, we discuss the ecological roles of predators and parasites in agroecosystems, and the various strategies used in biological control to employ entomophagous insects in order to regulate insect populations in agriculture.

3.2. THE ROLE AND IMPACT OF PREDATORS

Insects that prey upon other insects and spider mites occur in most orders but primarily in the orders Coleoptera, Odonata, Neuroptera, Hymenoptera, Diptera, and Hemiptera. Predatory insects feed on all host stages, i.e., egg, larval (or nymphal), pupal, and adult. From the standpoint of feeding habit, there are two kinds of predators, those with chewing mouth parts, e.g., lady beetles (Coccinellidae) and ground beetles (Carabidae), which simply chew up and bolt down their prey, and those with piercing mouth parts, which suck the juices from their prey, e.g., assassin bugs (Reduviidae), lacewing larvae (Chrysopidae) and hover fly larvae (Syrphidae). The sucking type of feeder often injects a powerful toxin that quickly immobilizes the prey. Many predators are agile, ferocious hunters, actively seeking their prey on the ground or on vegetation, as do beetles, lacewing larvae and mites, or catching it in flight, as do dragonflies and robber flies (Huffaker and Messenger, 1976).

Many species are predaceous in both larval and adult stages, although not necessarily on the same kinds of prey. Others are predaceous only as larvae, whereas the adults may feed on nectar, honeydew, etc., and prey for larvae by depositing eggs among the prey, because their larvae are sometimes incapable of finding it on their own (DeBach and Rossen, 1991; Beckage *et al.,* 1993).

The importance of predators in naturally occurring biological control is crucial. This role has been highlighted by the world-wide eruption of spider mites in many cropping systems in the wake of widespread use of chemical insecticides which mainly resulted from the elimination of predators of spider mites by the insecticides (van den Bosch and Messenger, 1973). Tetranychid mite pest species were abundant in commercial apple orchards due to the elimination of predator populations by pesticide sprays and due to better nutritional vigour of commercial apple trees, factors known to stimulate phytophagous mite buildup (Croft, 1990).

Predator species richness in particular agroecosystems can be impressive. For example, Whitcomb and Bell (1964) reported 602 species of predaceous arthropods in Arkansas cotton fields and about 1,000 species of predatory species in Florida soybean fields (Whitcomb, 1981). Such diversity can exert major regulatory pressures on herbivores and DeBach (1964) considered indigenous predators to act as a sort of balance wheel in the "pest-natural enemy complex", tending to feed upon whatever pest is present in abundance. Even in situations where predators may be incapable of achieving natural control below economic levels, they slow down the rate of increase of potential pests or infestations when more host-specific natural enemies are ineffective. In California's San Joaquin Valley cotton fields, predators are more important in restraining lepidopterous pests (i.e., bollworm, cabbage looper, beet armyworm, etc.) than are parasites (van den Bosch and Messenger, 1973). In Canada, researchers found that in insecticide-free apple orchards, five species of predaceous mirids accounted for 43.5-68.3 per cent mortality of codling moth eggs and in Maine, researchers found a correlation between predation and aphid declines in potatoes (Croft, 1990).

Among the most neglected and less understood predators, spiders can have a strong stabilizing influence on prey. Spiders rely on a complex assemblage of prey. The result is a spider community that is diverse and that maintains a fairly constant numerical representation-one that should exert considerable control on associated prey populations without extinction of these prey. Spiders thus serve as buffers that limit the initial exponential growth of given prey populations (Riechert and Lockley, 1984). In apple orchards in Israel, larval populations of the pest, *Spodoptera littoralis* (Boisduval), did not develop to damaging proportions on trees occupied by spiders, whereas significant damage was observed on trees from which spiders had been removed. Further experimentation revealed that spider activity was responsible for a 98 per cent reduction in larval densities. The reduction resulted both from spider consumption of prey (64% of the larvae present) and from larval abandonment of branches occupied by spiders (34%). In the absence of spiders, larvae abandoned branches with a frequency of only 1.4 per cent. In another study, the presence of micryphantids (sheet line weavers of the family Linyphiidae) in experimental plots resulted in significantly less leaf damage by the tobacco cutworm, *Spodoptera litura* (Fabricius), than was observed in plots from which the spiders had been removed. Here the primary predatory effect was one of causing the larvae to abandon plants occupied by spiders (Riechert and Lockley, 1984).

In agricultural systems, predator population can be increased by releasing them directly into the fields such as in the case of *Chrysoperla carnea* (Stephens), various coccinellids, *Geocoris, Nabis* and phytoseiid mites, or by providing supplementary food (i.e., sucrose solutions, yeast products, pollen, etc.) to retain, arrest or attract specific predator species to the fields (Hagen *et al.*, 1976). Another approach is to enhance resources and environmental opportunities for predators through complex cropping designs in time and space (Altieri, 1994).

3.3. PARASITOIDS: BIOLOGICAL CHARACTERISTICS, ROLE AND IMPACT

Most insects parasitic upon other insects are *protelean parasites,* i.e., they are parasitic only in their immature (larval) stages and lead free lives as adults. They usually consume all or most of the host's body and then pupate, either within or external to the host. The adult parasite emerges from the pupa and starts the next generation anew by actively searching for hosts in which to oviposit. Most adult parasites require food such as honeydew, nectar or pollen and many feed on their host's body fluids, as mentioned earlier. Others require free water as adults (DeBach and Rossen, 1991).

Parasites may be categorized as *ectoparasites,* feeding externally upon the host, and as *endoparasites,* developing internally within the host. Parasites may have one generation (*univoltine*) to one generation of the host or two or more generations (*multivoltine*) to one of the host. Life cycles are commonly short, ranging from 10 days to 4 weeks or so in summer but correspondingly longer in cold weather. The main groups of parasites utilized in biological control of insect pests are the Hymenoptera (mostly wasps of the superfamilies Chalcidoidea, Ichneumonoidea and Proctotrupoidea) and Diptera (flies, especially of the family Tachinidae).

Research on the diversity of parasitic Hymenoptera in agroecosystems has concentrated mostly on the study of parasitoid complexes attacking particular native as well as exotic pest species. Some pest species support a large number of parasitoid species, such as the Hessian fly, *Mayetiola destructor* (Say); the wheat-stem sawfly, *Cephus pygmeus* (Linnaeus); the coconut beetle, *Promecotheca coeruleipennis* Blanchard; the bean gall sawfly, *Pontania proxima;* and the coffee leaf miner, *Perileucoptera coffeella* (Guerin-Meneville). Different crops support particular herbivore species, which, in turn, are attacked by one or several parasitoid species (Table 3.1), although such associations may change according to geographical location, management intensity and crop arrangements (Waage and Greathead, 1986).

Table 3.1. Species richness of parasitoid complexes associated with different insect pests in a range of annual cropping systems

Crop system	Pest species	No. of parasitoid species	Location
1	2	3	4
Alfalfa	*Colias eurytheme* Boisduval	2	California
	Spodoptera exigua (Hubner)	11	California
	S. praefica	13	California
	Helicoverpa zea (Boddie)	13	California
Cassava	*Erinnyis ello* (Linnaeus)	4	Brazil
	Jatrophobia brasiliensis	4	Colombia
	Saissetia sp.	2	West Indies, Peru, Cuba
Cotton	*S. exigua*	11	California
	Trichoplusia ni (Hubner)	11	California
	H. zea	14	California
	Bucculatrix thurberiella Busck	3	California
	Estigmene acraea (Drury)	3	California
	S. praefica	13	California
Potato	*Myzus persicae* (Sulzer)	7	Maine, USA
	Aulacorthum solani (Kaltenbach)	5	Maine, USA
	Aphis nasturtii Kaltenbach	5	Maine, USA
Rice	*Nephotettix* spp.	3	Philippines
	Chilo suppressalis (Walker)	5	Philippines
Sorghum	*Schizaphis graminum* (Rondani)	3	USA
Soybeans	*Plathypena scabra* (Fabricius)	14	Missouri, USA
	Chrysodeixis includens (Walker)	12	Louisiana, USA
Tabacco	*Heliothis virescens* (Fabricius)	2	North California, USA

Source: Altieri *et al.* (1993)

The relative complexity of Hymenoptera parasitoid communities associated with different cropping systems is determined by biological, environmental and management factors. In large-scale monocultures, diversity is suppressed by pesticides, vegetational simplification and other environmental disturbances. In less disturbed agroecosystems, in addition to the absence of pesticides, parasitoid diversity seems related to crop diversity, ground cover, weeds and native vegetation adjacent to crops. In fact, the few studies conducted on this topic indicate that the vegetational settings associated with particular crops influence the kind, abundance and time of arrival of parasitoids (Waage, 1992; Waage and Greathead, 1986).

In many cases, only one or two species of such complexes prove vital in the natural biological control of key insect pests. For example, in California's alfalfa fields, the braconid wasp, *Apanteles medicaginis* plays a key role in regulating the numbers of the alfalfa caterpillar, *Colias eurytheme* Boisduval. Apparently, this pristine butterfly-wasp system moved from native clovers into the new and artificial irrigated alfalfa fields. Similarly in North Carolina's tobacco fields, *Campoletis flavicincta* (Ashmead) exerts a high parasitization rate on the budworm, *Heliothis virescens* (Fabricius), in early summer prior to flowering when plants are most susceptible to budworm injury. After flowering and on post-harvest sucker tobacco, parasitization by *C. flavicincta* declines and the action of *Cardiochiles nigriceps* Viereck becomes an important budworm mortality factor (Huffaker and Messenger, 1976). In other cases, it is a combination of several parasitoid species that exerts regulation on a specific insect pest (Ehler and Miller, 1978).

3.4. STRATEGIES OF BIOLOGICAL CONTROL

3.4.1. Classical Biological Control

Classical biological control is the regulation of a pest population by exotic natural enemies (parasites, predators, pathogens, etc.) that are imported for this purpose. Usually, the target species (pest) is an exotic that has reached higher population density in the new environment because of more favourable conditions than in its area of indigeneity (Rosen *et al.*, 1994). It involves the introduction of preferable host-specific, self-reproducing, density-dependent, hostseeking exotic natural enemies adapted to an exotic introduced pest, usually resulting in permanent control (Caltagirone, 1981).

Biological control agents, because they are often carefully selected to be those best adapted to their hosts, usually spontaneously spread throughout much of the host's range, affecting widespread control at relatively little cost. Caltagirone (1981) described 12 cases of successful classical biological control projects in which target pest species were reduced to a non-pest status by introduced natural enemies (Table 3.2). Other case studies which describe the various steps (discovery, introduction and evaluation) involved in the development of a classical biocontrol programme are described in van Driesche and Bellows (1996).

Since the successful control of the cottony-cushion scale, *Icerya purchasi* Maskell, in California with the imported Vedalia beetle, *Rodolia cardinalis* (Mulsant), from Australia in 1888, hundreds of attempts at classical biological control have taken place throughout the world. The biological control campaign conducted against the black scale of citrus, *Sissetia oleae* (Olivier) in California, included about 42 different natural enemy introductions from Africa, Mexico, Pakistan, China, Brazil,

Table 3.2. Successful examples of classical biological control

Exotic pest	Introduced natural enemy	Cropping system
Aleurocanthus woglumi Ashby (Citrus blackfly)	*Eretmocerus serius* (Parasite)	Citrus
Antonina graminis (Maskell) (Rhodegrass mealybug)	*Anagyrus antoninae* (Parasite)	Grasses
Aonidiella aurantii (Maskell) (California red scale)	*Aphytis* spp. (Parasites)	Citrus
Chromaphis juglandicola (Kaltenbach) (Walnut aphid)	*Trioxys pallidus* (Haliday) (Parasite)	Walnut
Nezara viridula (Linnaeus) (Green stink bug)	*Trissolcus basalis* (Wollaston) (Parasite)	Vegetable - field crops
Operophtera brumata (Linneaus) (Winter moth)	*Cyzenis albicans* (Fallen) and *Agrypon flaveolatum* (Gravenhorst) (Parasites)	Oaks, apples
Oryctes rhinoceros (Linneaus) (Rhinoceros beetle)	*Rabdionvirus oryctes* (Baculovirus)	Coconut palm, oil palm
Parlatoria oleae (Colvee) (Olive scale)	*Aphytis maculicornis* and *Coccophagoides utilis* (Parasites)	Olive
Quadraspidiotus perniciosus (Comstock) (San Jose scale)	*Encarsia perniciosi* (Tower) (Parasite)	Stone and pome fruits
Tetranychus urticae Koch (Two spotted spider mite)	*Phytoseiulus persimilis* Athias-Henriot (Predator)	Greenhouse
Therioaphis trifolii (Monell) (Spotted alfalfa aphid)	*Praon exsoletum, Trioaxys complanatus* Quilis Perez and *Aphelinus asychis* Walker (Parasites)	Alfalfa
Trialeurodes vaporariorum (Westwood) (Greenhouse whitefly)	*Encarsia formosa* Gahan (Parasite)	Greenhouse

Source: Caltagirone (1981)

Argentina and Taiwan (Luck, 1981). Of all the importation efforts into the USA in the last 80 years, it is estimated that about 128 introduced natural enemies resulted in some degree of biological control (Luck, 1981). The economic savings of such introductions have been substantial. The combined savings attributed to the California agricultural industry from seven major classical biological control programmes conducted between 1928 and 1979 were estimated at about US$ 320 million, not accounting for inflation and/or discount (van den Bosch *et al.*, 1982).

All classical biological control projects, by definition, involve procurement of exotic natural enemies. In the majority of the cases, exploration is conducted in the presumed area of origin of the target species. After foreign exploration is completed, entomophagous insects must be introduced into the host country, where they are subjected to quarantine. Following quarantine, most natural enemies are mass cultured to allow release of sufficient numbers of the species at particular colonization sites, providing material for colonization in a variety of environments in a region, and allowing for repeated colonizations over time if required (van den Bosch and Messenger, 1973).

The historical record shows that 34 per cent of attempts to colonize natural enemies have succeeded (Hall and Ehler, 1979). Such relatively low rates of field establishment can be due to factors such as inappropriate selection of the natural enemy, climatic dissimilarity between originating and release locations, negative features of the crop plants and/or the receiving agroecosystem. Once establishment of the natural enemy is documented, the regulating effect of specific enemies on the target host population is evaluated, including an assessment of economic costs and social benefits.

3.4.2. Augmentative Biological Control

This strategy involves the mass propagation and periodic release of exotic or native natural enemies that may multiply during the growing season but are not expected to become a permanent part of the ecosystem (Batra, 1982). Augmentative release may by made with either short-or long-term expectations depending upon the target pest, the species of natural enemy and the crop involved. The mass culture and dissemination of natural enemies was a popular method in the former Soviet Union and in China where the socioeconomic structure, including collectivization of agriculture, integration of research and production, and a large, well organized labour force permitted the successful mass culture and widespread release of augmentative control agents. Recent political and socio-economic changes which embrace capitalist modes of production in those regions changed such scenarios. Cuba is the only country experiencing a massive revival in augmentative biological control since the collapse of the Soviet bloc in 1989. The island has suffered an 80 per cent decrease of fertilizer and pesticide imports and in order to ensure food security Cuban researchers and farmers have launched a massive biological control project. By the end of 1994, some 222 centres for production of entomopathogens and entomophagous insects (CREEs) have been created (Rosset and Benjamin, 1993). In such centres they produce massive amounts of *Trichogramma* spp., *Beauveria bassiana* (78 metric tonnes), *Bacillus thuringiensis* (1,312 tonnes), *Verticillium lecanii* (196 tonnes) and *Metarhizium anisopliae* (142 tonnes) for the control of various pests in several major crops of the island (Table 3.3).

Table 3.3. Biological organisms for the control of insect pests in Cuba

Organism	Crop	Target pest
Bacillus thuringiensis Berliner	Collards, tomatoes, watercress, pepper	*Pieris* sp., *Heliothis* and *Spodoptera*
	Cassava	*Erynnis* sp.
	Yucca	*Spodoptera*
	Sweet potato	*Spodoptera*
	Potato, corn, tobacco	*Heliothis*
Beauveria bassiana (Balsamo) Vuillemin	Banana	*Cosmopolites sordidus* (Germar)
	Sweet potato, rice, citrus	Curculionidae (Weevils)
Metarhizium anisopliae (Metschnikoff) Sorokin	Grasses	Cercopidae (Spittle bug)
	Rice, citrus	Curculionidae
Paecilomyces lilacinus	Guava	*Meloidogyne*
	Coffee	*Meloidogyne*
	Banana	*Meloidogyne*, *Radopholus similis*
Verticillium lecanii (Zimmerman) Viegas	Tomatoes	Whiteflies
	Pepper	Whiteflies
	Cucumber, squash	Whiteflies
	Potato	Whiteflies
	Beans	Whiteflies
Pheidole megacephala (Fabricius)	Sweet potato	Weevil
Trichogramma sp.	Grasses	*Mocis* sp.
	Cassava	*Erynnis* sp.
	Sugarcane	Sugarcane borer

Source: Rosset and Benjamin (1993)

In the USA, success of augmentative pest control depends on the total number of individuals released (Ables and Ridgeway, 1981). Among the common agents commercially available for releases are *Trichogramma* wasps; the lacewing, *Chrysoperla carnea* (Stephens); and several insect pathogens, i.e., *Bacillus* spp., *Beauveria bassiana* and various nuclear polyhedrosis viruses. There are several potential natural enemy candidates for augmentation against *Heliothis* spp. in numerous crops. Examples include *C. carnea, Trichogramma* spp., *Glabromicroplitis croceipes* (Cresson) and *Campoletis sonorensis* (Cameron). Also, aphids in numerous crops have a range of parasites (i.e., *Praon* spp., *Lysiphlebus* spp., *Aphidius* spp., *Diaeretiella* spp. and others) subject to mass rearing and release (Huffaker and Messenger, 1976). Selected examples of entomophages with potential for augmentation in the USA are given in Table 3.4 (Ables and Ridgeway, 1981).

Table 3.4. Potential natural enemy candidates for augmentation through rearing in the United States

Augmentation natural enemy candidate	Target pest(s)	Cropping system(s)
1	2	3
Acari		
Phytoseiulus spp.	*Tetrancychus urticae* Koch	Strawberries
	Tetranchus sp.	Glasshouse crops
Typhlodromus spp.	*Tetranychus mcdanieli* McGregor	Apples
	Steneotarsonemus pallidus	Strawberries
Coleoptera		
Coccinella spp.	Aphids	Vegetables, fruits and nuts
Cryptolaemus montrouzieri Mulsant	Mealybugs	Citrus
Stethorus picipes	*Oligonychus punicae* (Hirst)	Avocado, other fruits and vegetables
Diptera		
Eucelatoria spp.	*Heliothis* spp.	Numerous crops
Lixophaga diatraeae (Townsend)	*Diatraea* spp.	Sugarcane
Voria ruralis (Fallen)	*Trichoplusia ni* (Hubner)	Cole crops, many other crops
Hemiptera		
Jalysus spinosus (Say)	*Heliothis virescens* (Fabricius), *Manduca* spp.	Tobacco
Hymenoptera		
Apanteles melanoscelus (Ratzeburg)	*Lymantria dispar* (Linnaeus)	Forest and shade trees
Apanteles rebecula Marshall	*Pieris rapae* (Linnaeus)	Cabbage
Aphidius smithi Sharma & Subba Rao	*Acyrthosiphon pisum* (Harris)	Peas, other vegetables
Aphytis melinus DeBach	California red scale	Citrus
Bracon kirkpatricki (Wilkinson)	*Pectinophora gossypiella* (Saunders)	Cotton
Bracon mellitor Say	*Anthonomus grandis* Boheman	Cotton
Campoletis sonorensis (Cameron)	*Heliothis* spp.	Numerous crops
Chelonus blackburni	*P. gossypiella*	Cotton

Table 3.4 contd...

1	2	3
Cameron		
Diaeretiella spp.	Aphids	Cole crops
Encarsia formosa Gahan	Whiteflies	Greenhouse crops
Glabromicropolitis croceipes (Cresson)	*Heliothis* spp.	Numerous crops
Lysiphlebus spp.	Aphids	Numerous crops
Macrocentrus ancylivorus Rohwer	*Cydia molesta* (Busck)	Peaches
Pediobius foveolatus (J.C. Crawford)	*Epilachna* spp.	Soybean, legume, vegetables, squash
Praon spp.	Aphids	Numerous crops
Trichogramma spp.	*Heliothis*	Numerous food and fiber crops
	Plusiine noctuids	Numerous food and fiber crops
	Pieris spp.	Cole crops
	Manduca spp.	Tobacco and tomatoes
	Ostrinia nubilalis (Hubner)	Corn
	Cydia pomonella (Linnaeus)	Apples, other fruits
Neuroptera		
Chrysoperla carnea (Stephens)	*Heliothis* spp.	Numerous food & fiber crops
	Pseudococcus spp.	Pears and other fruits
	Trichoplusia ni (Hubner)	Cabbage
	Aphids	Potatoes

Source: Ables and Ridgeway (1991)

In cotton, data show that 50,000 to 100,000 *Trichogramma* spp. must be released per acre at a 2-5 day interval during peak *Heliothis* spp. egg-laying period to significantly increase parasitization and obtain control. Release of up to 28,000 *Lysiphlebus testaceipes* (Cresson) per acre did not reduce greenbugs below economic threshold levels under normal monoculture conditions in the Texas high plains. However, recent developments in the use of semiochemicals (i.e., kairomones) point out to great possibilities for increasing field behaviour and efficiency of several parasitoids under monoculture situations (Nordlund *et al.*, 1981). The greatest utility of kairomones so far appears to be for aggregating and/or retaining released parasites in target locations (Hoy, 1994; Hoy and Herzog, 1985).

Augmentative control can be cost-effective. Several corporations are commercially rearing and marketing a number of genera of parasitic wasps, the aphid predator, *C. carnea*, and the insect pathogens, *Bacillus thuringiensis, B. popilliae, Beauveria bassiana,* and several nuclear polyhedrosis viruses. Treatments in which insects are used cost from $24.70 to $29.60 per hectare on orchards and $133 to $2,398 per hectare in greenhouses (Batra, 1982). Today, costs remain competitive.

3.4.3. Conservation and Habitat Management

This approach emphasizes management of agroecosystem in order to provide a general environment conducive to the conservation and enhancement of a complex natural enemy biota. By improving the availability of food, shelter and other environmental resources within and outside the crop field, the possibilities of increasing the populations and effective predatory and parasite behaviour of beneficial arthropods through habitat management are great (Huffaker and Messenger, 1976). Small changes in agricultural practices can cause substantial increases in natural enemy populations during critical periods of the growing season. Some practices may simply involve the withdrawal of chemical pesticides or avoiding disturbing practices such as ploughing and cultivating. Total removal of pesticides can restore parasitoid diversity and lead to renewed biological control of specific pests. Within two years, virtually all banana insect pests in Golfito, Costa Rica dropped to below economic threshold levels, due to enhanced parasitization and predation, after stopping insecticide (dieldrin and carbaryl) sprays. Similarly, in California's walnut orchards, natural biological control of the frosted scale and the calico scale was soon achieved by encyrtid parasitoids after removal of DDT sprays (Croft, 1990).

At times, it is necessary to provide supplementary resources. For example, erection of artificial nesting structures for *Polistes annularis* (Linnaeus) has increased predation against *Alabama argillacea* (Hubner) in cotton and *Manduca sexta* (Johannsen) in tobacco. The addition of subsidiary food (e.g., mixtures of hydrolyzate, sugar and water) multiplied six-fold the oviposition of *C. carnea* and increased the abundance of Syrphidae, Coccinellidae and Malachiidae in alfalfa and cotton plots. To improve survival and reproduction of beneficial insects within an agroecosystem, it is often desirable to have subeconomic, fluctuating populations of alternate prey permanently present in the crop (Huffaker and Messenger, 1976). For example, the relative abundance of aphids on cabbage in South Africa can be a determining factor upon the effectiveness of general predators against larvae of the diamond-back moth, *Plutella xylostella* (Linnaeus). Addition of host populations proved effective in controlling *Pieris rapae* (Linnaeus) in cabbage. The continuous release of fertile *Pieris* butterflies increased the pest population nearly tenfold above normal spring populations, enabling the parasites, *Trichogramma evanescens* Westwood and *Apanteles rebecula* Marshall to increase early and maintain themselves at an effective level throughout the season (van den Bosch and Messenger, 1973).

It is widely believed that agroecosystem diversity is associated with long-term stability of populations, presumably because a variety of parasites, predators, and competitors is always available to suppress population growth of potential pest species. Dispersal of food plants among other nonhost plants may make migration, host, and mate location, and consequently exponential growth of phytophages or pathogens, more difficult (Andow, 1991). Plant diversification of agroecosystems can result in increased environmental opportunities for natural enemies and, consequently, improved biological pest control. The various vegetational designs available in the form of polycultures, weed diversified crop systems, cover crops and living mulches and their effects on pest populations and associated natural enemies have been extensively reviewed (Altieri, 1994 and references therein). Factors involved in pest regulation in diversified agroecosystems include: increased parasitoid/predator populations, available alternative prey/hosts for natural enemies, decreased colonization and reproduction of pests, feeding inhibition or chemical repellency from non-host plants, prevention of movement and emigration and optimum synchrony between pests and natural enemies.

Research has shown that by adding plant diversity to existing annual monocultures, it is possible to exert changes in habitat diversity which in turn favour natural enemy abundance and effectiveness. This information can be used to design mixed cropping systems that enhance predator and parasitoid diversity and abundance, thus resulting in lower pest loads than in monocultures. In general, it is accepted that in polycultural agroecosystems there is an increased abundance of arthropod predators and parasitoids due to enhanced availability of alternate prey, nectar sources and suitable microhabitats (Altieri, 1994). Table 3.5 provides various examples of observed pest reduction in polycultures.

Increasing within-field plant diversity can enhance biological control. Considerable work in the former USSR was devoted to the use of nectar-bearing plants within orchards as a source of adult food for entomophagous insects to increase their effectiveness. Field experiments by Russians in the North Caucasus showed that the growing of *Phacelia* spp. in orchards greatly increased the parasitization of *Quadraspidiotus perniciosus* (Comstock) by its parasite *Aphytis proclia* (Walker). Three successive plantings of *Phacelia* flowers in orchards increased parasitization in about 70 per cent. These same plants have been

	B. brassicae	control
Cabbage intercropped with white and red clover	*Delia radicum* (Linnaeus), cabbage aphids and imported cabbage butterfly, *Pieris rapae* (Linnaeus)	Interference with colonization and increase of ground beetles
Cassava intercropped with cowpeas	Whiteflies, *Aleurotrachelus socialis* and *Trialeurocles variabilis*	Changes in plant vigour and increased abundance of natural enemies
Corn intercropped with fava beans and squash	Aphids, *Tetranychus urticae* Koch and *Macrodactylus* sp.	Enhanced abundance of predators
Corn intercropped with sweet potatoes	Leaf beetles, *Diabrotica* spp. and leafhoppers, *Agallia lingula*	Increase in parasitic wasps
Corn intercropped with forage cowpea	Boll weevil, *Anthomonus grandis* Boheman	Population increase of parasitic wasps, *Eurytoma* sp.
Intercropping cotton with sorghum or maize	Corn earworm, *Helicoverpa zea* (Boddie)	Increased abundance of predators
Strip cropping of cotton and alfalfa	Plant bugs, *Lygus hesperus* Knight and *L. elisus* Van Duzee	Prevention of emigration and synchrony in the relationship between pests and natural enemies
Peaches intercropped with strawberries	Strawberry leafroller, *Ancylis comptana* (Froelich) and Oriental fruit moth, *Cydia molesta* (Busck)	Population increase of parasites, *Macrocentrus ancylivora* Rohwer, *Bracon gelechiae* Ashmead and *Lixophaga variabilis*
Peanut intercropped with maize	Corn borer, *Ostrinia furnacalis* (Guenee)	Abundance of spider, *Lycosa* sp.
Seasame intercropped with cotton	*Heliothis* spp.	Increase of beneficial insects and trap cropping

Source : Altieri (1994)

shown to increase the abundance of the wasp, *Aphelinus mali* (Haldeman) for the control of apple aphids and improve the activity of *Trichogramma* spp. (Hagen *et al.*, 1976).

Manipulation of wild vegetation adjacent to crop fields can also be used to promote biological control, since the survival and activity of natural enemies often depends upon the resources provided by the vegetation around crop fields (Fig. 3.1). Hedgerows and other landscape features have received significant attention in Europe regarding their effects on arthropod distribution and abundance in adjacent crop fields (Fry, 1995). There is wide acceptance of the importance of wild vegetation field margins as reservoirs of natural enemies of crop pests (van Emden, 1965). Habitats may be important as overwintering sites for predators or they may provide increased resources such as pollen and nectar for parasitoids and predators from flowering plants. Many studies have documented the movement of beneficial arthropods from margins into crops and higher biological control is usually observed in crop rows close to wild vegetation edges than in rows in the centre of the fields (Altieri, 1994). Studies of tachinid and ichneumonid parasites attacking *Mamestra brassicae* (Linnaeus) and *P. xylostella* were conducted near Moscow and the data show that parasite efficiency was substantially higher in cabbage fields when they were grown near flowering umbelliferous plants (Huffaker and Messenger, 1976).

In California, the egg parasite, *Anagrus epos* Girault was effective in controlling the grape leafhopper, *Erythroneura elegantula* Osborn in vineyards adjacent to wild blackberries which harbour a non-economic leafhopper, *Dikrella cruentata,* whose eggs serve as the only overwintering resource for *Anagrus*. Recent studies have shown that prune trees planted next to vineyards also allow early season buildup of *A. epos* and early benefits of parasitism are promoted in vineyards with prune trees planted upwind from the vineyard (Flint and Roberts, 1988). Researchers now recommend that as many prune trees as possible should always be planted upwind from the vineyard. Also in California, parasitization of the alfalfa caterpillar, *Colias eurytheme* Boisduval, by *Apanteles medicaginis* was far greater in California's San Joaquin Valley where weeds were in bloom along irrigation canals in contrast to areas where the weeds were destroyed (DeBach, 1964).

In Norway's apple orchards, numbers of the key pest, *Argyresthia conjugella* Zeller are largely dependent on the amount of available food, i.e., the number of berries of the wild shrub, *Sorbus aucuparia* that develop each year. Since only one larva develops in a single berry, the number of *Argyresthia* can never be higher than the total number of berries. Thus, in years when *Sorbus* has no berries in a certain area, no *Argyresthia* larvae are produced, and consequently there will be no parasites (the braconid, *Microgaster politus*) in this area. Entomologists

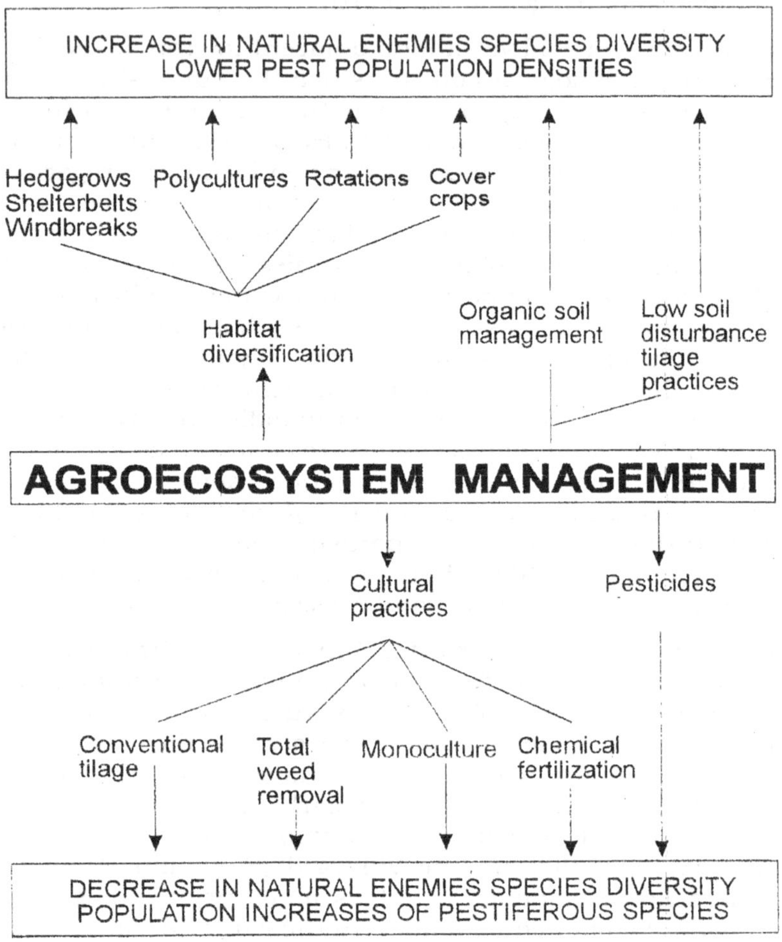

Fig. 3.1. Effects of agroecosystem management and associated cultural practices on the biodiversity of natural enemies and the abundance of insect pests

have suggested planting of *Sorbus* which produces an abundant and regular crop every year. *Argyresthia* always finds enough food to maintain a reasonably high population level. Under such conditions *Microgaster* and other natural enemies will also operate and reproduce sufficiently every year to regulate their host below the level where *Argyresthia* is forced to emigrate. Hence, apple avoids infestation (Edland, 1995).

3.5. CONCLUSIONS

Biological control through importation, augmentation and/or conservation of natural enemies can provide long-term regulation of pest species provided that in target agroecosystems proper cultural management (i.e., avoidance of disruptive agricultural practices and diversification of cropping systems) is adapted to foster an environment conducive to further the abundance and efficiency of predators and parasites. Under such conditions biological control can potentially become a self-perpetuating strategy, providing control at low cost and with none or minimal environmental impacts (Flint and Roberts, 1988).

Large scale commercial agriculture involving crops that have a major complex of pests is initially likely to require the integration of chemical and cultural pest control methods along with the use of natural enemies. In such cases, the conversion to a production system totally dependent on biological control will require a stepwise process of agroecological conversion including the efficient use of pesticides (IPM) and input substitution (the replacement of insecticides for botanical or microbial insecticides). The process ends with the re-design of a diversified farming system that provides the environmental conditions for natural enemies, thus allowing the agroecosystem to sponsor its own natural protection against pests (Altieri, 1994).

Diversified cropping systems, such as those based on intercropping and agroforestry or cover cropping of orchards, have been target of much research recently. This is largely based on the newly emerging evidence that these systems are more sustainable and more resource-conserving (Vandermeer, 1995). Much of these attributes are connected to the higher levels of functional biodiversity (including natural enemies) associated with such complex farming systems. The key is to identify the type of biodiversity that is desirable to maintain and/or enhance in order to carry out ecological services, and then to determine the best practices that will encourage the desired biodiversity components. As shown in Fig. 3.1, there are many agricultural practices and designs that have the potential to enhance functional biodiversity, and others that negatively affect it. The idea is to apply the best management practices in order to enhance and/or regenerate the kind of biodiversity that can subsidize the sustainability of agroecosystems through improved biological pest control.

REFERENCES

Ables, J.R. and Ridgeway, R.L. 1981. Augmentation of entomophagous arthropods to control insect pests and mites. *In*: G. Papavizas (ed.). *Biological Control in Crop Production.* Allanheld, Osmun Pub., London, UK, pp. 273-305.

Altieri, M.A. 1994. *Biodiversity and Pest Management in Agroecosystems.* Haworth Press, New York, USA. 185 pp.

Altieri, M.A., Cure, J.A. and Garcia, M.A. 1993. The role and enhancement of parasitic hymenoptera biodiversity in agro–ecosystems. *In:* J.La Salle and I.D. Gauld (eds). *Hymenoptera and Biodiversity.* CAB International, Wallingford, UK, pp. 257-275.

Andow, D.A. 1991. Vegetational diversity and arthropod population response. *A. Rev. Ent.* **36**: 561-586.

Batra, S.W.T. 1982. Biological control in agroecosystems. *Science* **215**: 134-139.

Beckage, N.E., Thompson, S.N. and Frederici, B.A. (eds). 1993. *Parasites and Pathogens of Insects, Vol. 1: Parasites.* Academic Press, London, UK.

Caltagirone, C.E. 1981. Landmark examples in classical biological control. *A. Rev. Ent.* **26**: 213-232.

Croft, B.A. 1990. *Arthropod Biological Control Agents and Pesticides.* John Wiley & Sons, New York, USA.

DeBach, P. 1964. *Biological Control of Insect Pests and Weeds.* Reinhold, New York, USA.

DeBach, P. and Rosen, D. 1991. *Biological Control by Natural Enemies.* Cambridge University Press, Cambridge, UK.

Edland, T. 1995. Integrated pest management in fruit orchards. *In:* H.M.T. Hokkanen and J.M. Lynch (eds). *Biological Control: Benefits and Risks.* Cambridge University Press, Cambridge, UK, pp. 97-105.

Ehler, L.E. and Miller, J.C. 1978. Biological control in temporary agroecosystems. *Entomophaga* **23**: 213-232.

Flint, M.L. and Roberts, P.A. 1988. Using crop diversity to manage pest problems: some California examples. *Am. J. Alter. Agric.* **3**: 164-167.

Fry, G. 1995. Landscape ecology of insect movement in arable ecosystem. *In:* D.M. Glen, M.P. Greaves and H.M. Anderson (eds). *Ecology and Integrated Farming Systems.* John Wiley & Sons, Bristol, UK, pp. 97-105.

Hagen, K.S., Bombosch, S. and McMurtry, J.A. 1976. The biology and impact of predators. *In:* C.B. Huffaker and P.S. Messenger (eds). *Theory and Practice of Biological Control.* Academic Press, New York, USA, pp. 93-142.

Hall, R.W. and Ehler, L.E. 1979. Rate of establishment of natural enemies in classical biological control. *Bull. ent. Soc. Am.* **25**: 280-282.

Hoy, M.A. 1994. Parasitoids and predators in management of arthropod pests. *In:* R.K. Metcalf and W.H. Luckmann (eds). *Introduction to Insect Pest Management.* John Wiley and Sons, New York, pp. 129-198.

Hoy, M.A. and Herzog, D.C. 1985. *Biological Control in Agricultural IPM Systems.* Academic Press, Orlando.

Huffaker, C.B. and Messenger, P.S. (eds). 1976. *Theory and Practice of Biological Control.* Academic Press, New York, USA.

Luck, R.F. 1981. Parasitic insects introduced as biological control agents for arthropod pests. *In:* D. Pimentel (ed.). *Handbook of Pest Management in Agriculture. Vol. II.* CRC Press, Boca Raton, Florida, USA, pp. 125-284.

Nordlund, D.A., Jones, R.L. and Lewis W.J. 1981. *Semiochemicals: Their Role in Pest Control.* John Wiley & Sons, New York, USA.

Papavizas, G.C. 1981. *Biological Control in Crop Production.* Beltsville Symposia in Agricultural Research, Allanheld, Osmun Pub., London, UK.

Price, P.W. 1981. Semiochemicals in evolutionary time. *In*: D.A. Nordlund, R.L. Jones and W.J. Lewis (eds). *Semiochemicals: Their Role in Pest Control.* John Wiley & Sons, New York, USA, pp. 251-279.

Riechert, S.E. and Lockley, T. 1984. Spiders as biological control agents. *A. Rev. Ent.* **29**: 294-320.

Rosen, D., Bennett, F.D. and Capinera, J.L. 1994. *Pest Management in the Tropics: Biological Control-A Florida Perspective.* Intercept, Andover, USA.

Rosset, P. and Benjamin, M. 1993. *Two Steps Backward, One Step Forward: Cuba's Nationwide Experiment with Organic Agriculture.* Global Exchange, San Francisco, USA.

van den Bosch, R. and Messenger, P.S. 1973. *Biological Control.* Intext Educational Publishers, New York, USA.

van den Bosch. R., Messenger, P.S. and Gutierrez, A.P. 1982. *An Introduction to Biological Control.* Plenum Press, New York, USA 247 pp.

van Emden, H.F. 1965. The role of uncultivated land in the biology of crop pests and beneficial insects. *Scient. Hort.* **17**: 121-136.

Vandermeer, J. 1995. The ecological basis of alternative agriculture. *A. Rev. Ecol. Syst.* **26**: 201-224.

van Driesche, R.G. and Bellows, T.S. 1996. *Biological Control.* Chapman & Hall, New York, USA.

Waage, J. 1992. Biological control in the year 2000. *In:* A.A.S.A. Kadir and H.S. Barlow (eds). *Pest Management and the Environment in 2000.* CAB International, Wallingford, UK, pp. 329-340.

Waage, J. and Greathead, D. 1986. *Insect Parasitoids.* Academic Press, London, UK.

Whitcomb, W.H. 1981. The use of predators in insect control. *In:* D. Pimentel (ed.). *CRC Handbook of Pest Management in Agriculture. Vol. II.* CRC Press, Boca Raton, Florida, USA, pp. 105-124.

Whitcomb, W.H. and Bell, K.O. 1964. Predaceous insects, spiders and mites in Arkansas cotton fields. *Arkansas Agric. Exp. Sta. Bull.* **690**: 1-84.

4

MICROBIAL CONTROL OF INSECTS : USE OF BACTERIAL INSECTICIDES

HYUN-KU LEE, HOSUN CHEONG AND SARJEET S. GILL[1]
Department of Entomology
University of California, Riverside, CA 92521-127, USA

4.1. INTRODUCTION

The extensive use of synthetic insecticides has caused significant problems including detrimental effects on human health and the environment, and the development of insect resistance to these insecticides. Consequently, it is advisable to use alternative approaches for insect pest management. One strategy that has significant promise is the use of microbial agents for insect control, and this method is relatively safe for the environment and can be integrated into pest management approaches.

Various microbial agents can be used for insect pest management, however, the most extensively used to date are bacterial preparations, and are the focus of this review. Insect viruses also have the potential for use, and new strategies using recombinant viruses are likely to see commercial applications in the near future. Excellent reviews, dealing with insect viruses, which the reader should refer are available. Additionally, fungi and nematodes can be used for insect control. However, their use is relatively restricted.

Of the many bacteria that affect insects, including *Pseudomonas, Clostridium* and *Serratia* to name a few, only preparations containing *Bacillus* spp. are commercially marketed at this jucture. Two species in particular, *B. thuringiensis* Berliner and *B. sphaericus* Neide, are of greatest interest. Other bacilli including *B. popilliae* and *B. lentimorbus* are of lesser significance.

Of these two bacilli, only *B. thuringiensis* is used for the control of numerous agricultural pests, particularly of lepidopteran and coleopteran pests. Both *B. thuringiensis* and *B. sphaericus* are effective against dipterans, particularly insect vectors of human diseases such as encephalitis, filariasis, yellow fever, and malaria (Table 4.1). Today, these diseases still threaten a major portion of the population in developing countries resulting in millions of human infections and deaths each year (Gibbons, 1992).

---

[1] Author for correspondence

In this review, brief discussions of both *B. thuringiensis* and *B. sphaericus* is provided. The focus is on the classification of the strains, mechanism of action of the toxins and the development of insect resistance of these toxins that may affect their future use.

Table 4.1. Human diseases vectored by different mosquito species

Disease	Causative agent	Mosquito species
Malaria	*Plasmodium* spp.	*Anopheles* spp.
Filariasis	*Wuchereria bancrofti* *Brugia malayi*	*Culex pipiens* Linnaeus, some species of *Aedes, Mansonia* and *Anopheles*
Yellow fever, dengu and encephalitis	Arboviruses	Some species of *Aedes* and *Culex*

4.2. *BACILLUS THURINGIENSIS* BERLINER

4.2.1. Classification

B. thuringiensis was first isolated in 1902 from diseased larvae of the silkworm, *Bombyx mori* (Linnaeus), in Japan by Ishiwata. However, the name *Bacillus thuringiensis* was first given in 1915 by Berliner who isolated another *B. thuringiensis* from the Mediterranean flour moth, *Ephestia kuehniella* Zeller in Thuringia, Germany (Luthy and Ebersold, 1981).

The parasporal inclusions in sporulating cultures of *B. thuringiensis* was first described by Hannay in 1953 who later demonstrated that these inclusions were proteinaceous and might have insecticidal entity (Hannay and Fitz-James, 1955). Heimpel and Angus (1959) later demonstrated that the parasporal inclusions were responsible for the insecticidal activity. Earlier classification of *B. thuringiensis* serotypes was based on the flagella (H) agglutination that depended on the existence of type-specific H antigens (de Barjac and Bonnefoi, 1962). This identification method is still used for the classification of numerous serotypes. Many serotypes have been classified, and more are continuously being identified. However, the insecticidal activity of a *B. thuringiensis* serotype is not dependent on the serotype but rather on the insecticidal proteins produced by these stains. To classify these insecticidal proteins or toxins Hofte and Whiteley (1989) developed a classification system that is based on both the host specificity of these toxins and on their structural similarities (Table 4.2). They proposed five major classes of insecticidal crystal proteins, that is, CryI as lepidopteran-specific toxin, CryII as lepidopteran-and dipteran-specific toxin, CryIII as coleopteran-specific toxin, CryIV as dipteran-specific toxin, and additionally, CytA as cytolytic and hemolytic toxin. With the discovery of numerous new toxins this classification system became onerous, and a new classification system based entirely on the protein structure was proposed recently (Table 4.2). Although this classification system is not officially accepted at this juncture, it is increasingly used for classification of new toxins. The reader is advised to use the Internet to review the latest toxin classification (see the Web address in Table 4.2).

Table 4.2. Classification of insecticidal proteins from *B. thuringiensis*

Old name	New name	Host range	Mass (kDa)
CryIA(a)	Cry1Aa	Lepidopteran	133.5
CryIA(b)	Cry1Ab	Lepidopteran	131.0
CryIA(c)	Cry1Ac	Lepidopteran	133.3
CryIB	Cry1Ba	Lepidopteran	138.0
CryIC(a)	Cry1Ca	Lepidopteran	134.8
CryIC(b)	Cry1Cb	Lepidopteran	134.0
CryID	Cry1Da	Lepidopteran	132.5
CryIF	Cry1Fa	Lepidopteran	133.6
CryIG	Cry9Aa	Lepidopteran	129.7
CryIIA	Cry2Aa	Lepidopteran/Dipteran	70.9
CryIIB	Cry2Ab	Lepidopteran	70.8
CryIIIA	Cry3Aa	Coleopteran	73.1
CryIIIB	Cry3Ba	Coleopteran	74.2
CryIIIC	Cry7Aa	Coleopteran	74.4
CryIIID	Cry3Ca	Coleopteran	73.8
CryIVA	Cry4Aa	Dipteran	134.4
CryIVB	Cry4Ba	Dipteran	127.8
CryIVC	Cry10Aa	Dipteran	77.8
CryIVD	Cry11Aa	Dipteran	72.4
CryV	Cry1Ib	Colepteran/Lepidopteran	81.2
CytA	Cyt1Aa	Dipteran/Cytolytic	27.4
CytB	Cyt2Aa	Dipteran/Cytolytic	29.2

***Source*:** Modified after Hofte and Whiteley (1989) and http://epunix. biols. susx. ac. uk/Home/Neil-Crickmore/Bt/index.html (see this Web site for additional details)

Although *B. thuringiensis* possesses other toxins such as α-β-and γ-exotoxins and the vegetative insecticidal protein (Vip3A) (Estruch *et al.*, 1996), this review concentrates mainly on the parasporal δ-endotoxin of *B. thuringiensis*. This review addresses the δ-endotoxin structure and function, its mode of action, and the use of *B. thuringiensis* for insect control.

4.2.2. Toxin Structure

Two different protein types, Cry and Cyt toxins, are observed in parasporal bodies of many *B. thuringiesis* strains. In general, most Cry toxins, with the possible exception of toxins in Cry2, Cry3, Cry6, Cry10, Cry11, Cry15 and Cry19 are synthesized as large, inactive protoxins which are activated by proteolysis in the insect gut (Gill *et al.*, 1992). A number of these protoxins, for example Cry1Ac, Cry4A, and Cry4B, are of 130 to 140-kDa in molecular weight. Proteins of this size are predominant in many *B. thuringiensis* strains, while other toxins belonging to Cry2A, Cry3A, Cry10A and Cry11A appear to be natural C-terminal truncations of the larger protoxins, and these have molecular weights in the range of 65 to 80-kDa

Activation of 130-140-kDa protoxins occurs by removal of the C-terminal half and ca. 25-30 residues from the N-terminal end of these proteins, resulting in the formation of active toxins of 65-70-kDa. Although naturally truncated, the smaller Cry2A and Cry3A toxins also require proteolytic activation, and are processed to active 60-65-kDa toxins by the removal of a short carboxyl stretch (Cry2A) or an amino-proximal portion (Cry3A) (Aronson, 1993).

In general, the C-terminal part of the larger 130-kDa proteins (Cry1A, Cry4A, and Cry4B) is not essential for toxicity, but is a highly conserved domain of these crystal proteins (Hofte and Whiteley, 1989).

All of the active N-terminal fragments, the protease-resistant core, which is formed by alkaline mediated proteolysis in the insect midgut are likely to have a similar three-domain structure as that found with the Cry3A and Cry1Aa toxins (Grochulski *et al.*, 1995; Li *et al.*, 1991). The generalized structure for most Cry toxins will thus consist of three domains. For example in Cry3A domain I, from amino acid residue 58-290, is amphipathic and hydrophobic consisting of a seven α-helix bundle in which the central helix, α helix 5, is completely surrounded by six outer helices. Domain II, from residues 291-500, contains three antiparallel β-sheets, while domain III, from residues, 291-500, contains three antiparallel β-sheets, while domain III, from residues 501-644 at the C-terminus, is a sandwich of two antiparallel β-sheets. The hydrophobic α-helices of domain I are possibly involved in inserting the toxin into the insect midgut cell membrane, and span the cell membrane to form a pore. Domain II contains the variable loops which are likely responsible for initial cellular recognition, and hence the binding of the toxin to its receptor in the midgut of the target insect. Domain III is apparently thought to be important for stability of the active toxin by protecting the toxin from further proteolysis in the insect gut (Li *et al.*, 1991; Carrol *et al.*, 1989). More recent evidence suggests that this domain is also involved in toxicity (Lee *et al.*, 1995; Chen *et al.*, 1993). In concert with the mechanism of action of the three-domain toxins (Parker and Pattus, 1993; Choe *et al.*, 1992; Allured *et al.*, 1986) it is probable that domain III of *B. thuringiensis* is a critical domain involved in toxicity.

The Cyt proteins are significantly different, both in their structure and their biological activities from the Cry proteins. Presently, five *cyt* genes have been sequenced: three Cyt1 proteins, Cyt1Aa, Cyt1Ab and CytBa have been isolated from subsp. *israelensis, medellin* and *neoleoensis,* respectively (Waalwijk *et al.*, 1985), and two Cyt2 proteins, Cyt2A and Cyt2B isolated from subsp. *kyushuensis* and *jegathesan,* respectively (Koni and Ellar, 1993); http:/epunix. biols susx. ac. uk/ Home Neil_Crickmore/Bt/index.html). The Cty1Aa and Cyt2Aa proteins are of 27.4-kDa and 29.2-kDa, respectively in size. The Cyt1Aa protein is proteolytically activated by both N-and C-terminal cleavage to a protease-resistant protein of 24-kDa, which exhibits lytic activity of erythrocytes as well as toxicity to larval mosquitoes (Al-yahyaee and Ellar, 1995; Gill *et al.*, 1987; Armstrong *et al.*, 1985).

Recently, the crystal structure of one toxin, the Cyt1Aa was elucidated (Li *et al.*, 1996). The three-dimensional CryB structure shows that the protein has a single domain with a α/β architecture. There are two outer layers of α-helix hairpins that wrap around β-sheets. In the protoxin form the protein is a dimer formed by the interwining of their N-terminal $\beta1$ sheets, in addition to other contacts between the two protein molecules. Proteolytic processing, both at the N-and C-termini, is essential for activation. Proteolytic processing (A1-yahyaee and Ellar, 1995; Gill *et al.*, 1987; Armstrong *et al.*, 1985) cleaves the interwining N-terminal $\beta1$ arms and the C-terminal αF-helix releasing the active toxin as a monomer and uncovering the three-layered core (Li *et al.*, 1996).

4.2.3. Mechanism of Action

The gut epithelium is the primary target tissue for the *B. thuringiensis delta*-endotoxin action. Very early studies (Heimpel and Angus, 1959), identified three insect types exhibiting different symptoms upon ingestion of *B. thuringiensis.* Type I insects, being most sensitive die within 1-7 hours after toxin ingestion suffering general body and gut paralysis. The mortality resulting from an increase in hemolymph pH is caused by the invasion of the alkaline midgut contents. Gut epithelium changes occur within 10-30 min depending on the toxin concentration (Luthy and Ebersold, 1981). Gut paralysis also occurs in type II insects which die in 2-7 days but no pH change is detected in the hemolymph. In contrast, type III insects die mainly from septicemia after spore germination occurs in the midgut (Heimpel and Angus, 1959).

The basic histopathology of *B. thuringiensis* toxins on dipteran and coleopteran larvae is similar to that described for leipdopteran insects (Bravo *et al.*, 1992a,b; Singh *et al.*, 1986a,b). In both lepidopteran and coleopteran insects, the histopathological changes include disruption of disintegration of the cells (Bravo *et al.*, 1992a). However, in the coleopteran larvae, the Cry3A toxins accumulate in the posterior part of the apical microvilli of the midgut epithelial cells, whereas in the lepidopteran larvae, Cry1Ab toxins accumulate in the anterior part (Bravo *et al.*, 1992a,b). On the other hand, in the mosquito larvae, the *delta*-endotoxin mostly destroys the midgut epithelium in the regions of the gastric caeca and posterior stomach (Singh *et al.*, 1986a,b). The midgut apical brush border in the regions of the gastric caeca and posterior stomach are also primary target sites for the Cry11A and Cry1Aa toxins in a mosquito species, *Anopheles gambiae* Giles (Ravoahangimalala *et al.*, 1993).

In general, the insect midgut proteases play a major role in the activation of *B. thuringiensis* toxins. Activation usually involves the removal of a small number of N-terminal amino acid residues as well as the cleavage of C-terminal half. N-terminal processing is observed in all Cry toxins including the natural truncated Cry toxins, while the latter is observed in all the larger Cry protoxins (Gill *et al.*, 1992). The processing at the C-terminus is predicated by the presence of a conserved region at position 791-805 which is very near the C-terminus of the naturally truncated toxins as well as the larger protoxins after activation (Hodgman and Ellar, 1990). The cleavage sites at the N-terminus are variable. For naturally truncated Cry3A and Cry2A proteins, the cleaved sites are at positions 179 and 184, respectively (Nicholls *et al.*, 1989; Carroll *et al.*, 1989). On the other hand, amino acid analysis of the activated Cry1Ac toxin showed that the N- and C-terminal residues are Ile^{29} and Lys^{623}, respectively (Bietlot *et al.*, 1989), while for Cry1Ab, the activated toxin N-and C-termini appear to be Ile^{29} and Arg^{601}, respectively (Hofte *et al.*, 1986). In cry 11A protein, the alkali-solubilized toxin is first cleaved between Thr^{347} and Phe^{348}, and between Phe^{348} and Tyr^{349} by mosquito larval midgut proteases, generating a 40-kDa N-terminal fragment and a 32.5-kDa C-terminal fragment. Afterward, the 40-kDa and 32.5-kDa fragments are cleaved further to 30-kDa and 31-kDa peptides, respectively (Dai and Gill, 1993). Interestingly, the N-amd C-terminal moities of Cry11A toxin individually do not have any or negligible mosquitocidal activity. However, when these two fragments are produced in the same bacterial cell, and hence packaged together, they are mosquitocidal (Dai and Gill, 1993).

In the case of Cyt toxins, the intact 27-kDa Cyt1Aa protein has little cytolytic activity *in vitro*, but if this protein is activated to 24-kDa molecule by both N- and C-terminal proteolysis, its activity is increased (A1-yahyaee and Ellar, 1995; Gill *et al.*, 1987; Armstrong *et al.*, 1985). Similarly, the intact 29-kDa Cyt2Aa protein is completely inactive *in vitro* until this protein is processed to a smaller molecule of 22-23-kDa by proteolysis (Koni and Ellar, 1994).

Solution of the Cry3A crystal structure demonstrated that domain II of the toxin, which contains three antiparallel β-sheets around a hydrophobic core, is likely responsible for initial recognition and binding of the Cry toxins to its binding sites on the cell membrane of the target insect midgut (Li *et al.*, 1991). The binding regions predicted from this structure are highly variable in many *B. thuringiensis* toxins, and lie between positions 322 and 560 which lies predominantly in domain II. For instance, a *B. mori* receptor binding domain on Cry1Aa toxin includes the N-terminal portion of the hypervariable region, amino acids 332-450 (Lee *et al.*, 1992; Ge *et al.*, 1989). Additional studies have demonstrated that determinants of insect specificity are located within domain II of the toxin (Ge *et al.*, 1991). More recent studies support the role of this domain with other toxins. For example, a site-directed mutagenesis shows that two surface-exposed loops (Gly^{317}-Phe^{320} and Gln^{374}-Pro^{377}) are important determinants of both the activity and specificity of the Cry1C toxin (Smith and Ellar, 1994).

Earlier studies showed that either the binding affinity, Kd, or the binding site concentration, B^{max} is generally correlated to the toxicity. For instance, the Cry1Aa toxin binds saturably to the *B. mori* larval brush border membrane vesicles (BBMV) with high affinity, whereas the Cry1Ac toxin, which is 200 times less toxic to *B. mori* than Cry1Aa toxin, shows much lower affinity binding (Lee *et al.*, 1992). In contrast, in *Heliothis virescens* (Fabricius), the toxic activities of three toxins, Cry1Aa, Cry1Ab, and Cry1Ac, are highly dependent on the binding site concentration on the BBMV rather than the affinity, reflecting the differences in the *in vivo* toxicity (Van Rie *et al.*, 1989). However, the high affinity or high concentration does not always result in high toxicity. For instance, in *Lymantria dispar* (Linnaeus), the *in vitro* binding affinity for the Cry1Ac toxin is much greater than the affinity for the Cry1Ab toxin, although Cry1Ac is almost 400 times less toxic to this lepidopteran insect than the Cry1Ab toxin (Wolfersberger, 1990). However, recent studies demonstrate that irreversible binding kinetics are a better correlate to toxicity than just binding studies, and that the toxicity of the Cry1Ab and Cry1Ac toxins is also correlated with irreversible binding kinetics (Liang *et al.*, 1995). Consequently, a given toxin may have more than one class of binding site in a single insect, and more than one type of toxin may compete for binding to a single site (see review by Gill *et al.*, 1992 for additional information).

The Cyt toxins interact not only with dipteran epithelial cells but also with lepidopteran and mammalian cells (Chow *et al.*, 1989; Gill and Hornung, 1987; Thomas and Ellar, 1983). Initial binding of the Cyt toxins occurs with unsaturated phospholipids of the plasma membrane (Gill *et al.*, 1987; Thomas and Ellar, 1983).

In general, the irreversible binding of *B. thuringiensis delta*-endotoxin to the insect epithelial cell membrane results from membrane insertion and transmembrane pore formation. This event is then followed by an increase in cell membrane permeability which leads eventually to cell lysis, disruption of gut integrity and finally to the death of the insect from starvation or septicemia (Gill *et al.*, 1992). In addition, the insertion of lepidopteran-specific *delta*-endotoxins increases the K^+ conductance of columnar cell apical membranes in the midgut of susceptible insects and subsequently inhibiting K^+-dependent amino acid transport also (Sacchi *et al.*, 1986).

Although it is unambiguous that activated *delta*-endotoxins insert into the midgut cell membranes and increase the K^+ conductance, the mechanism of the toxic action, however, has not yet been elucidated. Early studies proposed an hypothesis that the *delta*-endotoxin itself may form K^+ channels or open pre-existing K^+ channels in the midgut columnar cell apical membrane (Crawford and Harvey, 1988). However, this hypothesis has not been confirmed although an effort has been made utilizing Ba^{2+} and Ca^{2+}, which are universal K^+ channel blockers, to prevent the *delta*-endotoxin inhibition of the midgut K^+ homeostasis (Wolfesberger, 1989; Crawford and Harvey, 1988).

In artificial planar lipid bilayers, Cry1Ac and Cry3a toxins as well as Cyt1Aa toxin have been shown to form cation-selective channels. The cation-selective channels formed by Cyt1Aa toxin are permeable to K^+ and Na^+ but not to N-methylglucamine or Cl^- (Knowles *et al.*, 1989), while the channels formed by Cry1Ac and Cry3A toxins are highly selective with a similar ability to select K^+ over Cl^-, but the channel formation is observed only at high pH (Slatin *et al.*, 1990). However, artificial planar lipid bilayers with the absence of cell membrane proteins may not appropriately demonstrate the membrane interaction of Cry toxins which bind to a midgut glycoprotein receptor, although it may be applicable to the Cyt toxins which interact with phospholipids as binding sites (Gill *et al.*, 1987; Thomas and Ellar, 1983). The toxin pore has been estimated in a radius of 0.6-1.0 nm, and these appear to result ultimately in cell lysis (Gill *et al.*, 1992; Knowles and Ellar, 1987). This model assumes that the cell lysis is dependent ultimately on the formation of pores.

Consequently, although the Cry and Cyt toxins recognize different binding sites; glycoproteins and unsaturated phospholipids, respectively, their pores, however, can probably function with a common mechanism, leading to osmotic imbalance which results from the influx of water and cations or other small molecules and ultimately to cell lysis and death.

Several studies have revealed that the Cry1Ac toxin receptor in lepidopteran insects is a 120kDa aminopeptidase N (Valaitis *et al.*, 1995; Cowles *et al.*, 1995; Knight *et al.*, 1994; Sangadala *et al.*, 1994).

Binding of the toxin to this purified 120kDa glycoprotein is inhibited by GalNAc, but not by GlcNAc (Cowles *et al.*, 1995; Knight *et al.*, 1994). In addition, a Cry 1Ac-sensitive phosphatase of 65kDa has also been isolated as a complex with the 120kDa aminopeptidase N, and when reconstituted into phospholipid vesicles these proteins have increased toxin binding and enhanced toxin-induced $^{86}Rb^{+}$ release up to 1000-fold (Sangadala *et al.*, 1994).

Recently, two of the aminopeptidase N's have been cloned, and the deduced amino acid sequence has revealed significant similarity to the zinc-dependent aminopeptidase gene family (Gill *et al.*, 1995; Knight *et al.*, 1995). In *Manduca sexta* (Johannsen) the 120kDa aminopeptidase N, which binds the Cry1Ac toxin, is anchored to the membrane via a glycosyl-phosphatidylinositol (GPI) anchor (Garczynski and Adang, 1995). The C-terminal sequence of both *M. sexta* and *H. virescens* aminopeptidase NcDNAs contains putative anchor signals (Gill *et al.*, 1995; Knight *et al.*, 1995). In contrast, the putative receptor in *M. sexta* for the Cry1Ab toxin is a 210kDa protein (Vadlamudi *et al.*, 1993). Interestingly, this protein is a novel cadherin-like glycoprotein, which binds Cry1Ab toxin in ligand blotting experiments when it is expressed in either COS-7 or human embryonic 293 cells (Vadlamudi *et al.*, 1995).

Early studies proposed that the Cyt toxins initially interact with unsaturated phospholipids which are likely to be the receptors of Cyt toxins. For instance, in an experiment with lipid dispersions and multilamellar liposomes, the Cyt1Aa toxin bound to phsphatidylcholine, sphingomyelin and phosphatidylethanolamine (Thomas and Ellar, 1983). Furthermore, the protease- treated Cyt1Aa toxin was inactivated by aqueous suspensions of dioleolylphosphatidylcholine, indicating the involvement of phospholipids in the cytolytic action (Gill *et al.*, 1987). In particular, acyl unsaturation of the *syn*-2 position of these phospholipids is essential for thc initial binding of Cyt1Aa toxin (Gill *et al.*, 1987).

Recent studies propose that at least in several insects, the receptor-like interaction with Cyt toxin is involved *in vivo*. For instance, in an immunological localization study, Cyt1Aa toxin bounds only to the gastric caeca and posterior stomach of the *A. gambiae* larvae (Ravoahangimalala *et al.*, 1993).

4.3. *BACILLUS SPHAERICUS* NEIDE

Another mosquitocidal microbial agent, *B. sphaericus*, has attracted particular attention due to its longer environmental persistence and excellent control over *Culex, Psorophora,* and *Mansonia* mosquito larvae (Silapanuntakul *et al.*, 1983; Yap, 1990). This aerobic, rodshaped, gram-variable bacterium first described by Neide in 1904, produces terminally or subterminally swollen spherical spores (Davidson, 1981).

4.3.1. Classification and Biological Recycling

Since the discovery of the first mosquitocidal strain, strain K (Kellen *et al.*, 1965), more than 300 strains have been found. In 1973, a strain designated SSII-1, which was 10,000 times as effective as strain K, was isolated. Since then, a number of mosquito larvicidal strains with far greater toxicity have been isolated from different parts of the world. Among the most utilized toxic isolates, strain 1593 was discovered in Indonesia; strain MR4, now designated 2297, was found in Sri Lanka; strain 2362 was isolated in Nigeria; and IAB59 was found in West Africa (Berry *et al.*, 1991). Of these, strains 1593 and 2362 demonstrate the highest levels of toxicity. Strain 2362 has attracted particular attention because this isolate is capable of producing the binary larvicidal toxin at higher temperatures than other toxin strains of *B. sphaericus*. This may be an important factor for production of these control agents in most tropical areas (Yousten, 1984b). Isolation and screening continues in the quest for strains of *B. sphaericus* with increased toxicity and broader host ranges (Thiery *et al.*, 1992).

Despite the significant amount of heterogeneity among strains of *B. sphaericus*, the mosquito pathogenic strains of *B. sphaericus* were grouped together (Group IIA) with relatively high DNA homologies (Krych *et al.*, 1980). This was confirmed by Alexander and Priest (1990) in an extensive study using numerical analysis based on 88 variable features including DNA homology data. These authors suggested that the insect pathogenic strains be given species status since the phenotypic and genetic-features were sufficiently distinct to justify speciation.

Attempts have been made to differentiate entomopathogenic strains of *B. sphaericus* based on their toxicity to mosquito larvae and to develop techniques for rapid identification of toxic strains among new isolates. Serological classification of mosquito pathogenic strains is based on antibody reactivity with H (flagellar) antigens. Group H5a5b which includes strains 1593 and 2362 contains only strains with the high toxicity. Group H25 includes the toxic strain 2297 (de Barjac *et al.*, 1985), and group H6 contains the larvicidal strain 1AB59 and recent isolates from Ghana (Thiery *et al.*, 1992). To this list, two more serotypes, H3 and H48, were recently added (Thiery *et al.*, 1992). Bacteriophage typing has also been used to distinguish pathogenic subgroups among the strains of *B. sphaericus*. Groups 3 and 4 contained the majority of entomopathogenic strains of *B. sphaericus*. Groups 3 and 4 contained the majority of entomopathogenic strains (Yousten, 1984a). However, none of these methods provides a fast and reliable identification for new isolates of *B. sphaericus*. As the number of new isolates climbs at an ever increasing rate, more and more exceptions to long established classification methods are being discovered.

Toxicity was found to be correlated with the sporulation process. Asporogenous mutants blocked at early stages of sporulation are generally unable to produce detectable levels of the *B. sphaericus* binary toxin when incubated under conditions that induce sporulation in wild type cells (Baumann *et al.,* 1991). The absence of toxicity in asporogenous mutants was correlated to the absence of crystal formation in the exosporium when the cells were induced to attempt sporulation (Charles *et al.*, 1988). These crystals were produced during sporulation in the wild type cells and found to be proteinaceous in nature. This material accumulated as polyhedral body within the exosporium adjacent to the spore (Lacey and Undeen, 1986; Yousten and Davidson, 1982; Yousten, 1984b).

Field tests have been performed using two of the most potent strains of *B. sphaericus,* strains 2362 and 1593, in a variety of environments targeting different species and genera of mosquito larvae. Generally, both strains are highly effective against larvae of *Culex* and *Psorophora* (Yap, 1990). However, these strains exhibit variable activity against *Anopheles* and *Aedes* larvae, *Simulium* (black fly) larvae, and other insects (Davidson, 1981; de Barjac *et al.*, 1985; Lacey and Undeen, 1986; Priest, 1992; Yousten, 1984b). Both strains show excellent control over *Culex tarsalis* Coquillett, but little control over *Anopheles pseudopunctipenis franciscanus* McCracken (Davidson *et al.*, 1984).

One important factor in the field application of bioinsecticides is environmental persistence. Persistence of mosquito larvicidal activity using *B. sphaericus* strains and *B. thuringiensis* subsp. *israelensis* has been tested in field experiments (Silapanuntakul *et al.*, 1983; Mulla *et al.*, 1984; Mulligan *et al.*, 1980). Depending on the experimental settings, *B. sphaericus* exhibited residual activity ranging from two days to four weeks. Overall persistence was much longer in *B. sphaericus* than in *B. thuringiensis* subsp. *israelensis*. During field tests, recycling of *B. sphaericus* spores was observed (Davidson *et al.*, 1984; Seyler *et al.*, 1991). Recycling occurs through the germination and multiplication of bacteria in the dead larvae killed by toxic spore consumption. Upon exhaustion of nutrients, these vegetative cells undergo sporulation and are released when the larval cadaver disintegrates (Charles and Nicolas, 1986). Whether the recycled spores play a role in persistence of *B. sphaericus* is controversial. Recently, the spores of *B. sphaericus* were also found to recycle in the gut of non-biting midges (Karch *et al.*, 1990).

Various safety tests have been performed on non-target organisms (Mulla *et al.*, 1984; Singer, 1980) including mammals (Shadduck *et al.*,

1980) and non-target invertebrates in aquatic environments (Lacey and Mulla, 1990) and no adverse effects have been observed.

In spite of the recycling and persistence capabilities of *B. sphaericus*, the spore formulations that are used in field trials have been reported to settle down rapidly from the surface water of mosquito breeding sites to the bottom mud (Davidson, 1981; Mulla *et al.*, 1984; Mulligan *et al.*, 1980). Since it is probable that some larvae will survive the initial spore treatment and that adults from untreated breeding sites may oviposit in the treated habitat, repeated applications of spores are required to control the larval population. In addition, *B. sphaericus* spores are sensitive to ultraviolet light. In field and laboratory studies, exposure of both spores and vegetative cells to ultraviolet light dramatically reduced both cell viability and toxicity (Burke *et al.*, 1983).

There have been several attempts to transfer the larvicidal toxin activity to plankton organisms. Cyanobacterium has been reported as a model system for the delivery of the insecticidal crystal proteins of *B. thuringiensis* subsp. *israelensis* and *B. sphaericus* (Chungjatupornchai, 1990). *Caulobacter crescentus* was also used to express the *B. sphaericus* larvicidal toxins (Thanabalu *et al.*, 1992b). However, this research was performed utilizing strain of these aquatic microorganisms that have been manipulated in laboratories.

The use of habitat-specific microorganisms to deliver the insecticidal toxins may alleviate many of the problems associated with the original toxin-producing organisms. For the control of mosquito species such as *Anopheles* and *Culex*, which breed in the surface or subsurface level of aquatic environments, these microorganisms expressing larvicidal toxins may find important value. Another group of mosquitoes that is far from being controlled is *Mansonia* spp. due to their specific environment of breeding area. Even though these mosquito species are susceptible to a variety of microbial insecticides such as *B. thuringiensis* subsp. *israelensis* and *B. sphaericus* in laboratory setting, it is almost impossible for these insecticides to reach the larval breeding area in field. Investigation and isolation of natural organisms that co-exist in the specific environment may lead to useful delivery systems for a variety of insecticidal agents.

4.3.2. Toxin Structure

In attempts to define the toxic components from the crystals of *B. sphaericus* 2362 spores, the 43-, 63-, 98-, 110-, and 125-kDa proteins were identified utilizing SDS-PAGE (Baumann *et al.*, 1985). However, when Bowditch *et al.* (1989) cloned the 125-kDa protein gene from *B. sphaericus* 2362, it was found that the 125-kDa protein was the precursor of the cell wall S-layer protein, and that the 98-, 110-, and 125-kDa proteins observed from previous experiments were not parasporal crystal proteins, but contaminating derivatives of this cell wall protein.

Identification of toxic components prompted a number of research groups to clone and sequence the toxin genes from larvicidal strains (Hindley and Berry, 1987: Baumann *et al.*, 1987, 1988; Arapinis *et al.*, 1988; Berry *et al.*, 1989). Sequencing of the cloned DNA revealed a high degree of conservation among the toxin genes of different strains and identified two major open reading frames corresponding to the proteins of 51-and 42-kDa individually translated from a single bicistronic transcription unit. When bioassayed against susceptible mosquito larvae, the recombinant-made 42-kDa protein did not exert toxicity against mosquito larvae unless accompanied by the 51-kDa protein (Baumann *et al.*, 1987, 1988; Broadwell *et al.*, 1990a). Confirmation that the 51- and 42-kDa proteins are not toxic alone but are toxic in combination further supports the existence of a binary toxin (Broadwell *et al.*, 1990b; Davidson *et al.*, 1990). However, recent data on toxicity for tissue culture-grown cells of *Culex* suggest that only the 42-kDa or its processed 39-kDa protein is required for full toxicity to tissue culture-grown cells (Baumann and Baumann, 1991). These two different modes of action against larvae and tissue culture-grown cells by *B. sphaericus* larvicide need to be further elucidated. The 42-and 51-kDa proteins were cloned and over expressed as inclusion bodies in a non-mosquitocidal strain of *B. thuringiensis* subsp. *thuringiensis* under the promoter of cyt1Aa gene of *B. thuringiensis* subsp. *israelensis* (Nicolas *et al.*, 1993). The inclusion bodies consisting of 42-kDa protein showed toxicity to *Culex pipiens* Linnaeus larvae by itself. In contrast, the 51-kDa protein alone was not toxic, but showed synergistic effect when added to 42-kDa protein by increasing larvicidal activity of 42-kDa protein. These results indicated that the 42-kDa protein acts as a single toxin protein both in tissue culture cells and in larvae. The role of the 51-kDa protein is still speculative.

In addition to the crystal toxin proteins, a 100-kDa (Mtx1) soluble cytoplasmic toxin purified from *B. sphaericus* SSII-1 was cloned by Thanabalu *et al.* (1991). This gene was found to be widely distributed among both high- and low-toxicity *B. sphaericus* strains and was active against both *Culex* and *Aedes aegypti* (Linnaeus) mosquito larvae. Furthermore, the toxicity of the purified Mtx1 protein readily expressed in *E. coli* was found to be comparable to that of spore crystal proteins (51- and 42-kDa proteins) indicating that the 100-kDa Mtx1 protein is either expressed in very low quantity or has very low stability in *B. sphaericus* cells (Thanabalu *et al.*, 1992a). Further study revealed that the N-terminal domain contained ADP-ribosylation activity and the C-terminal toxic domain contained the cell binding function active on cultured *Culex quinquefasciatus* Say cell lines. However, both regions were necessary for toxicity to mosquito larvae (Thanabaly *et al.*, 1993).

More recently, two additionally toxins, Mtx2 and Mtx3, have been cloned from *B. sphaericus* (Liu *et al.*, 1996; Thanabalu and Porter, 1996). These mosquitocidal toxins, 31.8 and 35.8kDa, respectively, are homologous to the cytolytic toxins from *Clostridium perfringens* and *Pseudomonas aeruginosa.*

4.3.4. Mechanism of Action

Ultrastructural events as well as gross symptoms of intoxication were observed after ingestion of a high concentration of *B. sphaericus* cells or spores (Davidson, 1979, 1981; Charles, 1987; Singh and Gill, 1988). The midgut swelled to lie against the body wall, throwing the peritrophic membrane with its food contents into zigzag folds. The midgut cells separated from one another at their bases. Unlike the event occurring in *Aedes* larvae treated with *B. thuringiensis* subsp. *israelensis* crystals, the ultrastructural studies using electron microscopy revealed neither general cell swelling nor complete breakdown of the midgut epithelium indicating that the cellular mode of action of *B. sphaericus* toxin is different from that of the *B. thuringiensis* subsp. *israelensis* toxin. Davidson (1986) observed the cytopathic effect of purified toxin from *B. sphaericus* 1593 and 2362 on cultured *Culex* cells. Intoxicated cells rounded up, became swollen and phase-dark, and then detached from support. These swollen cells did not exclude trypan blue and eventually lysed.

Both *in vivo* and *in vitro* processing and activation of *B. sphaericus* crystal proteins have been reported (Baumann *et al.*, 1985; Broadwell and Baumann, 1987; Davidson *et al.*, 1987a; Broadwell *et al.*, 1990c). When the 42-kDa toxin was treated with mosquito gut proteases, the treated material was shown to increase the toxicity to culture grown cells of *Culex* by 54-fold (Broadwell and Baumann, 1987). Davidson *et al.* (1987a) demonstrated that the toxicity of the 42-kDa toxin increased when this material was treated with either mosquito gut enzymes or known proteolytic enzymes, such as trypsin and α-chymotrypsin. Products of *in vivo* digestion of the inclusions containing only 42-kDa protein were found to be primarily 39-kDa peptides, reduced in size by 3-4-kDa; whereas inclusions with the 51-kDa protein were converted to a protein of 44-kDa, reduced in size by 7-9-kDa (Broadwell *et al.*, 1990c). Overall, these data suggest that the activation of "protoxin" is achieved via two processing steps. First, a *B. sphaericus* protease removes 4 amino acids from the N-terminus of the 42-kDa protein (Baumann *et al.*, 1988); and, second, larval gut proteases remove 6 amino acids from the N-terminus and 20 amino acids from the C-terminus of the 42-kDa protein.

Processing of 100-kDa Mtx1 protein was studied by Thanabula *et al.* (1992a). The 100-kDa Mtx1 protein was found to be processed to

97-kDa by removal of the signal peptide in *B. sphaericus* cells and then acted on by larval gut extracts or trypsin to produce 27- and 70-kDa proteins.

To delineate the minimal size of the proteins required for toxicity to mosquito larvae, truncated proteins of both 51.4- and 41.9-kDa toxins were produced using a series of deletion mutants (Broadwell *et al.*, 1990a; Clark and Baumann, 1990; Oei *et al.*, 1990). The fragments which could be deleted without loss of toxicity were reported as follows: 15 amino acids from the N-terminal and 17 amino acids from the C-terminus of the 42-kDa toxin from strain 1593; 10 amino acids from the N-terminal and 17 amino acids from the C-terminus of the 42-kDa toxin from strain 2362; and 21 amino acids from the N-terminus and 53 amino acids from the C-terminus of the 51-kDa protein from the strain 2362. The most extensive study was done by Oei *et al.* (1990). In their study, a series of mutants was constructed with deletions in both the N- and C-termini of both the 51.4-kDa and 41.9-kDa toxin proteins of *B. sphaericus* 2297. They successfully removed 39 amino acids from the N-terminus and 54 amino acids from the C-terminus of the 51.4-kDa protein and 7 amino acids from the C-terminus and 17 amino acids from the N-terminus of the 41.9-kDa protein without loss of toxicity.

Binding assays were performed in an attempt to define the site of action on target mosquito gut by *B. sphaericus* toxin (Davidson *et al.*, 1987b; Davidson, 1988). Toxin labeled with fluorescein isothiocyanate (FITC) was localized in the gastric-caecum and the posterior midgut of larvae fed FITC-toxin. The toxin was found to bind strongly to susceptible *Culex* cells and to be internalized by receptor-mediated endocystosis, but far less binding to cells of *Aedes* species found.

At the molecular level, neither the mode of action nor the nature of the interaction between the 51- and 42-kDa proteins or their derivatives is understood. Using the rediolabelled 51- and 42-kDa proteins, Davidson *et al.* (1990) demonstrated that when the 42-kDa protein was fed up to 24 hours after the 51-kDa protein, the larvae would be killed. But, the toxicity could not be rescued when the order of feeding was reversed. An active form of the 51-kDa protein remained associated with the midgut tissue for up to 24 hours and was available for interaction with the 42-kDa protein. However, in the absence of the 51-kDa protein, the 42-kDa protein may bind to another site on the membrane, be internalized by endocytosis, or be enzymatically inactivated. In the mosquito *C. quinquefasciatus* these two toxins also bind specifically to a midgut cell membrane receptor with an affinity, K_d of 20nM and a maximal binding capacity B_{max} of 7 pmol/mg of membrane protein (Nielsen-Leroux and Charles, 1992). In mosquitoes resistant to these toxins there was no toxin binding at all (Nielsen-Leroux *et al.*, 1995).

One limitation of the *B. sphaericus* larvicide is that it has a limited host range. The sensitivity to *B. sphaericus* toxins among mosquito species varies markedly. Generally, *Culex* is highly sensitive to this toxin, whereas *A. aegypti* is relatively resistant. Susceptible species of *Anopheles* are intermediate in their sensitivity. The differences in susceptibility among these mosquito populations could be explained by the rate of toxin ingestion or dissolution, by the presence or absence of toxin activation enzyme in larval gut, by the presence or absence of necessary proteolytic enzymes in mosquito gut, or by the presence or absence of appropriate receptor sites on larval midgut cells. All of these factors have been examined (Davidson *et al.*, 1987a; Broadwell and Baumann, 1987). It was found that the differences in receptor binding affinity and/or the site-of-action might account for differences in the species specificity of *B. sphaericus* toxin as clearly shown by binding assay on three species of mosquito larvae (Davidson, 1989). In an attempt to broaden the host range, Bourgouin *et al.* (1990) cloned the toxin genes encoding the 51- and 42-kDa proteins from *B. sphaericus* 1593 into *B. thuringiensis* subsp. *israelensis*. Trisrisook *et al.* (1990) cloned the *B. thuringiensis* subsp. *israelensis* gene encoding the 130-kDa δ-endotoxin into *B. sphaericus* 1593 and 2362. Both transformants retained high toxicity against three species of mosquito larvae.

4.4. INSECT RESISTANCE TO BACTERIAL INSECTICIDES

The Indian meal moth, *Plodia interpunctella* (Hubner), when reared on diet treated with *B. thuringiensis* var. *kurstaki* -HD1 (Btk-HD1, Dipel) in the laboratory developed nearly 30-fold resistance in two generations and after 15 generations this resistance level reached a plateau, 100 times higher than that in the untreated insects (McGaughey, 1985).

Tabashnik *et al.* (1990) reported that a field population of the diamondback moth, *Plutella xylostella* (Linnaeus), treated with Btk-HD1 (Dipel) which contains Cry1A and Cry2 toxins showed 25-33 times less mortality than susceptible laboratory colonies. Reversal of 25- to 2800-fold resistance to Btk-HD1 occurred when exposure to the toxins was stopped for many generations (Tabashnik *et al.*, 1994b). Moreover, within 4 years, *P. xylostella* selected with Btk-HD1 (Dipel) exhibited over 200-fold cross-resistance to Cry1F toxin from *B. thuringiensis* subsp. *aizawai* (Tabashnik *et al.*, 1994a). Similarly, a laboratory selected strain of *H. virescens* resistant to Cry1Ac has also exhibited cross-resistance to other *B. thuringiensis* toxins that differ significantly in structure and activity (Gould *et al.*, 1992).

A number of early studies showed that primary mechanism of insect resistance to *B. thuringiensis* toxins is the reduced binding affinity of the toxins to the brush border membrane of the midgut epithelium (Lee *et al.*, 1995a; Ferre *et al.*, 1991; Van Rie *et al.*, 1990) rather than altered midgut proteolytic activity (Johnson *et al.*, 1990). For example, HD-1 resistant *P. interpunctella* had a 50-fold reduction in affinity (Van Rie *et al.*, 1990) and HD-1 resistant *P. xylostella* did not bind to the brush border membrane of the midgut epithelial cells, either because of strongly reduced binding affinity or because of the complete absence of the receptor molecule (Ferre *et al.*, 1991). Interestingly, both *P. interpunctella* and *P. xylostella* strains resistant to Btk-HD1 (Dipel) which contains Cry1A and Cry2 retained high sensitivity to Cry1Ba or Cry1Ca toxin.

In contrast, *H. virescens* resistant to Cry1Ab or Cry1Ac showed resistance to Cry1Ba and Cry1Ca toxins as well as to Cry1Aa, Cry1Ab, and Cry2A toxins (Gould *et al.*, 1992), implying that the mechanism of resistance is not always related to altered receptors; rather, post-binding events can be affected, leading to insect resistance.

Resistance development in mosquitoes to *B. thuringiensis* has not been observed in the field. However, laboratory selection has resulted in the development of resistance to individual toxins. Hence, selection of a *C. quinquefasciatus* colony, that had previously been exposed for 98-generations to crude *B. thuringiensis* subsp. *israelensis* powder, with just one toxin, Cry11A, resulted in a rapid development of Cry11A resistance, 85-fold, after just three generations (Dai, 1992). In contrast, when a previously unselected field collected population was exposed to a single toxin, the development of resistance in *C. quinquefasciatus* was much slower, 15-fold resistance after 22 generations (Dai, 1992). One factor in the slow development of resistance when the intact parasporal inclusion is used compared to the rapid development of resistance when single toxin is used is the presence of multiple toxins with varying modes of action in the intact parasporal inclusion. A key toxin is the Cyt1Aa cytolytic toxin that can overcome Cyt11A resistance in *C. quinquefasciatus* (Gill, unpublished).

In contrast field resistance to *B. sphaericus* has been observed both in Brazil (Silva-Filha *et al.*, 1995) and in India (Rao *et al.*, 1995). The mechanism of this resistance is not known. However, high level resistance observed in laboratory selected *C. quinquefasciatus* is due to decreased receptor binding in the midgut (Nielsen-Leroux *et al.*, 1995).

4.5. COMMERCIALIZATION

Dulmage (1970) reported an isolate of subsp. *kurstaki*, that later was produced by Abbott Laboratories as the first major commercial *B. thuringiensis* product called Dipel. To date several million kilograms of *kurstaki* HD-1 based products are produced annually in the United States with usage registered for nearly 30 crops and against over 100 pest insect species worldwide (Table 4.3).

Table 4.3. Examples of commercial *B. thuringiensis* insecticides

Commercial name	Producer	*B. thuringiensis* strain	Target insect pest
1	2	3	4
		Lepidoptera	
Agree	Ciba Geigy	*kurstaki* HD-1 + *aizawai*	Vegetable/Crop pests + *Spodoptera* spp.
Biobit	Novo-Nordisk	*kurstaki* HD-1	Vegetable/Crop pests
Certan	Sandoz	*aizawai*	*Galleria mellonella* (Linnaeus)
Condor	Ecogen	*kurstaki* HD-1	Forest pests
Cutlass	Ecogen	*kurstaki* HD-1	Vegetable/Crop pests
Cybout	Cyanamid	*kurstaki* HD-1	Vegetable/Crop pests
Dipel	Abbott	*kurstaki* HD-1	Vegetable/Crop pests
Foray	Novo-Nordisk	*kurstaki* HD-1	Forest pests
Javelin	Sandoz	*kurstaki* NRD-12	*Spodoptera* spp.
MVP	Mycogen	*kurstaki* HD-1 in *Pseudomonas*	Vegetable pests
Thuricide	Sandoz	*kurstaki* HD-1	Vegetable/Crop pests
		Coleoptera	
Di Terra	Abbott	*tenebrionis*	Vegetable pests
Foil	Ecogen	*tenebrionis*	*Ostrinia nubilalis* (Hubner)
M-One	Mycogen	*tenebrionis*	*Leptinotarsa decemlineata* (Say)
M-Trak	Mycogen	*tenebrionis*	Vegetable pests
Novodor	Novo-Nordisk	*tenebrionis*	Vegetable pests
Trident	Sandoz	*tenebrionis*	Vegetable pests
Trident II	Sandoz	*tenebrionis*	Vegetable pests
		Diptera	
Acrobe	Cyanamid	*israelensis*	Medical pests
Skeetal	Novo-Nordisk	*israelensis*	Medical pests
Teknar	Zoecon/Sandoz	*israelensis*	Medical pests
Vectobac	Abbott	*israelensis*	Medical pests

Subsequently, a number of *B. thuringiensis* screening trials allowed the identification of strains that possessed proteins that helped expand the host range among leipdopteran, coleopteran and dipteran insect pests. For example, Cry1Ac toxin produced by *kurstaki* HD-73 isolate shows superior activity against *Heliothis* spp. (Adang *et al.*, 1985), while kurstaki NRD-12 isolate is more toxic to the spruce budworm and gypsy moth than *kurstaki* HD-1 (Dubois *et al.*, 1988). Similarly, some isolates of subsp. *aizawai* are active against *Galleria mellonella* (Linnaeus) and *Spodotera* spp. larvae than HD-1, and subsp. *tenebrionis* is active against coleopteran pests (Huger and Krieg, 1989). Also, the discovery of subsp. *israelensis* (Goldbert and Margalit, 1977) led to the control of mosquito and blackfly larvae. It is now a major commercial product for the control of these human disease vectors worldwise.

ACKNOWLEDGEMENT

The original work reported herein was supported in part by grants from the NIH ES03298 and USDA 96-35302-3820.

REFERENCES

Adang, M.J., Staver, M.J., Rocheleau, T.A., Leighton, J., Barker, R.F. and Thompson, D.V. 1985. Characterized full-length and truncated plasmid clones of the crystal protein of *Bacillus thuringiensis* subsp. *kurstaki* HD-73 and their toxicity to *Manduca sexta*. *Gene* **36**:289-300.

Al-Yahyaee, S.A.S. and Ellar, D.J. 1995. Maximal toxicity of cloned CytA δ-endotoxin from *Bacillus thuringiensus* subsp. *israelensis* requires proteolytic processing from both the N- and C-termini. *Microbiology* **141**:3141-3148.

Alexander, B. and Priest, F.G. 1990. Numerical classification and identification of *B. sphaericus* including some strains pathogenic for mosquito larvae. *J. Gen. Microbiol.* **136**: 367-376.

Allured, V.S., Collier, R.J., Carroll, S.F. and McKay, D.B. 1986. Structure of exotoxin A of *Pseudomonas aeruginosa* at 3.0-Angstrom resolution. *Proc. natl. Acad. Sci.*, USA **83**: 1320-1324.

Arapinis, C., De La Torre, F. and Szulmajster, J. 1988. Nucleotide and deduced amino acid sequence of the *B. sphaericus* 1593M gene encoding a 51.4kDa polypetide which acts synergistically with the 42 kD protein for expression of the larvicidal toxin. *Nucleic Acids Res.* **16**: 7731.

Armstrong, J.L., Rohrmann, G.L. and Beaudreau, G.S. 1985. Delta endotoxin of *Bacillus thuringiensis* subsp. *israelensis*. *J. Bacteriol.* **161**: 39-46.

Aronson, A.I. 1993. The two faces of *Bacillus thuringiensis*: insecticidal proteins and post-exponential survival. *Mol. Microbiol.* **7**: 489-496.

Baumann, L. and Baumann, P. 1991. Effects of components of the *B. sphaericus* toxin on mosquito larvae and mosquito-derived tissue culture-grown cells. *Curr. Microbiol.* **23**: 51-57.

Baumann, L., Broadwell, A.H. and Baumann, P. 1988. Sequence analysis of the mosquitocidal toxin genes encoding 51.4-and 41.9-kilodalton proteins from *B. sphaericus* 2362 and 2297. *J. Bacteriol.* **170**: 2045-2050.

Baumann, P., Baumann, L., Bowditch, R.D. and Broadwell, A.H. 1987. Cloning of the gene for the larvicidal toxin of *B. sphaericus* 2362: evidence for a family of related sequences. *J. Bacteriol.* **169**: 4061-4067.

Baumann, P., Clark, M.A., Baumann, L. and Broadwell, A.H. 1991. *Bacillus sphaericus* as a mosquito pathogen: properties of the organism and its toxins. *Microbiol. Rev.* **55**: 425-436.

Baumann, P., Unterman, B.M., Baumann, L., Broadwell, A.H., Abbene, S.J. and Bowditch, R.D. 1985. Purification of the larvicidal toxin of *B. sphaericus* and evidence for high-molecular-weight precursors. *J. Bacteriol.* **163**: 738-747.

Berry, C., Hindley, J. and Oei, C. 1991. The *Bacillus sphaericus* toxins and their potential for biotechnical development. *In*: *Biotechnology for Biological Control of Pests and Vectors.* CRC Press, Boca Raton, USA, pp. 35-51.

Berry, C., Jackson-Yap, J., Oei, C. and Hindley, J. 1989. Nucleotide sequence of two toxin genes from *B. sphaericus* IAB59: sequence comparisons between five highly toxinogenic strains. *Nucleic Acids Res.* **17**: 7516.

Bietlot, H., Carey, P.R., Choma, C., Kaplan, H., Lessard, T. and Pozsgay, M. 1989. Facile preparation and characterization of the toxin from *Bacillus thuringiensis* var. *kurstaki. Biochem. J.* **260**: 87-91.

Bourgouin, C., Delecluse, A., De La Torre, F. and Szulmajster, J. 1990. Transfer of the toxin protein genes of *B. sphaericus* into *B. thuringiensis* subsp. *israelensis* and their expression. *Appl. Envir. Microbiol.* **56**: 340-344.

Bowditch, R.D., Baumann, P. and Yousten, A.A. 1989. Cloning and sequencing of the gene encoding a 125-kilodalton surface-layer protein from *B. sphaericus* 2362 and of a related cryptic gene. *J. Bacteriol.* **171**: 4178-4188.

Bravo, A., Jansens, S. and Peferoen, M. 1992a. Immunocytochemical localization of *Bacillus thuringiensis* insecticidal crystal proteins in intoxicated insects. *J. Invert. Pathol.* **60**: 237-246.

Bravo, A., Hendrickx, K., Jansens, S. and Peferoen, M. 1992b. Immuno cytochemical analysis of specific binding of *Bacillus thuringiensis* insecticidal crystal proteins to lepidopteran and coleopteran midgut membranes. *J. Invert. Pathol.* **60**: 247-253.

Broadwell, A.H. and Baumann, P. 1987. Proteolysis in the gut of mosquito larvae results in further activation of the *B. sphaericus* toxin. *Appl. Envir. Microbiol.* **53**: 1333-1337.

Broadwell, A.H., Baumann, L. and Baumann, P. 1990a. The 42- and 51-kilodalton mosquitocidal proteins of *B. sphaericus* 2362: construction of recombinants with enhanced expression and *in vivo* studies of processing and toxicity. *J. Bacteriol.* **172:** 2217-2223.

Broadwell, A.H., Baumann, L. and Baumann, P. 1990b. Larvicidal properties of the 42 and 51 kilodalton *B. sphaericus* proteins expressed in different bacterial hosts: evidence for a binary toxin. *Curr. Microbiol.* **21**: 361-366.

Broadwell, A.H., Clark, M.A., Baumann, L. and Baumann, P. 1990c. Construction by site-directed mutagenesis of a 39-kilodalton mosquitocidal protein similar to the larva-processed toxin of *B. sphaericus* 2362. *J. Bacteriol.* **172**: 4032-4036.

Burke, W.F., Jr., McDonald, K.O. and Davidson, E.W. 1983. Effect of UV light on spore viability and mosquito larvicidal activity of *B. sphaericus* 1593. *Appl. Envir. Microbiol.* **46**: 954-956.

Carroll, J., Li, J. and Ellar, D.J. 1989. Proteolytic processing of a coleopteran-specific δ-endotoxin produced by *Bacillus thuringiensis* var. *tenebrionis*. *Biochem. J.* **261**:99-105.

Charles, J.-F. 1987. Ultrastructural midgut events in culicidae larvae fed with *B. sphaericus* 2297 spore/crystal complex. *Ann. Microbiol.* **138**: 471-484.

Charles, J.-F. and Nicolas, L. 1986. Recycling of *B. sphaericus* 2362 in mosquito larvae: a laboratory study. *Ann. Microbiol.* **137**B: 101-111.

Charles, J.-F., Kalfon, A., Bourgouin, C. and De Barjac, H. 1988. *Bacillus sphaericus* asporogenous mutants: morphology, protein pattern and larvicidal activity. *Ann. Microbiol.* **139**: 243-259.

Chen, X.J., Lee, M.K. and Dean, D.H. 1993. Site-directed mutations in a highly conserved region of *Bacillus thuringiensis* delta-endotoxin affect inhibiton of short circuit current across *Bombyx mori* midguts. *Proc. natl. Acad. Sci.*, USA **90**: 9041-9045.

Choe, S., Bennett, M.J., Fujii, G., Curmi, P.M.G., Kantardjieff, K.A., Collier, R.J. and Eisenberg,D. 1992. The crystal structure of diphtheria toxin. *Nature* **357**: 216-222.

Chow, E., Singh, G.J.P. and Gill, S.S. 1989. Binding and aggregation of the 25-kilodalton toxin of *Bacillus thuringiensis* subsp. *israelensis* to cell membranes and alteration by monoclonal antibodies and amino acid modifiers. *Appl. Envir. Microbiol.* **55**: 2779-2788.

Chungjatupornchai, W. 1990. Expression of the mosquitocidal-protein genes of *B. thuringiensis sulesp. israelensis* and the herbicide-resistance gene *bar* in *Synechocystis* PCC6803. *Curr. Microbiol.* **21**: 283-288.

Clark, M.A. and Baumann, P. 1990. Deletion analysis of the 51=kilodalton protein of the *B. sphaericus* 2362 binary mosquitocidal toxin: construction of derivatives equivalent to the larva-processed toxin. *J. Bacteriol.* **172**: 6759-6763.

Cowles, E.A., Yunovitz, H., Charles, J.-F. and Gill, S.S. 1995. Comparison of toxin overlay and solid-phase binding assays to identify diverse CryIA(c) toxin binding proteins in *Heliothis virescens* midgut. *Appl. Envir. Microbiol.* **61**: 2738-2744.

Crawford, D.N. and Harvey, W.R. 1998. Barium and calcium block *Bacillus thuringiensis* subspecies *kurstaki* delta-endotoxin inhibition of potassium current across isolated midgut of larval *Manduca sexta. J. exp. Biol.* **137**: 277-286.

Dai, S.-M. 1992. *Bacillus thuringiensis* subsp. *israelensis* CryIVD toxin: Resistance development, proteolysis and binding characteristics in *Culex quinquefasciatus*. Ph.D. Dissertation, University of California, Riverside, CA. pp. IV-176.

Dai, S.-M. and Gill, S.S. 1993. *In vitro* and *in vivo* proteolysis of the *Bacillus thuringiensis* subsp. *israelensis* CryIVD protein by *Culex quinquefasciatus* larval midgut proteases. *Insect Biochem. Mol. Biol.* **23**: 273-283.

Davidson, E.W. 1979. Ultrastructure of midgut events in the pathogenesis of *B. sphaericus* strain SSII-1 infections of *Culex pipiens quinquefasciatus* larvae. *Can. J. Microbiol.* **25**: 178-184.

Davidson, E.W. 1981. A review of the pathology of bacilli infecting mosquitoes, including an ultrastructural study of larvae fed *B. sphaericus* 1593 spores. *Develop. Indus. Microbiol.* **22**: 69-81.

Davidson, E.W. 1986. Effects of *B. sphaericus* 1593 and 2362 spore/crystal toxin on cultured mosquito cells. *J. Invert. Pathol.* **47**: 21-31.

Davidson, E.W. 1988. Binding of the *B. sphaericus* (Eubacteriales: Bacillaceae) toxin to midgut cells of mosquito (Diptera: Culicidae) larvae: relationship to host range. *J. Med. Ent.* **25**: 151-157.

Davidson, E.W. 1989. Variation in binding of *B. sphaericus* toxin and wheat germ agglutinin to larval midgut cells of six species of mosquitoes. *J. Invert. Pathol.* **53**: 251-259.

Davidson, E.W., Bieber, A.L., Meyer, M. and Shellabarger, C. 1987a. Enzymatic activation of the *B. sphaericus* mosquito larvicidal toxin. *J. Invert. Pathol.* **50**: 40-44.

Davidson, E.W., Shellabarger, C., Meyer, M. and Bieber, A.A. 1987b. Binding of the *B. sphaericus* mosquito larvicidal toxin to cultured insect cells. *Can. J. Microbiol.* **33**: 982-989.

Davidson, E.W., Oei, C., Meyer, M., Bieber, A.L., Hindley, J. and Berry, C. 1990. Interaction of the *B. sphaericus* mosquito larvicidal proteins. *Can. J. Microbiol.* **36**: 870-878.

Davidson, E.W. Urbina, M., Payne, J., Mulla, M.S., Darwazeh, H., Dulmage, H.T. and Correa, J.A. 1984. Fate of *B. sphaericus* 1593 and 2362 spores used as larvicides in the aquatic environment. *Appl. Envir. Microbiol.* **47**: 125-129.

de Barjac, H. and Bonnefoi, A. 1962. Essai de classification biochimique et serologique de 24 souches de *Bacillus* du type *B. thuringiensis. Entomophaga* **8**: 5-31.

de Barjac, H., Larget-Thiery, I., Dumanoir, V.C. and Ripouteau, H. 1985. Serological classification of *B. sphaericus* strains on the basis of toxicity of mosquito larvae. *Appl. Microbiol. Biotechnol.* **21**: 85-90.

Dubois, N.R., Reardon, R.C. and Kolodny-Hirsch, D.M. 1988. Field efficacy of the NRD-12 strain of *Bacillus thuringiensis* against gypsy moth (Lepidoptera: Lymantriidae). *J. econ. Ent.* **81**: 1672-1677.

Dulmage, H.T. 1970. Insecticidal activity of HD-1, a new isolate of *Bacillus thuringiensis* var. *alesti. J. Invert. Pathol.* **15**: 232-239.

Estruch, J.J., Warren, G.W., Mullins, M.A., Nye, G.J., Graig, J.A. and Koziel, M.G. 1996. Vip3A, a novel *Bacillus thuringiensis* vegetative insecticidal protein with a wide spectrum of activities against lepidopteran insects. *Proc. natl. Acad. Sci.*, USA **88**: 5389-5394.

Ferre, J., Real, M.D., Van Rie, J., Jansens, S. and Peferoen, M. 1991. Resistance to the *Bacillus thuringiensis* bioinsecticide in a field population of *Plutella xylostella* is due to a change in a midgut membrane receptor. *Proc. natl. Acad. Sci.,* USA **88**: 5119-5123.

Garczynski, S.F. and Adang, M.J. 1995. *Bacillus thuringiensis* CryIA(c) δ-endotoxin binding aminopeptidase in the *Manduca sexta* midgut has a glycosyl-phosphatidylinositol anchor. *Insect Biochem. Mol. Biol.* **25**: 409-415.

Ge, A.Z., Shivarova, N.I. and Dean, D.H. 1989. Location of the *Bombyx mori* specificity domain on a *Bacillus thuringiensis* δ-endotoxin protein. *Proc. natl. Acad. Sci.,* USA **86**: 4037-4041.

Ge, A.Z., Rivers, D., Milne, R. and Dean, D.H. 1991. Functional domains of *Bacillus thuringiensis* insecticidal crystal proteins. Refinement of *Heliothis virescenes* and *Trichoplusia ni* specificity domains on CryI(c). *J. Biol. Chem.* **266**: 17954-17958.

Gibbons. A. 1992. Researcher fret over neglect of 600 million patients. *Science* **256**: 1135.

Gill, S.S. and Hornung, J.M. 1987. Cytolytic activity of *Bacillus thuringiensis* proteins to insect and mammalian cell lines. *J. Invert. Pathol.* **50**: 16-25.

Gill, S.S., Cowles, E.A. and Francis, V. 1995. Identification, isolation and cloning of a *Bacillus thuringiensis* CryIAc toxin binding protein from the midgut of the lepidopteran insect *Heliothis virescens*. *J. Biol. Chem.* **270**: 27277-27282.

Gill, S.S., Cowles, E.A. and Pietrantonio, P. 1992. The mode of action of *Bacillus thuringiensis* endotoxins. *A. Rev. Ent.* **37**: 615-636.

Gill, S.S., Singh, G.J.P. and Hornung, J.M. 1987. Cell membrane interaction of *Bacillus thuringiensis* subsp. *israelensis* cytolytic toxins. *Infect. Immun.* **55**: 1300-1308.

Goldberg, L.J. and Margalit, J. 1977. A bacterial spore demonstrating rapid larvicidal activity against *Anopheles sergentii, Uranotaenia unguiculata, Culex univitattus, Aedes aegypti* and *Culex pipiens*. *Mosq. News* **37**: 355-358.

Gould, F., Martinez-Ramirez, A., Anderson, A., Ferre, J., Silva, F.J. and Moar, W.J. 1992. Broad-spectrum resistance to *Bacillus thuringiensis* toxins in *Heliothis virescens*. *Proc. natl. Acad. Sci.*, USA **89**: 7986-7990.

Grochulski, P., Masson, L., Borisova, S., Pusztai-Carey, M., Schwartz, J.-L., Brousseau, R. and Cygler, M. 1995. *Bacillus thuringiensis* CryIA(a) insecticidal toxin: Crystal structure and channel formation. *J. Mol. Biol.* **254**: 447-464.

Hannay, C.L. and Fitz-James, P.C. 1995. The protein crystals of *Bacillus thuringiensis* Berliner. *Can. J. Microbiol.* **1**: 694-710.

Heimpel, A.M. and Angus, T.A. 1959. The site of action of crystalliferous bacteria in lepidoptera larvae. *J. Insect Pathol.* **1**: 152-170.

Hindley, J. and Berry, C. 1987. Identification, cloning and sequence analysis of the *B. sphaericus* 1593 41.9 kD larvicidal toxin gene. *Mol. Microbiol.* **1**: 187-194.

Hodgman, T.C. and Ellar, D.J. 1990. Models for the structure and function of the *Bacillus thuringiensis* δ-endotoxins determined by computational analysis. *J. DNA Sequ. Map.* **1**: 97-106.

Hofte, H. and Whiteley, H.R. 1989. Insecticidal crystal proteins of *Bacillus thuringiensis*. *Microbiol. Rev.* **53**: 242-255.

Hofte, H., Greve, H., Seurinck, J., Jansens, S., Mahillon, J., Ampe, C., Vandekerdkhove, J., Vanderbruggen, H., Montagu, M., Zabeau, M. and Vaeck, M. 1986. Structural and functional analysis of a cloned delta endotoxin of *Bacillus thuringiensis* Berliner 1715. *Eur. J. Biochem.* **161**: 273-280.

Huger, A.M. and Krieg, A. 1989. Two types of parasporal crystals produced by the coleopteran-toxin strain BI256-82 of *Bacillus thuringiensis* ssp. *tenebrionis*. *J. appl. Ent.* **108**: 490-497.

Johnson, D.E., Brookhart, G.L., Kramer, K.J., Barnett, B.D. and McGaughey, W.H. 1990. Resistance to *Bacillus thuringiensis* by the Indian meal moth, *Plodia interpunctella:* Comparison of midgut proteinases from susceptible and resistant larvae. *J. Invert. Pathol.* **55**: 235-244.

Karch, S., Monteney, N., Julien, J.L., Sinegre, G. and Co, J. 1990. Control of *Culex pipiens* by *Bacillus sphaericus* and role of nontarget arthropods in its recycling. *J. Am. Mosq. Cont. Assoc.* **6**: 47-54.

Kellen, W.R., Clark, T.B., Lindegren, J.E., Ho. B.C., Rogoff, M.H. and Singer, S. 1965. *B. sphaericus* Neide as a pathogen of mosquitoes. *J. Invert. Pathol.* **7**: 442-448.

Knight, P.J.K., Crickmore, N. and Ellar, D.J. 1994. The receptor for *Bacillus thuringiensis* CryIA(c) δ-endotoxin in the brush border membrane of the lepidopteran *Manduca sexta* is aminopeptidase *N*. *Mol. Microbiol.* **11**: 429-436.

Knight, P.J.K., Knowles, B.H. and Ellar, D.J. 1995. Molecular cloning of an insect aminopeptidase N that serves as a receptor for *Bacillus thuringiensis* CryA(c) toxin. *J. Biol. Chem.* **270**: 17765-17770.

Knowles, B.H. and Ellar, D.J. 1987. Colloid-osmotic lysis is a general feature of the mechanism of action of *Bacillus thuringiensis* δ-endotoxins with different insect specificity. *Biochem. Biophys. Acta* **924**: 509-518.

Knowles, B.H., Blatt, M.R., Tester, M., Horsnell, J.M., Carroll, J., Menestrina, G. and Ellar, D.J. 1989. A cytolytic delta-endotoxin from *Bacillus thuringiensis* var. *israelensis* forms cation-selective channels in planar lipid bilayers. *FEBS Lett.* **244**: 259-262.

Koni, P.A. and Ellar, D.J. 1993. Cloning and characterization of a novel *Bacillus thuringiensis* cytolytic delta-endotoxin. *J. Mol. Biol.* **229**: 319-327.

Koni, P.A. and Ellar, D.J. 1994. Biochemical characterization of *Bacillus thuringiensis* cytolytic α-endotoxins. *Microbiology* **140**: 1869-1880.

Krych, V.K., Johnson, J.L. and Yousten, A.A. 1980. Deoxyribonucleic acid homologies among strains of *B. sphaericus*. *Int. J. Syst. Bacteriol.* **30**: 476-484.

Lacey, L. and Mulla, M. 1990. Safety of *Bacillus thuringiensis* and *Bacillus sphaericus* to nontarget organisms in the aquatic environment. *In*: M. Laird, L. Lacey and E.W. Davidson (eds). *Safety of Microbial Insecticides*. CRC Press, Boca Raton, USA, pp. 169-188.

Lacey, L.A. and Undeen, A.H. 1986. Microbial control of black flies and mosquitoes. *A. Rev. Ent.* **31**: 265-296.

Lee, M.K., Milne, R.E., Ge, A.Z. and Dean, D.H. 1992. Location of a *Bombyx mori* receptor binding region on a *Bacillus thuringiensis* δ-endotoxin. *J. Biol. Chem.* **267**: 3115-3121.

Lee, M.K., Rajamohan, F., Gould, F. and Dean, D.H. 1995. Resistance to *Bacillus thuringiensis* CryIA δ-endotoxins in a laboratory-selected *Heliothis virescens* strain is related to an alteration in one of the receptors. *Appl. Envir. Microbiol.* **61**: 3836-3842.

Li, J., Carroll, J. and Ellar, D.J. 1991. Crystal structure of insecticidal δ-endotoxin from *Bacillus thuringiensis* at 2.5 Å resolution. *Nature* **353**: 815-821.

Li, J., Koni, P.A. and Ellar, D.J. 1996. Structure of the mosquitocidal δ-endotoxin CytB from *Bacillus thuringiensis* sp. *kyushuensis* and implications for membrane pore formation. *J. Mol. Biol.* **257**: 129-152.

Liang, Y., Patel, S.S. and Dean, D.H. 1995. Irreversible binding kinetics of *Bacillus thuringiensis* CryIA δ-endotoxins to Gypsy moth brush border membrane vesicles is directly correlated to toxicity. *J. Biol. Chem.* **270**: 24719-24724.

Liu, J.W., Porter, A.G., Wee, B.Y. and Thanabalu, T. 1996. New gene from nine *Bacillus sphaericus* strains encoding highly conserved 35.8-kilodalton mosquitocial toxins. *Appl. Envir. Microbiol.* **62**: 2174-2176.

Luthy, P. and Ebersold, H.R. 1981. The entomocidal toxins of *Bacillus thuringiensis*. *Pharmac. Ther.* **13**: 257-283.

McGaughey, W.H. 1985. Insect resistance to the biological insecticide *Bacillus thuringiensis*. *Science* **229**: 193-195.

Mulla, M.S., Darwazeh, H.A., Davidson, E.W. and Dulmage, H.T. 1984. Efficacy and persistence of the microbial agent *Bacillus sphaericus* against mosquitoes in organically enriched habitats. *Mosq. News* **44**: 166-173.

Mulligan, F.S., Schaefer, C.H. and Wilder, W.H. 1980. Efficacy and persistence of *B. sphaericus* and *B. thuringiensis* H-14 against mosquitoes under laboratory and field conditions. *J. econ. Ent.* **73**: 684-688.

Nicholls, C.N., Ahmad, W. and Ellar, D.J. 1989. Evidence for two different types of insecticidal P2 toxins with dual specificity in *Bacillus thuringiensis* subspecies. *J. Bacteriol.* **171**: 5141-5147.

Nicolas, L., Nielsen-Leroux, C., Charles, J.F. and Delecluse, A. 1993. Respective role of the 42- and 51-kDa components of the *Bacillus sphaericus* toxin overexpressed in *Bacillus thuringiensis. FEMS Microbiol. Lett.* **106**: 275-280.

Nielsen-Leroux, C. and Charles, J.-F, 1992. Binding of *Bacillus sphaericus* binary toxin to a specific receptor on midgut brush-border membranes from mosquito larvae. *Eur. J. Biochem.* **210**: 585-590.

Nielsen-Leroux, C., Charles, J.-F., Thiery, I. and Georghiou, G.P. 1995. Resistance in a laboratory population of *Culex quinquefasciatus* (Diptera: Culicidae) to *Bacillus sphaericus* binary toxin is due to a change in the receptor on midgut brush-border membranes. *Eur. J. Biochem.* **228**: 206-210.

Oei, C., Hindley, J. and Berry, C. 1990. An analysis of the genes encoding the 51.4- and 41.9-kDa toxins of *B. sphaericus* 2297 by deletion mutagenesis: the construction of fusion proteins. *FEMS Microbiol. Lett.* **72**: 265-274.

Parker, M.W. and Pattus, F. 1993. Rendering a membrane protein soluble in water: a common packing motif in bacterial protein toxins. *TIBS*: 391-395.

Priest, F.G. 1992. Biological control of mosquitoes and other biting flies by *Bacillus sphaericus* and *Bacillus thuringiensis. J. appl. Bacteriol.* **72**: 357-369.

Rao, D.R., Mani, T.R., Rajendran, R. and Joseph, A.S. 1995. Development of a high-level of resistance to *Bacillus sphaericus* in a field population of *Culex quinquefasciatus* from Kochi, India. *J. Am. Mosq. Cont. Assoc.* **11**: 1-5.

Ravoahangimalala, O., Charles, J.-F. and Schoeller-Raccaud, J. 1993. Immunological localization of *Bacillus thuringiensis* serovar *israelensis* toxins in midgut cells of intoxicated *Anopheles gambiae* larvae (Diptera: Culicidae). *Res. Microbiol.* **144**: 271-278.

Sacchi, V.F., Parenti, P., Hanozet, G.M., Giordana, B., Luthy, P. and Wolfersberger, M.G. 1986. *Bacillus thuringiensis* toxin inhibits K^+-gradient-dependent amino acid transport across the brush order membrane of *Pieris brassicae* midgut cells. *FEBS Lett.* **204**: 213-218.

Sangadala, S., Walters, F.S., English, L.H. and Adang, M.J. 1994. A mixture of *Manduca sexta* aminopeptidase and phosphatase enhances *Bacillus thuringiensis* insecticidal CryIA(c) toxin binding and $^{86}Rb^+$-K^+ efflux *in vitro. J. Biol. Chem.* **269**: 10088-10092.

Seyler, R.W. Jr., Yousten, A.A. and Burke, W.F., Jr. 1991. Plasmid stability in *Bacillus sphaericus* 2362 during recycling in mosquito larvae. *J. Invert. Pathol.* **58**: 362-366.

Shadduck, J.A., Singer, S. and Lause, S. 1980. Lack of mammalian pathogenicity of entomocidal isolates of *Bacillus sphaericus. Envir. Ent.* **9**: 403-407.

Silapanuntakul, S., Pantuwatana, S., Bhumiratana, A. and Charoensiri, K. 1983. The comparative persistence of toxicity of *B. sphaericus* strain 1593 and *B. thuringiensis* serotype H-14 against mosquito larvae in different kinds of environments. *J. Invert. Pathol.* **42**: 387-392.

Silva-Filha, M.-H., Regis, L., Nielsen-Leroux, C. and Charles, J.-F. 1995. Low-level resistance to *Bacillus sphaericus* in a field-treated population of *Culex quinquefasciatus* (Diptera: Culicidae). *J. econ. Ent.* **88**: 525-530.

Singer, S. 1980. *Bacillus sphaericus* for the control of mosquitoes. *Biotechnol. Bioengg.* **22**: 1335-1355.

Singh, G.J.P. and Gill, S.S. 1988. An electron microscope study of the toxic action of *Bacillus sphaericus* in *Culex quinquefasciatus* larvae. *J. Invert. Pathol.* **52**: 237-247.

Singh, G.J.P., Schouest, L.P., Jr. and Gill, S.S. 1986a. The toxic action of *Bacillus thuringiensis* var. *israelensis* in *Aedes aegypti in vivo.* I. The relevance of midgut lesions to its poisoning syndrome. *Pestic. Biochem. Physiol.* **26**: 36-46.

Singh, G.J.P., Schouest, L.P., Jr. and Gill, S.S. 1986b. The toxic action of *Bacillus thuringiensis* var. *israelensis* in *Aedes aegypti in vivo.* II. The relevance of skeletal muscle system lesions to its poisoning syndrome. *Pestic. Biochem. Physiol.* **26**: 47-55.

Slatin, S.L., Abrams, C.K. and English, L.H. 1990. Delta-endotoxins from cation-selective channels in planar lipid bilayers. *Biochem. Biophys. Res. Commn.* **169**: 765-772.

Smith, G.P. and Ellar, D.J. 1994. Mutagensis of two surface-exposed loops of the *Bacillus thuringiensis* CryIC δ-endotoxin affects insecticidal specificity. *Biochem. J.* **302**: 611-616.

Tabashnik, B.E., Cushing, N.L., Finson, N. and Johnson, M.W. 1990. Field development of resistance to *Bacillus thuringiensis* in diamondback moth (Lepidoptera: Plutellidae). *J. econ. Ent.* **83**: 1671-1676.

Tabashnik, B.E., Finson, N., Johnson, M.W. and Heckel, D.G. 1994a. Cross-resistance to *Bacillus thuringiensis* toxin CryIF in the diamondback moth (*Plutella xylostella*). *Appl. Envir. Microbiol.* **60**: 4627-4629.

Tabashnik, B.E., Finson, N., Groeters, F.R., Moar, W.J., Johnson, M.W., Luo, K. and Adang, M.J. 1994b. Reversal of resistance to *Bacillus thuringiensis* in *Plutella xylostella*. *Proc. natl. Acad. Sci.*, USA **91**: 4120-4124.

Thanabalu, T. and Porter, A.G. 1996. A *Bacillus sphaericus* gene encoding a novel type of mosquitocidal toxin of 31.8kDa. *Gene* **170**: 85-89.

Thanabalu, T., Berry, C. and Hindley, J. 1993. Cytotoxicity and ADP-ribosylating activity of the mosquitocidal toxin from *Bacillus sphaericus* SSII-1: possible roles of the 27- and 70-kilodalton peptides. *J. Bacteriol.* **175**: 2314-2320.

Thanabalu, T., Hindley, J. and Berry, C. 1992a. Proteolytic processing of the mosquitocidal toxin from *Bacillus sphaericus* SSII-1. *J. Bacteriol.* **174**: 5051-5056.

Thanabalu, T., Hindley, J., Jackson-Yap, J. and Berry, C. 1991. Cloning, sequencing and expression of a gene encoding a 100-kilodalton mosquitocidal toxin from *B. sphaericus* SSII-1. *J. Bacteriol.* **173**: 2776-2785.

Thanabalu, T., Hindley, J., Brenner, S., Oei, C. and Berry, C. 1992b. Expression of the mosquitocidal toxins of *Bacillus sphaericus* and *Bacillus thuringiensis* subsp. *israelensis* by recombinant *Caulobacter crescentus*, a vehicle for biological control of aquatic insect larvae. *Appl. Envir. Microbiol.* **58**: 905-910.

Thiery, I., Ofori, J. and Cosmao-Dumanoir, V. 1992. New mosquitocidal strains from Ghana belonging to serotypes H3, H6 and H48 of *Bacillus sphaericus*. *Appl. Microbiol. Biotechnol.* **37**: 718-723.

Thomas, W.E. and Ellar, D.J. 1983. Mechanism of action of *Bacillus thuringiensis* var. *israelensis* insecticidal delta-endotoxin. *FEBS Lett.* **154**: 362-368.

Trisrisook, M., Pantuwatana, S., Bhumiratana, A. and Panbangred, W. 1990. Molecular cloning of the 130-kilodalton mosquitocidal δ-endotoxin gene of *B. thuringiensis* subsp. *israelensis* in *B. sphaericus*. *Appl. Envir. Microbiol.* **56**: 1710-1716.

Vadlamudi, R.K., Ji, T.H. and Bulla, L.A.J. 1993. A specific binding protein from *Manduca sexta* for the insecticidal toxin of *Bacillus thuringiensis* subsp. *berliner*. *J. Biol. Chem.* **268**: 12334-12340.

Vadlamudi, R.K., Weber, E., Ji, I., Ji, T.H. and Bulla, L.A., Jr. 1995. Cloning and expression of a receptor for an insecticidal toxin of *Bacillus thuringiensis*. *J. Biol. Chem.* **270**: 5490-5494.

Valaitis, A.P., Lee, M.K., Rajamohan, F. and Dean, D.H. 1995. Brush border membrane aminopeptidase-N in the midgut of the gypsy moth serves as the receptor for the CryIA(c) δ-endotoxin of *Bacillus thuringiensis*. *Insect Biochem. Mol. Biol.* **25**: 1101-1114.

Van Rie, J., Jansens, S., Hofte, H.S., Degheele, D. and Van Mellaert, H. 1989. Specificity of *Bacillus thuringiensis* δ-endoxins. Importance of specific receptors on the brush border membrane of the mid-gut of target insects. *Eur. J. Biochem.* **186**: 239-247.

Van Rie, J., McGaughey, W.H., Johnson, D.E., Barnett. B.D. and Van Mellaert, H. 1990. Mechanism of insect resistance to the microbial insecticide *Bacillus thuringiensis*. *Science* **247**: 72-74.

Waalwijk, C., Dullemans, A.M., Workum, M.E.S. and Visser, B. 1985. Molecular cloning and the nucleotide sequence of the Mr 28 000 crystal protein gene of *Bacillus thuringiensis* subsp. *israelensis*. *Nucleic Acids Res.* **13**: 8207-8217.

Wolfersberger, M.G. 1989. Neither barium nor calcium prevents the inhibition by *Bacillus thuringiensis* delta endotoxin of sodium or potassium gradient dependent amino acid accumulation by tobacco hornworm midgut brush border membrane vesicles. *Arch. Insect Biochem. Physiol.* **12**: 267-277.

Wolfersberger, M.G. 1990. The toxicity of two *Bacillus thuringiensis* delta-endotoxins to gypsy moth larvae is inversely related to the affinity of binding sites on midgut brush border membranes for the toxins. *Experientia* **46**: 474-477.

Yap, H.H. 1990. Field trials of *B. sphaericus* for mosquito control. *In*: H. de Barjac and D.J. Sutherland (eds). *Bacterial Control of Mosquitoes and Black Flies*. Rutgers University Press, New Brunswick, pp. 307-320.

Yousten, A.A. 1984a. Bacteriophage typing of mosquito pathogenic strains of *B. sphaericus*. *J. Invert. Pathol.* **43**: 124-125.

Yousten, A.A. 1984b. *Bacillus sphaericus*: microbiological factors related to its potential as a mosquito larvicide. *Adv. Biotechnol. Proc.* **3**: 315-343.

Yousten, A.A. and Davidson, E.W. 1982. Ultrastructural analysis of spores and parasporal crystals formed in *Bacillus sphaericus* 2297. *Appl. Envir. Microbiol.* **44**: 1449-1455.

5

HOST PLANT RESISTANCE TO INSECTS

Z.R. KHAN AND R.C. SAXENA

International Centre of Insect Physiology and Ecology
P.O. Box 30772, Nairobi, Kenya

5.1. INTRODUCTION

The increasing complexities of insect control in recent in years, and the current awareness of problems associated with chemical control of insect pests, such as resistance to insecticides, rising cost of insecticide application, destruction of beneficial insects and other fauna, accumulation of undesirable residues, environmental pollution, etc. have led to an increased emphasis on utilisation of plant resistance as a major tactic in pest management (Kogan, 1975; Pathak and Saxena, 1981; Saxena and Khan, 1991).

Breeding insect-resistant crop plants is now a major objective in many national and international crop improvement programmes and cultivation of resistant cultivars has become a key component in pest management programmes. The growing of insect-resistant crop varieties had led to substantial reductions in the use of insecticides and yet increased food production in many tropical developing countries of the world. The pest-resistant varieties have played an important role in the 'Green revolution' in South and Southeast Asia.

Plant resistance as an approach in pest management confers many advantages. This germplasm-based technology virtually wraps the seed in a genetic coat of armor against insect pests and offers the grower the advantage of genetically incorporated insect control for the cost of seed alone (Smith, 1989; Saxena and Khan, 1991). Crop varieties that are insect resistant have the unique advantage of providing inherent insect control which is genetically compatible with other methods of insect control.

Depending on its level, plant resistance can be used as a principal control method or as a supplement to other measures in insect management. Earlier concerns that varietal resistance is generally associated with poor quality and low yield have largely been overcome by incorporating resistance into high-yielding varieties of good quality (Pathak and Saxena, 1981; Saxena and Khan, 1991). Also, the cultivation of resistant plants is not subject to vagaries of weather, as are chemical and biological control measures. Host plant resistance is especially of great significance where crops are low in value, particularly in developing countries where crops are planted in small farms and farmers have limited access to capital, pesticides, and application equipment. Also, resistant varieties control insects even at low density whereas the use of insecticides is justifiable only when the pest density reaches the economic injury level. In some cases resistance developed in plants for one pest species may also be effective against several others. Varietal resistance should not be misconstrued as 'immunity', which is a rarity among crop plants. Varietal resistance is generally a relative quality; the degree of resistance is based on comparison to susceptible plants that are more severely damaged under similar test conditions.

This chapter discusses the salient aspects of insect resistance in crop plants and its role in insect pest control. It uses selected examples as appropriate illustrations, but is not intended to be an exhaustive review of the literature on varietal resistance.

5.2. HISTORICAL PERSPECTIVES

Although serious efforts to utilise insect resistance in crop plants have been made only during the last three decades or so, the natural resistance to insect pests has been on record from the earliest days of applied entomology. The early literature includes several significant examples of differences in susceptibility or resistance of crop cultivars to insect attacks. As early as 1792, Havens reported that 'Underhill' variety of wheat was resistant to the Hessian fly, *Mayetiola destructor* (Say). Another early report of insect resistance was the apple variety 'Winter Majetin' found to be resistant to the woolly apple aphid, *Eriosoma lanigerum* (Hausmann) (Lindley, 1831). The most spectacular early success in plant resistance, however, was the control of the grape phylloxera, *Viteus (Phylloxera) vitifoliae* (Fitch) in 1861 in France through the use of American rootstocks resistant to the pest (Howard, 1930). This discovery saved the grape and wine industry in many parts of Europe and is still a major means of control of this pest.

Despite this spectacular beginning, plant resistance attracted little attention of plant breeders and entomologists during the first half of this century. It was to the credit of R.H. Painter, who recognised the practical implications of host plant resistance and provided a scientific and systematic basis to its study and applications for pest management. The first review by Snelling on resistance of plants to insect attack appeared in 1941. The first book on insect resistance in crop plants was published by Painter in 1951. It gives a comprehensive overview of the contemporary literature on host plant resistance and discusses the general principles involved in plant resistance to insects. Although somewhat simplistic in approach, Painter's book still remains a classic and a standard reference for the study of resistance of crop plants to insects. Some notable reviews on resistance of plants to insects are by

Auclair (1957), Painter (1958, 1968), Beck (1965), van Emden (1966), Horber (1972), Maxwell *et al.* (1972), Maxwell (1977), Pathak and Saxena (1979, 1981), etc. The theoretical aspects of insect-plant relationships have been discussed in volumes edited by van Emden (1973), Jermy (1976), Chapman and Bernays (1978), Brattsten and Ahmad (1986), Juniper and Southwood (1986), Miller and Miller (1986), Labeyrie *et al.* (1987), Barbosa and Letourneau (1988) and Heinrichs (1988), and in books authored by Ahmad (1983), Bernays and Chapman (1994), etc. Also, books edited by Harris (1979), Maxwell and Jennings (1980), Mandal *et al.* (1991) and Dhaliwal and Dilawari (1993), and authored by Panda (1979), Singh (1986), Smith (1989) and Panda and Khush (1995) deal with theoretical and applied aspects of breeding crop plants resistant to insects. The recent book by Smith *et al.* (1994) is the first publication to elaborate various entomological techniques used in evaluating insect resistance in crop plants.

5.3. CLASSIFICATION OF RESISTANCE

Snelling (1941) included in plant resistance those characteristics that enable a plant to avoid, tolerate, or recover from attacks of insects under conditions that would cause injury to other plants of same species. Painter (1951) gave a more comprehensive definition of varietal resistance to insect pests as 'the relative amount of heritable qualities possessed by the plant which influences the ultimate degree of damage done by the insect. In practical agriculture, it represents the ability of a certain variety to produce a larger crop of good quality than do ordinary varieties at the same levels of insect populations'. Beck (1965) emphasized the process of establishment of insects on plants and considered resistance as collective heritable plant characteristics.

The influence of abiotic factors, resulting in *pseudoresistance,* should not be overlooked while studying insect resistance in plants. It may favour the plant or the insect unequally or unpredictably and thus affect the expression of resistance. Painter (1951) described the phenomenon of *pseudoresistance* due to *host evasion* — when the host passes through the most susceptible stage quickly or at a time when insect numbers are reduced, *induced resistance* — resulting from some condition of plant or environment such as a change in the amount of water, soil fertility, etc. and *escape* — lack of infestation of, or injury to the host plant because of transitory circumstances such as incomplete infestation. According to Painter (1951), although cultivars or crops showing pseudoresistance are important in economic entomology, they should be distinguished from cultivars that show resistance throughout the wide range of environments.

Painter (1951) established three broad categories of true resistance based on inherited characters:

Nonpreference. When a plant possesses characteristics that make it unattractive to insect pests for feeding, oviposition, or shelter.

Antibiosis. When the host plant adversely affects the biology and physiology, such as food assimilation, development, survival and egg production of insects feeding on it.

Tolerance. When the damage to the host plant is only slight in spite of its supporting an insect population of a size that would severely damage susceptible hosts.

Painter's categorisation of resistance was based mainly on field observations. Though widely adopted by plant breeders and entomologists working on insect resistance, these terms are arbitrary and not clearly separated. The modalities of resistance often show a high degree of interdependence and may be shared by common plant characters. Also, nonpreference has often been projected as a property of the plant, rather than as the insect's negative response or avoidance of the plant. To overcome this confusion, Kogan and Ortman (1978) proposed a new term *antixenosis*.

Antixenosis (derived from the Greek word *xenos* = guest) describes the unsuitability of a plant to serve as a host to an insect herbivore. The new term is parallel to antibiosis and conveys the message that the plant is unsuitable as a host. Both antixenosis and nonpreference denote the presence of morphological or chemical plant factors that adversely alter insect behaviour, resulting in poor establishment of the insect pest. Antibiosis is considered a major aspect of resistance, and under some circumstances, antixenosis can be equally important, particularly where brief infestations can cause severe plant damage. The example of the latter is the insect transmission of virus diseases.

Tolerance is sometimes misconstrued as a low or general resistance, but modern insect-plant co-evolutionary theories prove that tolerance is much different from antixenosis and antibiosis. Tolerance is essentially an adaptive mechanism for survival of a plant against herbivore pressure, which is largely independent of the herbivore response. Unlike antixenosis and antibiosis, tolerance involves only plant characteristics and is not a part of insect-plant interactions. Beck (1965) stressed this difference and excluded tolerance from his review of host plant resistance. Kogan (1976) also expressed similar views. However, tolerance often occurs in combination with antixenosis and antibiosis. According to Pathak (1970), tolerant plants do not inhibit an insect population and under certain conditions, may even be responsible for buildup of a population larger than on susceptible plants. On the other hand, tolerant plants, by supporting insect infestations for a longer period, provide the natural enemies an arena to survive and multiply. Frequently, tolerant varieties are valuable in a programme of insect pest management. Velusamy and Heinrichs (1986) have extensively reviewed the literature on tolerance in crop plants to insect pests. Several different techniques have been used for the quantitative measurement of insect tolerance in crop plants (Table 5.1). These techniques have been developed in an attempt to separate tolerance from antixenosis and antibiosis.

5.4. FACTORS AFFECTING INSECT-HOST PLANT INTERACTIONS

Plants are the primary producers and manufacturers of organic substrates in the trophic system. Plants defend themselves against insect herbivory by certain adaptations. In due course of time, the insects counteradapt to overcome plant resistance. This cycle of adaptation and counteradaptation between plants and insects has been going on for millions of years (Brues, 1924; Ehrlich and Raven, 1964) resulting in diverse insect-host plant co-evolutionary process. From an evolutionary point of view, resistant traits are preadaptive characteristics of a plant and plants with such genes withstand the selective pressure of insect populations, thus augmenting the chances of their own survival and reproduction.

According to Norris and Kogan (1980), interactions between insects and plants are complex phenomena and it is seldom that a single factor can be assigned as a basis for resistance in a genotype to any one species of insect. Single factor resistance is especially unlikely if a complex of pests is involved. Successful host plant utilization by insect species very much depends upon a 'proper-fit' between biological characteristics of insects and plants (Beck, 1965).

Most species of insect pests are well adapted and possess highly developed mechanisms for host selection according to their nutritional requirements. Even the so-called polyphagous insects can not feed on just any plant species, while the oligophagous and monophagous insects are further limited to utilise different species or varieties of plants. Thus, every plant species is inherently resistant to some insect herbivore (Norris and Kogan, 1980).

There have long been different opinions as to the fundamental basis of insect-host plant relationships. Brues (1920) suggested that host selection by insects is through botanical instinct which might be a complex of chemical and physical stimuli emanating from the plant. Fraenkel (1959, 1969) postulated that because all insects have similar nutritional requirements, host selection must be governed by the presence of secondary plant biochemicals, such as alkaloids, flavonoids, quinones, etc. The nutritional factors, such as sugars, amino acids, sterols, phospholipids, ascorbic acid, and vitamin B have been shown to stimulate feeding of some insect species (Beck, 1956; Beck and

Table 5.1. Selected examples of methods used to measure insect tolerance in crop plants

Crop	Pest	Method(s) of tolerance measurement
1	2	3
Alfalfa	*Hypera postica* (Gyllenhal)	Yield, plant damage
	Therioaphis maculata (Buckton)	Plant survival, yield, damage index
	Acyrthosiphon kondoi Shinji	Damage threshold
	Acyrthosiphon pisum (Harris)	Yield
Barley	*Delia flavibasis*	Seedling survival
	Schizaphis graminum (Rondani)	Seedling survival, damage
	Diuraphis noxia (Mordvilko)	Tolerance index, seedling survival, damage
	Sipha flava (Forbes)	Damage, seedling height
Corn	*Diabrotica virgifera* Le Conte	Root pull resistance
	Diabrotica longicornis (Say)	Root volume
	Blissus leucopterus (Say)	Seedling survival
	S. flava	Damage, seedling height
Cotton	*Lygus lineolaris* (Palisot de Beauvois)	Yield
	Earias vittella (Fabricius)	Plant growth
Cowpea	*Empoasca dolichi* Paoli	Plant damage
Muskmelon	*Phyllotreta cruciferae* Glover	Plant damage
Mustard	*Phyllotreta cruciferae* (Goeze)	Plant growth
Oats	*D. noxia*	Damage
	S. flava	Seedling height
	S. graminum	Damage
Okra	*Amrasca biguttula* (Ishida)	Plant damage
Rice	Stem borers	Plant damage
	Spodoptera frugiperda (J.E.Smith)	Plant dry weight reduction, plant growth, root volume, root weight
	Nilaparvata lugens (Stal)	Yield, plant damage, functional plant loss index (FPLI), photosynthetic activity, root volume
	Lissorhoptrus oryzophilus Kuschel	Yield, root volume
Small grains	*D. noxia*	Plant growth, seedling survival
Sorghum	*Atherigona varia* (Meigen)	Seedling survival
	S. graminum	Seedling height, tolerance

Table 5.1 contd...

1	2	3
		index, functional plant loss index (FPLI), damage, foliage loss
	S. flava	Seedling height
	D. noxia	Seedling height
	Blissus leucopterus (Say)	Seedling survival
	Chilo partellus (Swinhoe)	Yield
Sugarcane	*S. flava*	Leaf discolouration & photo-synthetic activity
Triticale	*D. noxia*	Plant height
	S. graminum	Plant damage
	S. flava	Damage, height
Wheat	*S. graminum*	Presence of plant growth hormones in honeydew
	S. flava	Damage, height
	D. noxia	Damage
Wheat grass	*Labops hesperius*	Yield
	D. noxia	Plant growth

Source. Smith *et al.* (1994)

Hanec, 1958; House, 1962; Mittler and Dadd, 1963; Auclair, 1969; Beck and Reese, 1976; Saxena, 1986; Bernays and Chapman, 1994). Kennedy and Booth (1951) and Kennedy *et al.* (1958) propounded the theory of 'dual discrimination', according to which host selection is due to two types of insect responses: (a) response to secondary plant biochemicals acting as flavor stimuli, and (b) nutritional stimuli which may or may not constitute required nutrients for the insect. Whittaker (1970) proposed the term *allelochemics* for secondary plant substances and defined these as non-nutritional chemicals produced by an organism of one species which affect growth, health, behaviour or population biology of another species. According to him, two classes of allelochemicals are of particular pertinence in insect-plant interactions: (ii) *allomones* — chemical factors that give an adaptive advantage to the producing organism (host plant), and (iii) *kairomones* — chemical factors that give an adaptive advantage to the receiving organism (insect).

According to Saxena (1969), differences in relative susceptibility or resistance of plants to insects are determined by factors which influence the process of establishment of insect populations on them. The extent of insect establishment on plants depends on an interaction between insect responses to various plant characteristics. Six main categories of insect responses have been considered important during insect establishment on plants (Saxena, 1969; Saxena *et al.*, 1974; Khan and Saxena, 1985b): (i) orientation, (ii) feeding, (iii) metabolism of ingested food, (iv) growth, (v) survival of adults and egg production, and (vi) oviposition. Saxena and Pathak (1977) identified egg hatchability as another important factor during establishment of insects, particularly those that deposit their eggs inside the plant tissue. Saxena and Kumar (1980) and Khan and Saxena (1986) reported that insect mating may also be influenced by plant characters. The degree to which an insect responds positively in each of the response categories determines the degree of its establishment on different plants. Thus for a realistic determination of the resistance of a plant to an insect, it is desirable to understand various behavioural and physiological responses of the insects during different phases of establishment on the plant.

5.5. CAUSES OF RESISTANCE

The interruption of any one or more of the above mentioned insect responses due to unfavourable plant characters renders some degree of plant resistance (Visser, 1983). Such interruptions may be caused by the biophysical or biochemical characteristics of a plant. Often such responses are due to several factors. However, the available information on the causes of resistance is broadly summarized as follows:

5.5.1. Biophysical Factors

The behaviour of insects in selecting a host plant for food and shelter is affected by a wide array of physical stimuli. There is ample evidence to suggest that plant morphological characters such as tissue toughness, pubescence, spines, or bristles act as barriers to normal feeding and oviposition by insects. Similarly, factors such as lignification of cell walls, the presence of silica, tissue sclerotisation, and arrangement and number of vascular bundles affect the feeding and establishment of insects on plants. The general shape, size and colour of a plant also contribute to its attractiveness to insects by influencing the long range perception of the plant. Some examples of biophysical factors contributing towards insect resistance in plants are as follows.

Trichomes. Trichomes, also known as hairs or pubescence, are one of the most important morphological parameters of plant resistance to insects. In numerous plant species, a negative correlation has been established between trichome density on the plant surface and insect feeding and oviposition (Gallun *et al.*, 1973; Levin, 1973; Webster, 1975; Hoxie *et al.*, 1975; Roberts and Foster, 1983; Khan *et al.*, 1986a; Tingey, 1991). Long and dense trichomes reportedly hinder normal feeding and oviposition by various insect pests. According to Norris and Kogan (1980), the role of plant pubescence has been studied in 58 insect species in at least 19 crops belonging to 11 plant families. Trichomes reportedly contribute to plant resistance against 32 insects and susceptibility to 14 insects. No trichome effect was found for five insect species and no consistent relationship was found with seven other species. Non-glandular, simple erect trichomes on plants deter insect feeding and oviposition and may also adversely affect insect egg hatchability and larval development (Johnson and Hollowell, 1935; Sikka *et al.*, 1966; Ringland and Everson, 1968; Schillinger, Jr. and Gallun, 1968; Gallun *et al.*, 1973; Hoxie *et al.*, 1975; Roberts *et al.*, 1979; Khan *et al.*, 1986a; Schultz and Coffelt, 1987). Hooked trichomes on the foliage of the green bean, *Phaseolus vulgaris* L., impale nymphs of the potato leafhopper, *Empoasca fabae* (Harris) and the aphid, *Aphis craccivora* Koch during movement on the bean plant leaf (Johnson, 1955; Pillemer and Tingey, 1976, 1978). Glandular trichomes present on leaves, stems and reproductive structures of several plant species adversely affect the feeding, growth, survival and oviposition of several insect pests (Shade *et al.*, 1979; Duffey and Isman, 1981; Johnson *et al.*, 1980; MacLean and Byers, 1983; Tingey and Laubengayer, 1986; Farrar, Jr. and Kennedy, 1987; Tingey, 1991). Upon contact, glandular trichomes release a viscous substance which accumulates on tarsi and mouthparts, entrapping small insects, and decreasing the feeding and oviposition of larger insects. Trichome exudates are composed of various aldehydes, alkanes, ester, ketones, sesquitepenes, etc. (Triebe *et al.*, 1981; Lin *et al.*, 1987). The biochemical aspects of glandular trichome mediated insect resistance were reviewed by Steffens and Walters (1991).

Silica. Soluble silica absorbed by plant roots is translocated and deposited in the tissues of a number of plants. Several studies have demonstrated a relationship between silica deposits in certain plant tissues and the resistance of these plants to insect attack (Ponnaiya, 1951; Djamin and Pathak, 1967; Hanifa *et al.*, 1973; Lanning *et al.*, 1980). Djamin and Pathak (1967) and Hanifa *et al.* (1973) demonstrated that the presence of greater amounts of silica in resistant plants interfered with the feeding and boring of insects and caused severe wear to the mandibles.

Toughness and thickness of tissues. The toughness and thickness of various plant tissues has been reported to be correlated with resistance to insect. Toughness of stems, foliage and pods of various crops adversely affects penetration and feeding by insects (Issac, 1939; Khanna *et al.*, 1947; Sands and Brancatini, 1991; Tanton, 1962; Agarwal, 1969; Chiang and Jackai, 1988; Springer *et al.*, 1990, Kang *et al.*, 1990). Stems thickened by increased layers of epidermal cells deter or limit the entrance of stem damaging insects of several crops (Patanakamjorn and Pathak, 1967; Wallace *et al.*, 1974, Martin *et al.*, 1975; Fiori and Dolan, 1981). A thick cortex layer in the stems of wild tomato, *Lycopersicon hirsutum* var. *typicum*, is correlated with resistance to the aphid, *Macrosiphum euphorbiae* (Thomas), because the proboscis is not sufficiently long to reach the phloem (Quiros *et al.*, 1977).

Surface waxes. The plant surface is protected against desiccation, insect feeding and diseases by a layer of surface waxes over the epicuticle. Chemically, plant waxes are esters formed by the linkage of a long chain fatty acid and an aliphatic alcohol. Foliar wax coatings play an important role in the resistance of some crop cultivars to insect attack when sense organs on the insect tarsi and mouthparts receive negative chemical and tactile stimuli from the leaf surface. The raspberry species, *Rubus phoenicolasius* has heavy wax secretions that serve as a resistance barrier against the raspberry beetle, *Byturus tomentosus* (DeGeer), and the rubus aphid, *Amphorophora rubi* (Kaltenbach) (Lupton, 1967). Wax blooms on the leaves of some cruciferous crops deter feeding of the cabbage flea beetle, *Phyllotreta albionica* and diamondback moth, *Plutella xylostella* (Linnaeus) (Anstey and Moore, 1954; Stoner, 1990; Eigenbrode and Shelton, 1990; Eigenbrode *et al.*, 1991). Wax blooms on the leaves of brussels sprouts consist of vertical rods and dendritic plates that interfere with adhesion of the tarsal setae of the mustard beetle, *Phaedon cochleariae* (Fabricius), to the leaf surface (Stork, 1980). In contrast, the waxy leaves of kale stimulate feeding of the cabbage aphid, *Brevicoryne brassicae* (Linnaeus), and the cabbage whitefly, *Aleyrodes brassicae* Walker (Thompson, 1963) more than glossy-leaved cultivars. Foliar surface waxes of other plants contain allelochemicals that also affect the behaviour of insects. Sulfur compounds from the surface waxes of onions stimulate oviposition by the leek moth, *Acrolepiopsis assectella* (Zeller) (Thibout *et al.*, 1982). Linear furanocoumarins from carrot foliage waxes have a similar effect on oviposition by the carrot fly, *Psilia rosae* (Fabricius) (Stadler and Buser, 1982).

Hydrocarbon-and carbonyl-containing fractions of the wax of rice cultivars resistant to the brown planthopper, *Nilaparvata lugens* (Stal), adversely alter hopper feeding behaviour by causing hoppers to move away from their preferred feeding site (Woodhead and Padgham, 1988). Wax on sorghum leaves deters feeding by the migratory locust, *Locusta migratoria migratorioides* (Reiche & Fairmaire) (Atkin and Hamilton, 1982). Wax from young plants is more deterrent than that from older leaves, apparently because of the wax "blooms" of younger, faster growing foliage.

Cuticle. The cuticle of plants is a heterogeneous membrane in which wax, pectin and cellulose in varying proportions are contained in a cutin framework (Holloway and Baker, 1968). Plant cuticle plays a significant role in the probing and oviposition behaviour of insect pests, and young plant cuticle is normally preferred over thick mature cuticle (Zettler *et al.*, 1969; Walker and Aitken, 1985; Walker, 1987, 1988).

Colour. Specific colour-related insect resistance in plants does exist, but genetic manipulation of plant colour usually has an effect on some fundamental physiological plant process (Norris and Kogan, 1980). Most species of phytophagous insects are attracted to leaves reflecting light within 500 to 600 nm range (yellow-green). Therefore, red cotton plants are less attractive to the boll weevil, *Anthonomus grandis* Boheman, than green plants (Stephens, 1957). The imported cabbage worm, *Pieris rapae* (Linnaeus), is less attracted to the foliage of red brussels sprout varieties than to green-leaved varieties (Dunn and Kempton, 1976). Some cucurbit cultivars with silver-coloured leaves reflect more blue and ultraviolet wavelengths of light than normal cultivars and are resistant to aphids (Shifriss, 1981). A purple pigment in the stem epidermis of wild soybean, *Glycine soja*, has been used as an indicator of soybean stem resistance to the bean fly, *Ophiomyia centrosematis* (de Meijere) (Chiang and Norris, 1984). Similarly, the colour intensity of leaf supernatants of birch trees has been used to determine the degree of resistance to oviposition by the birch leaf-miner, *Fenusa pusilla* (Lepeletier) (Fiori and Craig, 1987). Foliar colour influences oviposition by the cabbage root fly, *Delia radicum* (Linnaeus) (Roessingh and Stadler, 1990); onion fly, *Delia antiqua* (Meigen) (Harris and Miller, 1983); carrot fly, *Psila rosae* (Fabricius) (Stadler, 1977) and Hessian fly, *Mayetiola destructor* (Say) (Harris and Rose, 1990).

Shape and size. Shape and size perception is likely to elicit some generalized behavioural patterns by insects, but no resistance mechanisms among crop plants have been reported using these parameters as selection criteria (Norris and Kogan, 1980). Certain plant shapes and sizes may, however, be linked with other resistance factors. For example, the frego-bract and okra-leaf traits of cotton exhibit resistance to the boll weevil, *Anthonomus grandis* Boheman (Niles, 1980). The efficacy of frego-bract cotton in weevil resistance is due mainly to modifications of weevil behaviour (Mitchell *et al.*, 1973), whereas weevil mortality in squares under the canopy of okra-leaf cotton is increased because of desiccation (Reddy, 1974). Although other morphological and chemical factors may also be involved in resistance associated with shape and size, plant shape and size could be used as a selection criterion in insect resistance valuation (Prokopy and Bush, 1973; Roessingh and Stadler, 1990; Pittara and Katsoyannos, 1992).

5.5.3. Biochemical Factors

Plants remain surrounded in an external environment of vapours of volatile compounds which emanate from the outer layers of plant parts. These compounds serve as olfactory (attractant or repellent) cues for insects. The internal environment of plants consists of a large variety of chemicals, some of which are of nutritional value to insects and may also act as feeding stimulants. There may also be other non-nutritional chemicals, called 'allelochemicals' which greatly influence the behaviour and physiology of insects and are very important in imparting resistance or susceptibility to insects in plants. Allelochemicals, as stated earlier, are classified as 'allomones' and 'kairomones'. The allomones are associated with repellence, feeding deterrence, toxicity, or other adverse effects on insects. The kairomones may serve as attractants and feeding or ovipositional stimulants.

Plant chemicals conferring resistance against insect pests are diverse (Table 5.2) and include isoprenoids or terpenes (i.e., from hemiterpenes through polyterpenes), acetogenins (e.g., phloroglucinols, orsellinic acid derivatives, naphthalenes, naphthaloquinones, anthraquinones, etc.), acetogenous molecules (e.g., benzoquinones, flavonoids, certain coumarins, condensed tannins, stilbenes, DIMBOA), aromatic derivatives from shikimic acid, and acetate (e.g., morin, myristicin, sasamin, quercetin, etc.), alkaloids (e.g., nicotine, demissine, solanine, tomatine, etc.), protein inhibitors, non-protein amino acids and glycosides, etc. (Norris and Kogan, 1980; Frazier, 1986; Frazier and Hanson, 1986). These plant chemicals may have adverse effects on insect feeding and assimilation of food, thus reducing the probability for survival, particularly among species in which larval forms are incapable of locating a more suitable host. Other physiological disturbances in growth, metamorphosis, and reproduction are also caused by the occurrence of metabolic inhibitors in plant tissues.

The chemicals present in external and internal environments of the plants are perceived by the insect's sensory organs and are decoded by the central nervous system into an expression of 'suitable host', 'unsuitable host' or 'non-host'. For example, rice stem borer, *Chilo suppressalis* (Walker) moths can discriminate resistant rice varieties from susceptible ones by their odor. Pathak and Saxena (1980) reported that certain resistant rice varieties possessed a volatile ovipositional deterrent factor while the susceptible varieties contained some ovipositional attractant factor. The effect of rice plant odor on oviposition was so strong that the moths oviposited heavily even on resistant plants when they were sprayed with an extract of susceptible plants. Similarly, feeding by rice leafhoppers and planthoppers on susceptible plants was significantly reduced when volatile extracts from resistant plants were applied (Saxena and Okech, 1985; Khan and Saxena, 1986). Khan and Saxena (1985a) reported that the green leafhopper, *Nephotettix virescens* (Distant), shifted from phloem feeding to xylem drinking when allowed to feed on susceptible TN1 seedlings sprayed with steam distillate extracts of resistant ASD7 plants. Another classical example of the presence or absence of a plant biochemical affecting the resistance of the plant to pests are cucurbitacins. Cucurbitacins, tetracyclic triterpenoids, are feeding attractants to the spotted cucumber beetle, *Diabrotica undercimpunctata howardi* Barber. Cucurbitacin-free cucumbers are resistant because they are not preferred by the bettle,

Table 5.2. Selected examples of plant allelochemicals that inhibit insect feeding

Chemical class	Compound	Insect affected
1	2	3
Alkaloids	Isoboldine	*Spodoptera litura* (Fabricius)
	N-methyl flindersine	*Spodoptera exempta* (Walker) *Spodoptera littoralis* (Boisduval)
	Tomatine	*Pieris brassicae* (Linnaeus)
	Lycorine	*Schistocerca gregaria* (Forskall)
	Senkirkine	*Choristoneura fumiferana* (Clemens)
	Trewiastine	*Acalymma vittatum* (Fabricius)
	Quinine	*P. brassicae*
	Strychnine	*P. brassicae*
	2,5-dihydroxymethyl 3,4-dihydroxypyrro-liodone	*S. gregaria*
Phenylpropanoids	Cinnamonitrile	*P. brassicae*
	Isopimpinellin	*S. litura*
	Phlorizin	*Schizaphis graminum* (Rondani)
	R=2'-glucosyl procyanidin	*S. graminum*
	(-) -Vestitol	*Costelytra zealandica* (White)
	Chlorogenic acid	A. *vittatum* *P. brassicae*
	Phenylhepta trinyne	*Euxoa messoria* (Harris)
	Juglone	*Scolytus multistriatus* (Marsham)
Monoterpenes	Specionon	*C. fumiferana*
	β-Ocimene	*Atta cephalotes* (Linnaeus)
Sesquiterpenes	Shiromodiol diacetate	*Trimeresia miranda*
	Tulirionol	*Lymantria dispar* (Linnaeus)
	Warburganal	*S. exempta*
	Glaucolide	*Spodoptera ornithogalli* (Guenee)
Diterpenes	Clerodin	*S. litura, S. littoralis*
	Jugarin I	*S. gregaria*
	Kalmitoxin I	*L. dispar*
	Harrisonin	*S. exempta*
	Cucurbitacin E	*Phyllotreta nemorum* (Linnaeus)
	Cucurbitacin A	*Diabrotica undecimpunctata* Mannerheim
	Chaparrinone	*Heliothis virescens* (Fabricius)
	Withanolide E	*S. littoralis*

Source: Frazier (1986)

but they are susceptible to the two-spotted spider mite, *Tetranychus urticae* Koch, while those containing cucurbitacin were mite-resistant (DaCosta and Jones, 1971).

Knowledge of chemical defenses of plants against insect pests has mushroomed since the mid 1960s, with the majority of the research in this area being directed at identifying insect antifeedants and feeding deterrents. It is a clear indication that chemical-based resistance is a major factor in plants' total defence against herbivores. Unfortunately, its application in breeding programmes is still in its scientific infancy. There are tremendous advantages to the full elucidation of the allelochemical bases of insect resistance in plants (Sexena, 1986). The identification of the chemicals that confer insect resistance or susceptibility would greatly improve the ability to breed for insect resistant varieties. Such understanding will also open new avenues for the manipulation of insect behaviour for use in insect pest management programmes.

5.6. ABIOTIC STRESSES AND EXPRESSION OF RESISTANCE

All plants are dependent on their environment for growth and development. Changes in light, temperature, relative humidity, and soil nutrient conditions commonly affect the expression of insect resistance in crop plants. The following are selected examples of effects of abiotic stresses on the expression of insect resistance.

5.6.1. Light

Both the quantity and quality of light affect the expression of plant resistance to insects. Increased light quantity enhances resistance in tomato to the tobacco hornworm, *Manduca sexta* (Johannsen) (Kennedy *et al.*, 1981). Exposure of plants to continuous light, however, decreases resistance in soybeans to defoliation by the cabbage looper, *Trichoplusia ni* (Hubner) (Khan *et al.*, 1986). The growth of plants produced under reduced light intensity decreases insect resistance in wheat (Roberts and Tyrell, 1961), alfalfa (Shade *et al.*, 1975), sorghum (Woodhead, 1981), corn (Manuwoto and Scriber, 1985a), sugarbeet (Lowe, 1976) and soybean (Reynolds and Smith, 1985).

5.6.2. Temperature

Plant resistance to insects may not be expressed at abnormally low or high temperatures. Higher than normal ambient temperatures diminish the expression of resistance in several small grain crops. The relationship has been established in wheat resistance to the Hessian fly, *Mayetiola destructor* (Say) (Sosa, 1979; Tyler and Hatchett, 1983; Ratanatham and Gallun, 1986), greenbug resistance in barley (Salas and Corcuera, 1991) and sorghum (Wood and Starks, 1972), and Russian wheat aphid resistance in sorghum (Girma *et al.*, 1990).

Kindler and Staples (1970b) noted that alfalfa resistance to the spotted alfalfa aphid was expressed more fully at fluctuating temperatures than at a constant temperature equivalent to the mean of the fluctuating regime. Similar results were recorded by van de Klashorst and Tingey (1979) when evaluating potato plant material for resistance to the potato leafhopper, *Empoasca fabae* (Harris).

Reduced insect resistance occurs at temperatures 10 to 15^0 C below normal in alfafa (Karner and Manglitz, 1985; Schalk *et al.*, 1969). The amount of exudate from the glandular trichomes controlling two-spotted spider mite resistance in geraniums also decreases with increased temperature (Walters *et al.*, 1991). Salim and Saxena (1991b) reported that both high and low temperature stresses reduced rice plant growth and biomass, and influenced plant chemical composition. Food intake, growth, longevity, fecundity and population of the whitebacked planthopper, *Sogatella furcifera* (Horvath), increased significantly when resistant IR2035-117-3 rice plants were grown at low and high (vs. optimum) temperatures.

5.6.3. Soil Fertility and Salinity

Soil fertility, where plants are grown, affects the expression of insect resistance. Increasing levels of nitrogen fertilizer decrease armyworm resistance in maize (Manuwoto and Scriber, 1985b; Wiseman *et al.*, 1973), grasses (Change *et al.*, 1985; Gaynor and Hunt, 1983), peanuts (Leuck and Hammons, 1974), alfalfa (Kindler and Staples, 1970b), tomato (Barbour *et al.*, 1991), and sorghum (Archer *et al.*, 1990), and *Sogatella furcifera* (Horvath) resistance in rice (Salim and Saxena, 1991a), Increasing the amount of potassium and phosphorous fertilizer, however, increases spotted alfalfa aphid resistance in alfalfa (Kindler and Staples, 1970a), greenbug resistance in sorghum (Schweissing and Wilde, 1979), and fall armyworm resistance in pearl millet (Leuck, 1972). Increase in potassium also increased the level of resistance in rice to *S. furcifera* (Salim and Saxena, 1991a). The addition of silica or aluminum to the fertilization regime of rice plants increases resistance to *S. furcifera* (Salim and Saxena, 1992). Resistant rice plants grown under saline conditions were less susceptible to *S. furcifera* than unstressed control plants (Salim *et al.*, 1990).

5.6.4. Relative Humidity

Relative humidity is especially important to the process of measuring plant resistance to the insect pests of stored grain. The expression of resistance in stored grain sorghum to the lesser rice weevil, *Sitophilus oryzae* (Linnaeus), and the maize weevil, *Sitophilus zeamais* Motschulsky, at differing humidities has been documented by Russel (1966) and Rogers and Mills (1974).

5.7. GENETICS OF INSECT RESISTANCE

Information on the mode of inheritance and the number of genes involved in resistance of plants to insects, although not essential for breeding resistant plants, has great significance in identifying donors for resistance, developing isogenic lines, and breeding broad-based resistant varieties. Knowledge of the genetics of insect resistance is equally as important as of the mechanisms of resistance, especially if biotypes of insects are prone to develop on the crop plant under investigation. The completeness of the information available on the number of genes conferring resistance depends on the thoroughness with which the germplasm has been screened. Table 5.3 lists the genes involved in resistance to major insect pests in selected agricultural crops. The genetics of insect resistance has been studied since the early twentieth century, when Harlan (1916) demonstrated that resistance to the leafblister mite, *Eriophyes gossypii,* was a heritable trait in cotton.

Most of the known high level of resistance is either monogenic or oligogenic. Monogenic resistance is easy to incorporate in improved plant types. However, such resistance is often short-lived because with a single gene conditioning, only a single gene mutation in the insect is enough to overcome resistance. Erosion of monogenic resistance, conferred by *Bph 1* gene, in the rice variety IR26 within two years of its intensive cultivation in the Philippines and Indonesia in mid-seventies, by Biotype 2 of the brown planthopper, *Nilaparvata lugens* (Stal), serves as a classical example of the inherent weakness of monogenic type of resistance (Saxena and Barrion, 1983). This problem has, however, been overcome to some extent by sequential release of rice varieties IR36 and IR56 with new major genes, *bph 2* and *Bph3,* respectively. Nonetheless, there are also examples of major gene resistance remaining effective for a very long period. The jassid-resistant cotton varieties having a major resistance gene have been resistant for more than 50 years in Africa (Gallun and Khush, 1980). Similarly, some rice varieties with major genes have been resistant to green leafhopper, *Nephotettix* spp., for more than 40 years (Khush and Beachell, 1972).

Table 5.3. Selected examples of major agricultural crops in which genetics of resistance to insect pests has been investigated

Crop	Insect	Mode of inheritance
1	2	3
Alfalfa	Pea aphid,	
	Acyrthosiphon pisum (Harris)	Monogenic, dominant
	Spotted alfalfa aphid,	Lack of cytoplasmic
	Therioaphis trifolii (Monell)	inheritance

Table 5.3 contd..

1	2	3
Corn	Corn earworm, *Helicoverpa zea* (Boddie)	Polygenic
	Corn leaf aphid, *Rhopalosiphum maidis* (Fitch)	Polygenic
	European corn borer, *Ostrinia nubilalis* (Hubner)	Polygenic
	Western corn rootworm, *Diabrotica virgifera* Le Conte	Monogenic, recessive
	Stem borer, *Chilo partellus* (Swinhoe)	Polygenic
Cotton	Boll weevil, *Anthonomus grandis* Boheman	2 dominant genes
	Jassid, *Empoasca* sp.	2 dominant genes
	Thrips, *Thrips* sp.	3 pairs of dominant genes
	Tobacco budworm, *Heliothis virescens* (Fabricius)	2 dominant additive genes
Cowpea	Aphid, *Aphis craccivora* Koch	2 nonallelic dominant genes
	Pod borer, *Maruca testulalis* (Geyer)	Polygenic
Rice	Brown planthopper, *Nilaparvata lugens* (Stal)	Monogenic, 4 dominant & 5 recessive genes
	Whitebacked planthopper, *Sogatella furcifera* (Horvath)	5 dominant, one recessive gene
	Zigzag leafhopper, *Recilia dorsalis* (Motschulsky)	3 dominant genes
	Gall midge, *Orseolia oryzae* (Wood-Mason)	3 pairs of recessive genes
	Green leafhopper, *Nephotettix virescens* (Distant)	Monogenic, 6 dominant & 2 recessive genes
	Striped rice borer, *Chilo suppressalis* (Walker)	Polygenic
	Yellow stem borer, *Scirpophaga incertulas* (Walker)	2 duplicate recessive genes
Sorghum	Chinch bug, *Blissus leucopterus* (Say)	Monogenic, single dominant gene
	Greenbug, *Schizaphis graminum* (Rondani)	Monogenic, single
	Shootfly, *Atherigona soccata* Rondani	Polygenic
Wheat	Cereal leaf beetle,	Polygenic

Table 5.3 contd...

1	2	3
	Oulema melanopus (Linnaeus)	
	Greenbug, *S. graminum*	Monogenic
	Hessian fly, *Mayetiola destructor* (Say)	Dominant and recessive genes involved
	Sawfly, *Cephus cinctus* Norton	1 or more dominant, recessive or complementary genes

Source: Saxena and Khan (1991)

Polygenic resistance is moderate, but it is more stable and longer lasting than major gene resistance. This type of resistance is biotype-non-specific due to the involvement of several minor genes and there is less likelihood that biotypes will develop. The resistance against the European corn borer, *Ostrinia nubilalis* (Hubner); the fall armyworm, *Spodoptera frugiperda* (J.E. Smith) and the squash bug, *Anasa tristes,* is polygenic. No biotypes of any of these insects have been reported so far (Gallun and Khush, 1980).

5.8. INSECT BIOTYPES

The protective properties of insect resistant cultivars may be overcome by the development of resistance-breaking insect populations or biotypes, that possess an inherent genetic capability to overcome plant resistance (Smith, 1989). Insect biotypes are the populations that are capable of thriving on and damaging plants that are resistant to other populations of the same species. They represent dynamic biological and taxonomic units that exist in nature as products of evolution and are capable of defeating host plant resistance. They are often selected out of a general population by a cultivar developed for resistance and grown in areas where exposure to the insect is common. Recognition of pest biotypes has strengthened the co-evolutionary concept of host plant-insect interactions (Kogan, 1975). However, failure to recognise the existence of insect biotypes may lead to severe infestation of crop plants even on formerly resistant cultivars. The study of insect biotypes has become a significant part of insect resistance programmes and has provided tools for the analysis of insect plant relationships that serve as bases for breeding resistant plants (Saxena and Barrion, 1985, 1987).

Although biotype is a fairly recent term, the basic concept is more than one hundred years old (Saxena and Barrion, 1987). Walsh (1864) recognised phytophagic varieties among phytophagous insects. Cholodkovsky (1908) working with adelgids, used the term biological species for populations which he regarded as distinct species because of their differing biologies. Broadly considered, the term biotype is an intraspecific category referring to insect populations of similar genetic composition for a biological attribute. The biotype populations may be partially and temporarily *sympatric* (coincident), *allopatric* or *parapatric* with other compatible populations, but differ in one or more biological attributes. Some of the commonly known biotypes are listed in Table 5.4. Biotypes reportedly occur in 36 arthropod species belonging to 17 families of six orders (Saxena and Barrion, 1987). Almost 50 per cent of the known biotypes are encountered among aphids.

The actual mechanism in the development of pest biotypes can be best explained by the 'gene-for-gene' theory propounded by Flor (1956) and Person (1959) in order to interpret relationships in host-parasite systems, involving pathogenic fungi. Insect biotypes are more likely to develop on plants that have monogenic resistance than on plants with polygenic resistance. Also, the possibility of the development of new biotypes by immigration of a new population to an area or through mutation, chromosomal translocations, genetic recombinations, and other evolutionary phenomena can not be overlooked.

The identification of new biotypes is of utmost importance in accurately identifying genetic resources of resistance in crop plants. The methods used to determine differences vary considerably. In many insects, biotypes can be detected by the response of a group of differential host varieties to an insect population (Gallun *et al.*, 1961; Oka, 1978; Saxena and Barrion, 1985; Lidell and Schuster, 1990). In others, electrophoretic (Sogawa, 1978; Saxena and Mujer, 1984; Wellso *et al.*, 1988; Black *et al.*, 1990), morphometric (Kim *et al.*, 1967; Saxena and Rueda, 1982; Fargo *et al.*, 1986; Damsteegt and Voegtlin, 1990) or cytological (Mayo and Starks, 1972; Saxena and Barrion, 1983, 1985; Mayo *et al.*, 1988) methods may be used to detect differences between biotypes. Majority of insect biotype occurrences are concerned with insect virulence in different host varieties (Maxwell and Jennings, 1980).

5.9. FUTURE CONSIDERATIONS AND CONCLUSIONS

Deployment of host plant resistance will continue to be an important strategy in pest control in the foreseeable future. It is a key component of varietal improvement programmes. Several national and international crop improvement programmes devote a major effort toward identifying and incorporating genetic resistance to important insect pests into their breeding materials. However, the fact that continuous planting of crop varieties with monogenic resistance may lead to evolution of virulent biotypes, the future thrust should be to broaden the genetic base of resistance both in terms of major and minor genes. Deployment of varieties with diverse resistance backgrounds should become the

ected examples of biotypes of agricultural insect pests

Family	Species	Common name	Crop(s)	Biotypes (No.)
2	3	4	5	6
Curculionidae	*Phaedonia inclusa*	Soybean weevil	Soybean	1
	Sitophilus oryzae (Linnaeus)	Rice weevil	Corn, wheat	2
Agromyzidae	*Agromyza oryzae* (Munakata)	Rice leaf miner	Rice	2
Cecidomyiidae	*Mayetiola destructor* (Say)	Hessian fly	Wheat	10
	Orseolia oryzae (Wood-Mason)	Gall midge	Rice	4
Chloropidae	*Chlorops oryzae* Matsumura	Rice stem maggot	Rice	2
Tephritidae	*Rhagoletis cerasi* (Linnaeus)	Cherry fruit fly	Cherry	2
Aphididae	*Acyrthosiphon pisum* (Harris)	Pea aphid	Afalfa, broad bean, lucerne, pea	9
	Amphorophora rubi (Kaltenbach)	Raspberry aphid	Blackberry, dewberry, raspberry	6
	Aphis craccivora Koch	Cowpea aphid	Grountnut, bush sitao	5
	Aphis fabae Scopoli	Bean aphid	Broad bean	2
	Aphis gossypii Glover	Melon or cotton aphid	Pepper	2
	Aphis nasturtii Kaltenbach	Buckthorn aphid	Potato	2
	Aulacorthum solani (Kaltenbach)	Green-spotted potato aphid	Potato	1
	Brevicoryne brassicae (Linnaeus)	Cabbage aphid	Cabbage	2
	Chaetosiphon fragaefolii (Cockerell)	Strawberry aphid	Strawberry	2
	Dysaphis devecta (Walker)	Apple aphid	Apple	3
	Eriosoma lanigerum (Hausmann)	Woolly apple aphid	Apple	1
	Macrosiphum euphorbiae (Thomas)	Potato aphid	Potato	2

Table 5.4. contd...

			aphid		
		Schizaphis graminum (Rondani)	Greenbug	Barley	1
				Oat	4
				Sorghum	5
				Wheat	3
		Therioaphis trifolii (Monell)	Yellow clover aphid	Alfalfa	8
	Cicadellidae	*Muelerianella farmairei*	Leafhopper	—	1
		Nephotettix virescens (Distant)	Green leafhopper	Rice	2
	Coccidae	*Saissetia oleae* (Olivier)	Black scale	Citron melon	1
	Delphacidae	*Nilaparvata lugens* (Stal)	Brown planthopper	Rice	5
	Diaspididae	*Lepidosaphes ulmi* (Linnaeus)	Oystershell scale	Apple	2
	Phylloxeridae	*Viteus vitifoliae* (Fitch)	Grape phylloxera	Grapevine	2
Lepidoptera	Pyralidae	*Ostrinia nubilalis* (Hubner)	European corn borer	Corn	4
	Toricidae	*Cydia pomonella* (Linnaeus)	Walnut codling moth	Apple, plum, walnut	3
	Yponomeutidae	*Yponomeuta pedellus* (Linnaeus)	Small ermine moth	Apple, hawthorn	1
Thysanoptera	Thripidae	*Thrips tabaci* Lindeman	ion thrips	Tobacco, deadnettle	2

Source: Saxena and Barrion (1987)

cornerstone of the achievement of long-term stability against rapidly evolving pests and their biotypes. Several short-and long-term strategies are being deployed in breeding for insect resistance in crop varieties. These include sequential release of varieties with major genes, pyramiding of major genes, multiline varieties, development of horizontal resistance, wide hybridization and tissue culture techniques.

Efforts are underway to incorporate the *Bacillus thuringiensis* (*Bt*) gene into crop plants through genetic engineering. Although *Bt* toxins have an excellent safety track record in agriculture, their increased use in some crops has led to resistance within a relatively short time span (McGaughey, 1985). Although deployment strategies such as mixtures of *Bt* and non-*Bt* plants, and rotation of *Bt* and non-*Bt* plants have been proposed for pest resistance management, their practical value in agriculture is still risky as experienced recently by farmers in the USA growing genetically engineered cotton with *Bt* genes. While genetic engineering facilitates a reduction in the time span for producing "novel plants", more time must be allocated for carefully ascertaining the risks involved, whether they pertain to possible gene transfer to wild or weedy plants or the development of pest resistance.

On the other hand, varieties with relatively more durable resistance are expected to provide effective insect control for many years by slowing down the process of biotype development. Increased understanding of the causes of resistance, identification of chemicals that confer resistance or susceptibility, and the study of their inheritance would greatly improve breeding for durable insect resistance. If resistance involves more than one defense chemical, it may become possible to develop durable resistance, because insect pests are not likely to overcome sensitivity to several allelochemicals simultaneously.

There is a great need to develop improved crop varieties with multiple resistance to important insect pests and diseases. However, the success of such a programme largely depends upon the availability of sources of resistance, a well-planned breeding strategy, and a liberal exchange of breeding materials among participating national and international programmes. International cooperation is also essential in collecting and evaluating germplasm, studying insect biotypes and identifying diverse genes for resistance. Development of precise and creative methods to enhance natural variation within crop cultivars should pave the way for ecologically sound and economically rewarding pest management. Also, the buffering effect of plant resistance should merit adequate attention in the context of integrated pest management. Such an approach would lead to a greater appreciation of the importance of using even moderately resistant and pest tolerant crop cultivars and would favour the conservation of pests' natural enemies.

REFERENCES

Agarwal, R.A. 1969. Morphological characteristics of sugarcane and insect resistance. *Ent. exp. Appl.* **12**: 767-776.

Ahmad, S. 1983. *Herbivorous Insects: Host-Seeking Behaviour and Mechanisms.* Academic Press, Inc., Orlando, Florida, USA.

Anstey, T.H. and Moore, J.F. 1954. Inheritance of glossy foliage and cream petals in green sprouting broccoli. *J. Hered.* **45**: 39-41.

Archer, T.L., Onken, A.B., Bynum, E.D., Jr. and Patterson, G.C. 1990. Banks grass mite (*Oligonychus pratensis*) abundance of sorghum cultivars with different levels of nitrogen use and metabolism efficiency. *Exp. appl. Acarol.* **9**: 177.

Atkin, D.S.J. and Hamilton, R.J. 1982. The effects of plant waxes on insects. *J. Nat. Prod.* **45**: 694-696.

Auclair, J.L. 1957. Development of resistance of plants to insects. *A. Rep. Ent. Soc. Ontario* **88**: 7-17.

Auclair, J.L. 1989. Nutrition of plant-sucking insects on chemically defined diets. *Ent. exp. Appl.* **12**: 623-641.

Barbosa, P. and Letourneau, D.K. (eds). 1988. *Novel Aspects of Insect-Plant Interactions.* John Wiley & Sons, New York, USA.

Barbour, J.D., Farrar, R.R., Jr. and Kennedy, G.G. 1991. Interactions of fertilizer regime with host plant resistance in tomato. *Ent. exp. Appl.* **60**: 189.

Beck, S.D. 1956. Nutrition of the European corn borer. IV. Feeding reactions of first instar larvae. *Ann. Ent. Soc. Am.* **49**: 399-405.

Beck, S.D. 1965. Resistance of plants to insects. *A. Rev. Ent.* **10**: 207-232.

Beck, S.D. and Hanec, W. 1958. Effect of amino acids on feeding behaviour of the European corn borer *Pyrausta nubilalis* (Hübn.). *J. Insect Physiol.* **2**: 85-96.

Beck, S.D. and Reese, J.C. 1976. Insect-plant interactions: nutrition and metabolism. *In:* J.W. Wallace and R.l. Mansell (eds). *Recent Advances in Phytochemistry, Vol. 10.* Plenum Press, New York, pp. 41-92.

Bernays, E.A. and Chapman, R.F. 1994. *Host-Plant Selection by Phytophagous Insects.* Chapman & Hall, New York, USA.

Black IV, W.C., Hatchett, J.H. and Krchma, L.J. 1990. Allozyme variation among populations of the Hessian fly (*Mayetiola destructor*) in the United States. *J. Hered.* **81**: 331-337.

Brattsten, L.B. and Ahmed, S. (eds). 1986. *Molecular Aspects of Insect-Plant Associations.* Plenum Press, New York, USA.

Brues, C.T. 1920. The selection of food plants by insects with special reference to lepidopterous larvae. *Am. Nat.* **54**: 313-322.

Brues, C.T. 1924. The specificity of food plants in the evolution of phytophagous insects. *Am. Nat.* **58**: 127-144.

Chang, N.T., Wiseman, B.R., Lynch, R.E. and Habeck, D.H. 1985. Resistance to fall armyworm: influence of nitrogen fertilizer application on nonpreference and antibiosis in selected grasses. *J. agric. Ent.* **2**: 137-146.

Chapman, R.F. and Bernays, E.A. 1987. *Insect and Host Plant.* Proceedings of the 4th International Symposium, 4-9 June 1978, Grange, Slough, England, Nederlandse Entomologische Vereniging.

Chiang, H.S. and Jackai, L.E.N. 1988. Tough pod wall : a factor involved in cowpea resistance to pool sucking bugs. *Insect Sci. Applic.* **9**: 389-393.

Chiang, H.S. and Norris, D.M. 1984. "Purple stem", a new indicator of soybean stem resistance to bean flies (Diptera: Agromyzidae). *J. econ. Ent.* **77**: 121-125.

Cholodkovsky, N. 1908. Zur frage iiber die biologischen arten. *Biol. Centralbl.* **28**: 169-182.

DaCosta, C.P. and Jones, C.M. 1971. Cucumber beetle resistance and mite susceptibility controlled by bitter gene in *Cucumis sativus* L. *Science* **172**: 1145-1146.

Damsteegt, V.D. and Voegtlin, D.J. 1990. Morphological and biological variation among populations of *Aulacorthum solani* (Homoptera: Aphididae), the vector of soybean dwarf virus. *Ann. ent. Soc. Am.* **83**: 949-955.

Dhaliwal, G.S. and Dilawari, V.K. (eds). 1993. *Advances in Host Plant Resistance to Insects.* Kalyani Publishers, New Delhi, India.

Djamin, A. and Pathak, M.D. 1967. Role of silica in resistance to Asiatic rice borer, *Chilo suppressalis* (Walker), in rice varieties. *J. econ. Ent.* **60**: 347-351.

Duffey, S.S. and Isman, M.B. 1981. Inhibition of insect larval growth by phenolics in glandular trichomes of tomato leaves. *Experientia* **37**: 574.

Dunn, J.A. and Kempton, D.P.H. 1976. Varietal differences in the susceptibility of Brussel sprouts to lepidopterous pests. *Ann. appl. Biol.* **82**: 11-19.

Ehrlich, P.R. and Raven, P.H. 1964. Butterflies and plants: a study in coevolution. *Evolution* **18**: 585-608.

Eigenbrode, S.D. and Shelton, A.M. 1990. Behaviour of neonate diamondback moth larvae (Lepidoptera: Plutellidae) on glossy-leaved resistant *Brassica oleracea* L. *Envir. Ent.* **19**: 1566-1571.

Eigenbrode, S.D., Stoner, K.A., Shelton, A.M. and Kain, W.C. 1991. Characteristics of glossy leaf waxes associated with resistance to diamondback moth (Lepidoptera: Plutellidae) in *Brassica oleracea. J. econ. Ent.* **84**: 1609-1618.

Fargo, W.S., Inayatullah, C., Webster, J.A. and Holbert, D. 1986. Morphometric variation within apterous females of *Schizaphis graminum* biotypes. *Res. Popul. Ecol.* **28**: 163-172.

Farrar, R.R., Jr. and Kennedy, G.G. 1987. Growth, food consumption and mortality cf *Heliothis zea* larvae on foliage of the wild tomato *Lycopersicon hirsutum* f. *glabratum* and the cultivated tomato, *L. esculentum. Ent. exp. Appl.* **44**: 213-219.

Fiori, B.J. and Craig, D.W. 1987. Relationship between color intensity of leaf supernatants from resistant and susceptible birchtree and rate of oviposition by the birch leafminer (Hymenoptera: Tenthredinidae). *J. econ. Ent.* **80**: 1331-1333.

Fiori, B.J. and Dolan, D.D. 1981. Field tests for *Medicago* resistance against the potato leafhopper (Homoptera: Cicadellideae). *Can. Ent.* **113**: 1049-1053.

Flor, H.H. 1956. The complementary genetic systems in flax and flax rust. *Adv. Genet.* **8**: 29-54.

Fraenkel, G. 1959. The raison d'etre of secondary plant substances. *Science* **129**: 1466-1470.

Fraenkel, G. 1969. Evaluation of our thoughts on secondary plant substances. *Ent. exp. Appl.* **12**: 473-486.

Frazier, J.L. 1986. The perception of plant allelochemicals that inhibit feeding. *In*: L.B. Brattsten and S. Ahmad (eds). *Molecular Aspects of Insect-Plant Interactions.* Plenum Press, New York, pp. 1-42.

Frazier, J.L. and Hanson, F.E. 1986. Electrophysiological recording and analysis of insect chemosensory responses. *In*: J.R. Miller and T.A. Miller (eds). *Insect-Plant Interactions.* Springer-Verlag, New York, USA, pp. 285-330.

Gallun, R.L. and Khush, G.S. 1980. Genetic factors affecting expression and stability of resistance. *In*: F.G. Maxwell and P.R. Jennings (eds). *Breeding Plants Resistant to Insects.* John Wiley & Sons, New York, USA, pp. 63-85.

Gallun, R.L., Deay, H.O. and Cartwright, W.B. 1961. Four races of Hessian fly selected and developed from an Indiana population. *Purdue Univ. Res. Bull.* **732**: 8.

Gallun, R.L., Roberts, J.J., Finney, R.E. and Patterson, F.L. 1973. Leaf pubescence of field grown wheat: a deterrent to oviposition by the cereal leaf beetle. *J. Envir. Qual.* **2**: 333.

Gaynor, D.L. and Hunt, W.F. 1983. The relationship between nitrogen supply, endophytic fungus, and Argentine stem weevil resistance in ryegrass. *Proc. N.Z. Grassland Assoc.* **44**: 257-263.

Girma, M., Wilde, G. and Reese, J.C. 1990. Influence of temperature and plant growth stage on development, reproduction, life span, and intrinsic rate of increase of the Russian wheat aphid (Homoptera: Aphididae). *Envir. Ent.* **19**: 1438-1442.

Hanifa, A.M., Subramaniam, T.R. and Ponnaiya, B.W.X. 1973. Role of silica in resistance to the leafroller, *Cnaphalocrocis medinalis* Guenee, in rice. *Indian. J. expt. Biol.* **12**: 463-465.

Harlan, S.C. 1916. The inheritance of immunity of leaf blister mite (*Eriophyes gossypii* Banks) in cotton. *West Indian Bull.* **17**: 162-166.

Harris, M.K. 1979. *Biology and Breeding for Resistance to Arthropods and Pathogens in Agricultural Plants.* Proc. Int. Short Course in Host Plant Resistance. Texas A & M University, Texas, USA.

Harris, M.O. and Miller, J.R. 1983. Color stimuli and oviposition behaviour of the onion fly, *Delia antiqua* (Meigen) (Diptera: Anthoyiidae). *Ann. ent. Soc. Am.* **76**: 766.

Harris, M.O. and Rose, S. 1990. Chemical, color and tactile cues influencing oviposition behaviour of the Hessian fly (Diptera: Cecidomyiidae). *Envir. Ent.* **19**: 303-308.

Havens, J.N. 1792. Observations on Hessian fly. *Trans. N.Y. Soc. Agron. Pt.* **1**: 89-107.

Heinrichs, E.A. (ed.). 1998. *Plant Stress-Insect Interactions.* John Wiley & Sons, New York, USA.

Holloway, P.J. and Baker, E.A. 1968. Isolation of plant cuticles with zinc chloride-hydrochloric acid solution. *Pl. Physiol.* **43**: 1878-1879.

Horber, E. 1972. Plant resistance to insects. *Agric. Sci. Rev.* **10**: 1-18.

House, H.L. 1962. Insect nutrition. *A. Rev. Biochem.* **31**: 653-672.

Howard, L.O. 1930. *A History of Applied Entomology.* Smithsm, Misc. Collect, 34. Smithsonian Institute.

Hoxie, R.P., Wellso, S.G. and Webster, J.A. 1975. Cereal leaf bettle response to wheat trichome length and density. *Envir. Ent.* **4**: 365.

Issac, P.V. 1939. How mid-rib hardness affords resistance to the sugarcane top-borer *Scirpophaga nivella* F., in India. *Curr. Sci.* **8**: 211-212.

Jermy, T. 1976. *The Host Plant in Relation to Insect Behaviour and Reproduction.* Plenum Press, New York, USA.

Johnson, B. 1955. The injurious effects of the hooked epidermal hairs of French beans (*Phaseolus vulgaris* L.) on *Aphis craccivora* Koch. *Bull ent. Res.* **44**: 779-782.

Johnson, H.W. and Hollowell, E.A. 1935. Pubescent and glabrous characters of soybeans as related to resistance to injury by the potato leafhopper. *J. agric. Res.* **51**: 371-381.

Johnson, K.J.R., Sorensen, E.L. and Horber, E.K. 1980. Resistance in glandular-haired annual *Medicago* species to feeding by adult alfalfa weevils (*Hypera postica*). *Envir. Ent.* **9**: 133.

Juniper, B. and Southwood, R. (eds). 1986. *Insects and the Plant Surface.* Edward Arnold, London, UK.

Kang, M.S., Sosa, O., Jr. and Miller, J.D. 1990. Genetic variation and advance for rind hardness, flowering and sugar yield trials in sugarcane. *Field Crops Res.* **23**: 69-73.

Karner, M.A. and Manglitz, G.R. 1985. Effects of temperature and alfalfa cultivar on pea aphid (Homoptera: Aphididae) fecundity and feeding activity of convergent lady beetle (Coleoptera: Coccinelidae). *J. Kansas ent. Soc.* **58**: 131-136.

Kennedy, G.G., Yamamoto, R.T., Dimock, M.B., Williams, W.G. and Bordner, J. 1981. Effect of daylength and light intensity on 2-tridecanone levels and resistance in *Lycopersicon hirsutum* f. *glabratum* to *Manduca sexta. J. chem. Ecol.* **7**: 707-716.

Kennedy, J.S. and Booth, C.O. 1951. Host alteration in *Aphis fabae.* I. Feeding perferences and fecundity in relation to age and kind of leaves. *Ann. appl. Biol.* **38**: 25-64.

Kennedy, J.S., Lamb, K.P. and Booth, C.O. 1958. Responses of *Aphis fabae* Scop. to water shortage in host plants in pots. *Ent. exp. Appl.* **1**: 274-291.

Khan, Z.R. and Saxena, R.C. 1985a. Effect of steam distillate extract of a resistant rice variety on feeding behaviour of *Nephotettix virescens* (Homoptera: Cicadellidae). *J. econ. Ent.* **78**: 562-566.

Khan, Z.R. and Saxena, R.C. 1985b. Behavioural and physiological responses of *Sogatella furcifera* (Homoptera: Delphacidae) to selected resistant and susceptible rice cultivars. *J. econ. Ent.* **78:** 1280-1286.

Khan, Z.R. and Saxena, R.C. 1986. Effect of steam distillate extract of resistant and susceptible rice cultivars on behaviour of *Sogatella furcifera* (Homoptera: Delphacidae). *J. econ. Ent.* **79**: 928-935.

Khan, Z.R., Ward, J.T. and Norris, D.M. 1986a. Role of trichomes in soybean resistance to cabbage looper, *Trichoplusia ni*. *Ent. exp. Appl.* **42**: 109-117.

Khan, Z.R., Norris, D.M., Chiang, H.S., Weiss, N.E. and Oosterwyk, S.S. 1986b. Light-induced susceptibility in soybean to cabbage looper, *Trichoplusia ni* (Lepidoptera: Noctuidae). *Envir. Ent.* **15**: 803-808.

Khanna, K.L., Sharma, S.L. and Ramanathan, K.R. 1947. Studies in the association of plant characters and pest incidence: III. Hardness of leaf midrib and top borer infestation in sugarcane. *Indian J. Ent.* **9**: 115-137.

Khush, G.S. and Beachell, H.M. 1972. Breeding for disease and insect resistance at IRRI. *In: Rice Breeding*. International Rice Research Institute, Los Banos, Philippines, pp. 309-322.

Kim, K.C., Brown, Jr., B.W. and Cook, E.F. 1967. Morphometric differences among four biotypes of *Ostrinia nubilalis* (Lepidoptera: Pyralidae). *Ann. ent. Soc. Am.* **60**: 798-801.

Kindler, S.D. and Staples, R. 1970a. Nutrients and the reaction of two alfalfa clones to the spotted alfalfa aphid. *J. econ. Ent.* **63**: 938-1201.

Kindler, S.D. and Staples, R. 1970b. The influence of fluctuating and constant temperatures, photoperiod and soil moisture on the resistance of alfalfa to the spotted alfalfa aphid. *J. econ. Ent.* **63**: 1198-1201.

Kogan, M. 1975. Plant resistance in pest management. *In*: R.L. Metcalf and W. Luckman (eds). *Introduction to Insect Pest Management*. John Wiley & Sons, New York, USA, pp. 103-146.

Kogan, M. 1976. Resistance in soybean to insect pests. *In*: R.M. Goodman (ed.) *Proc. Conf. for Asia and Oceania*. Chiang-Mai, Thailand Univ. III, pp. 165-169.

Kogan, M. and Ortman, E.E. 1978. Antixenosis - A new term proposed to define Painter's 'nonpreference' modality of resistance. *Bull. ent. Soc. Am.* **24**: 175-176.

Labeyrie, V., Fabres, G. and Lachaise, D. (eds). 1987. *Insect-Plants*. Proceedings of the 6th International Symposium on Insect-Plant Relationships. Dr W. Junk Publishers, Dordrecht.

Lanning, F.C., Hopkins, T.L. and Loera, J.C. 1980. Silica and ash content and depositional patterns is tissues of mature *Zea mays* L. plants. *Ann. Bot.* **45**: 549-551.

Leuck, D.B. 1972. Induced fall armyworm resistance in pearl millet. *J. econ. Ent.* **65**: 1608-1611.

Leuck, D.B. and Hammons, R.O. 1974. Nutrients and growth media: influence on expression of resistance to the fall armyworm in the peanut. *J. econ. Ent.* **67**: 564.

Levin, D.A. 1973. The role of trichomes in plant defense. *Quart. Rev. Biol.* **48**: 3.

Lidell, M.C. and Schuster, M.F. 1990. Effectiveness of wheat genes for Hessian fly (Diptera: Cecidomyiidae) resistance in Texas. *J. econ. Ent.* **83**: 1135-1139.

Lin, S.Y.H., Trumble, J.T. and Kumamoto, J. 1987. Activity of volatile compounds in glandular trichomes of *Lycopersicon* species against two insect herbivores. *J. chem. Ecol.* **13**: 837-842.

Lindley, G. 1831. *A Guide to the Orchard and Kitchen Garden.* Longman, Rees, Orme, Brown and Green Publ. Co., London, UK.

Lowe, H.J.B. 1967. Interspecific difference in the biology of aphids (Homoptera: Aphididae) on leaves of *Vicia faba*. II. Growth and excretion. *Ent. exp. Appl.* **10**: 413-420.

Lupton, F.G.H. 1967. The use of resistant varieties in crop protection. *World Rev. Pest Cont.* **6**: 47-58.

MacLean, P.S. and Byers, R.A. 1983. Ovipositional preferences of the alfalfa blotch leafminer (Diptera: Agromyzidae) among some simple and glandular-haired *Medicago* species. *Envir. Ent.* **12**: 1083-1086.

Mandal, A.K., Ganguli, P.K. and Banerjee, S.P. (eds). 1991. *Advances in Plant Breeding.* CBS Publishers & Distributors, New Delhi, India.

Manuwoto, S. and Scriber, J.M. 1985a. Neonate larval survival of European corn borer, *Ostrinia nubilalis,* on high and low DIMBOA genotypes of maize: effects of light intensity and degree of insect inbreeding. *Agric. Ecosys. Envir.* **14**: 221-236.

Manuwoto, S. and Scriber, J.M. 1985b. Differencial effects of nitrogen fertilization of three corn genotypes on biomass and nitrogen utilization by the southern armyworm, *Spodoptera eridania. Agric. Ecosys. Envir.* **14**: 25-40.

Martin, G.A., Richard, C.A. and Hensley, S.D. 1975. Host resistance to *Diatraea saccharalis* (F.): Relationship of sugarcane node hardness to larval damage. *Envir. Ent.* **4**: 687-688.

Maxwell, F.G. 1977. Plant resistance to cotton insects. *Bull. ent. Soc. Am.* **23**: 199-203.

Maxwell, F.G. and Jennings, P.R. (eds). 1980. *Breeding Plants Resistant to Insects.* John Wiley & Sons, New York, USA.

Maxwell, F.G., Jenkins, J.N. and Parrtt, W.L. 1972. Resistance of plants to insects. *Adv. Agron.* **24**: 187-265.

Mayo, Z.B. and Starks, K.J. 1972. Sexuality of the greenbug, *Schizaphis graminum*, in Oklahoma. *Ann. ent. Soc. Am.* **65**: 671.

Mayo, Z.B., Starks, K.J., Banks, D.J. and Veal, R.A. 1988. Variation in chromosome length among five biotypes of greenbug (Homoptera: Aphididae). *Ann. ent. Soc. Am.* **81**: 128-131.

McGaughey, W.H. 1985. Insect resistance to the biological insecticide *Bacillus thuringiensis*. *Science* **229**: 193-195.

Miller, J.R. and Miller. T.A. (eds). *Insect-Plant Interaction*. Springer-Verlag, New York, USA.

Mitchell, H.C., Cross, W.H., McGovern, W.L. and Dawson, E.M. 1973. Behaviour of the boll weevil on frego bract cotton. *J. econ. Ent.* **66**: 677-680.

Mittler, T.E. and Dadd, R.H. 1963. Studies on the artificial feeding of the aphid. II. Relative survival, development and larviposition on different diets. *J. Insect Physiol.* **9**: 741-757.

Niles, G.A. 1980. Breeding cotton for resistance to insect pests. *In*: F.G. Maxwell and P.R. Jennings (eds). *Breeding Plants Resistant to Insects*. John Wiley & Sons, New York, pp. 337-369.

Norris, D.M. and Kogan, M. 1980. Biochemical and morphological bases of resistance. *In*: F.G. Maxwell and P.R. Jennings (eds). *Breeding Plants Resistant to Insects*. John Wiley & Sons, New York, USA, pp. 23-61.

Oka, I.N. 1978. Quick method for identifying brown planthopper biotypes in the field. *Int. Rice Res. Newsl.* **3**(6): 11-12.

Painter, R.H. 1951. *Insect Resistance in Crop Plants*. The MacMillan Company, New York, USA.

Painter, R.H. 1958. Resistance of plants to insects. *A Rev. Ent.* **49**: 546-548.

Painter, R.H. 1968. *Crops that Resist Insects Provide a Way to Increase World Food Supply*. Kansa State Agric. Exp Stn. Bull. No. 520, Kansas State University of Agriculture and Applied Science, Manhattan, Kansas, USA.

Panda, N. 1979. *Principles of Host-Plant Resistance to Insect Pests*. Allanheld Osmun and Universe Books, New York, USA.

Panda, N. and Khush, G.S. 1995. *Host Plant Resistance to Insects*. CAB International, Wallingford, UK.

Patanakamjorn, S. and Pathak, M.D. 1967. Varietal resistance of rice to Asiatic rice borer, *Chilo suppressalis* (Lepidoptera: Crambidae), and its association with various plant characters. *Ann. ent. Soc. Am.* **60**: 287-292.

Pathak, M.D. 1970. Genetics of plants in pest management. *In*: R.L. Rabb and F.E. Guthrie (eds). *Concepts of Pest Management.* North Carolina State University, Releigh, USA, pp. 357-386.

Pathak, M.D. and Saxena, R.C. 1979. Insect resistance. *In*: J. Sneep and A.J.T. Hendriksen (eds). *Plant Breeding Perspectives.* Centre for Agriculture Publishing Documentation, Wageningen, The Netherlands, pp. 270-285.

Pathak, M.D. and Saxena, R.C. 1980. Breeding approaches in rice. *In*: F.G. Maxwell and P.R. Jennings (eds). *Breeding Plants Resistant to Insects.* John Wiley & Sons, New York, USA, pp. 421-455.

Pathak, M.D. and Saxena, R.C. 1981. Insect resistance in crop plants. *In*: H. Smith (ed.). *Commentaries in Plant Science, Vol. 2.* Pergamon Press, New York, USA, pp. 61-81.

Person, C.O. 1959. Gene-for-gene relationship in host-parasite systems. *Can J. Bot.* **37**: 1101-1130.

Pillemer, E.A. and Tingey, W.M. 1976. Hooked trichomes: a physical plant barrier to a major agricultural pest. *Science* **193**: 482-484.

Pillemer, E.A. and Tingey, W.M. 1978. Hooked trichomes and resistance of *Phaseolus vulgaris* to *Empoasca fabae* (Harris). *Ent. exp. Appl.* **24**: 83-94.

Pittara, I.S. and Katsoyannos, B.I. 1992. Effect of shape, size and color on selection of oviposition sites by *Chaetorellia australis. Ent. exp. Appl.* **63**: 105-113.

Ponnaiya, B.W.X. 1951. Studies in the genus *Sorghum* : the cause of resistance in sorghum to the insect pest *Atherigona indica* M. *Madras Univ. J.,* Section B, **21**: 203-251.

Prokopy, R.J. and Bush, G.L. 1973. Ovipositional response to different sizes of artificial fruit by flies of *Rhagoletis pomonella* species group. *J. econ. Ent.* **66**: 927-929.

Quiros, C.F., Stevens, M.A., Rick, C.M. and Kok-Yokomi, M.L. 1977. Resistance in tomato to the pink form of the potato aphid (*Macrosiphum euphorbiae* Thomas): the role of anatomy, epidermal hairs, and foliage composition. *J. Am. Soc. Hort. Sci.* **102**: 166-171.

Ratanatham, S. and Gallun, R.L. 1986. Resistance to Hessian fly (Diptera: Cecidomyiidae) in wheat as affected by temperature and larval density. *Envir. Ent.* **9**: 443-448.

Reddy, P.S.C. 1974. Effects of three leaf shape genotypes of *Gossypium hirsutum* L. and row types on plant microclimate, boll weevil survival, boll rot and important agronomic characters. Ph.D. Dissertation, Louisiana State University, Baton Rouge, USA.

Reynolds, G.W. and Smith, C.M. 1985. Effects of leaf position, leaf wounding and plant age of two soybean genotypes on soybean looper (Lepidoptera: Noctuidae) growth. *Envir. Ent.* **14**: 475-478.

Ringland, K. and Everson, E.H. 1968. Leaf pubescence in common wheat, *Triticum aestivum* L., and resistance to the cereal leaf beetle, *Oulema melanopus* (L.). *Crop Sci.* **8**: 707-710.

Roberts, D.W.A. and Tyrrell, C. 1961. Sawfly resistance in wheat IV. Some effects of light intensity on resistance. *Can. J. Pl. Sci.* **41**: 457-465.

Roberts, J.J. and Foster, J.E. 1983. Effect of leaf pubescence in wheat on the bird cherry oat aphid (Homoptera: Aphididae). *J. econ. Ent.* **76**: 1320-1322.

Roberts, J.J., Gallun, R.L., Patterson, F.L. and Foster, J.E. 1979. Effects of wheat leaf pubescence on the Hessian fly. *J. econ. Ent.* **72**: 211-214.

Roessingh, R. and Stadler, E. 1990. Foliar form, color and surface characteristics influence oviposition behaviour in the cabbage root fly, *Delia radicum*. *Ent. exp. Appl.* **57**: 93-100.

Rogers, R.R. and Mills, R.B. 1974. Reaction of sorghum varieties to maize weevil infestation under relative humidities. *J. econ. Ent.* **67**: 692.

Russell, M.P. 1966. Effects of four sorghum varieties on the longevity of the lesser rice weevil. *Sitophilus oryzae* (L.). *J. stored Prod. Res.* **2**: 75-79.

Salas, M.L. and Corcuera, L.J. 1991. Effect of environment on gramine content in barley leaves and susceptibility to the aphid *Schizaphis graminum*. *Phytochemistry* **30**: 3237.

Salim, M. and Saxena, R.C. 1991a. Nutritional stresses and varietal resistance in rice: effects on whitebacked planthopper. *Crop Sci.* **31**: 797-805.

Salim, M. and Saxena, R.C. 1991b. Temperature stress and varietal resistance in rice: effects on whitebacked planthopper. *Crop Sci.* **31**: 1620-1625.

Salim, M. and Saxena, R.C. 1992. Iron, silica and aluminum stresses and varietal resistance in rice: effects on whitebacked planthopper. *Crop Sci.* **32**: 212-219.

Salim, M., Saxena, R.C. and Akbar, M. 1990. Salinity stress and varietal resistance in rice: whitebacked planthopper. *Crop Sci.* **30**: 654-659.

Sands, D.P.A. and Brancatini, V.A. 1991. A portable penetrometer for measuring leaf toughness in insect herbivory studies. *Proc. ent. Soc. Wash.* **93**: 786-788.

Saxena, K.N. 1969. Patterns of insect-plant relationships determining susceptibility or resistance of different plants to an insect. *Ent. exp. Appl.* **12**: 751-766.

Saxena, K.N. and Kumar, H. 1980. Interruption of acoustic communication and mating in a leafhopper and a planthopper by aerial sound vibrations picked up by plants. *Experientia* **36**: 933-937.

Saxena. K.N., Gandhi, J.R. and Saxena, R.C. 1974. Patterns of relationship between certain leafhoppers and plants. I. Responses to plants. *Ent. exp. Appl.* **17**: 303-318.

Saxena, R.C. 1986. Biochemical bases of insect resistance in rice varieties. *In*: M.B. Green and P. A. Hedin (eds). *Natural Resistance of Plants to Pests: Roles of Allelochemicals*. ACS Symp. Ser. 296, Am. Chem. Soc., Washington, DC, USA, pp. 142-159.

Saxena, R.C. and Barrion, A.A. 1983. Biotypes of the brown planthopper, *Nilaparvata lugens* (Stal). *Korean J. Pl. Prot.* **22**: 52-66.

Saxena, R.C. and Barrion, A.A. 1985. Biotypes of the brown planthopper, *Nilaparvata lugens* (Stal) and strategies in deployment of host plant resistance. *Insect Sci. Applic.* **6**: 271-289.

Saxena, R.C. and Barrion, A.A. 1987. Biotypes of insect pests of agricultural crops. *Insect Sci. Applic.* **8**: 453-458.

Saxena, R.C. and Khan, Z.R. 1991. Genetics of insect-host plant interactions: cencepts, old and new. *In*: A.K. Mandal, P.K. Ganguli and S.P. Banerjee (eds). *Advances in Plant Breeding, Vol. I.* CBS Publishers & Distributors, New Delhi, India, pp. 111-120.

Saxena, R.C. and Mujer, C.V. 1984. Detection of enzyme polymorphism among populations of brown planthopper (BPH) biotypes. *Int. Rice Res. Newsl.* **9**(3): 18-19.

Saxena, R.C. and Okech, S.H. 1985. Role of plant volatiles in resistance of selected rice varieties to the brown planthopper *Nilaparvata lugens* (Stal) (Homoptera: Delphacidae). *J. chem. Ecol.* **11**: 1601-1616.

Saxena, R.C. and Pathak, M.D. 1977. Factors affecting resistance of rice varieties to the brown planthopper, *N. lugens*. Paper presented at the 8th Conf. Pest Control Council Philippines, Bacolod City, Philippines, May 18-20, 1977.

Saxena, R.C. and Rueda, L.M. 1982. Morphological variations among three biotypes of brown planthopper *Nilaparvata lugens* in the Philippines. *Insect Sci. Applic.* **3**: 192-210.

Schalk, J.M., Kindler, S.D. and Manglitz, G.D. 1969. Temperature and preference of the spotted alfalfa aphid for resistant and susceptible alfalfa plants. *J. econ. Ent.* **62**: 1000-1003.

Schillinger, Jr. J.A. and Gallun, R.L. 1968. Leaf pubescence of wheat as a deterrent to the cereal leaf beetle, *Oulema melanopus*. *Ann. ent. Soc. Am.* **61**: 900-905.

Schultz, P.B. and Coffelt, M.A. 1987. Oviposition and nymphal survival of the hawthorn lace bug (Hemiptera: Tingidae) on selected species of *Cotoneaster* (Rosaceae). *Envir. Ent.* **16**: 365-367.

Schweissing. F.C. and Wilde, G. 1979. Temperature and plant nutrient effects on resistance of seedling sorghum to the greenbug. *J. econ. Ent.* **72**: 20-23.

Shade, R.E., Doskocil, M.J. and Maxon, N.P. 1979. Potato leafhopper resistance in glandular-haired alfalfa species. *Crop Sci.* **19**: 287-289.

Shade, R.E., Thompson, T.E. and Campbell, W.R. 1975. An alfalfa weevil larval resistance mechanism detected in *Medicago*. *J. econ. Ent.* **72**: 20-23.

Shifriss, O. 1981. Do *Cucurbita* plants with silvery leaves escape virus infection? *Cucurbit Gen. Coop. Rep.* **4**: 42-45.

Sikka, S.M., Sahni, V.M. and Butani, D.K. 1966. Studies on jassid resistance in relation to hairiness of cotton leaves. *Euphytica* **15**: 383-388.

Singh. D.P. 1986. *Breeding for Resistance to Diseases and Insect Pests*. Springer-Verlag, New York, USA.

Smith, C.M. 1989. *Plant Resistance to Insects: A Fundamental Approach*. John Wiley & Sons, New York, USA.

Smith, C.M., Khan, Z.R. and Pathak, M.D. 1994. *Techniques for Evaluating Insect Resistance in Crop Plants*. CRC-Lewis Press, Boca Raton, Florida, USA.

Snelling, R.O. 1941. Resistance of plants to insect attack. *Bot. Rev.* **7**: 543-586.

Sogawa, K. 1978. Electrophoretic variations in esterase among biotypes of the brown planthopper. *Int. Rice Res. Newsl.* **3**: 8-9.

Sosa, O., Jr. 1979. Hessian fly: Resistance of wheat as affected by temperature and duration of exposure. *Envir. Ent.* **8**: 280-281.

Springer, T.L., Kindler, S.D. and Sorensen, E.L. 1990. Comparison of podwall characteristics with seed damage and resistance to the alfalfa seed chalcid (Hymenoptera: Eurytomidae) in *Medicago* species. *Envir. Ent.* **19**: 1614-1617.

Stadler, E. 1977. Host selection and chemoreception in the carrot rustfly (*Psila rosae* F., Dipt. Psilidae): Extraction and isolation of oviposition stimulants and their perception by the female. *In*: *Comportement des Insects et Millieu Trophique*. C.N.R.S. Edition, 493 pp.

Stadler, E. and Buser, H.R. 1982. Oviposition stimulation for the carrot fly in the surface wax of carrot leaves. *In*: J.H. Visser and A.K. Minks (eds). *Proceedings of the 5th International Symposium of Insect-Plant Relationships*. Wageningen Centre for Agricultural Publishing and Documentation, Wageningen, The Netherlands. 464 pp.

Steffens, J.C. and Walters, D.S. 1991. Biochemical aspects of glandular trichome mediated insect resistance in the Solanaceae. *In*: P.A. Hedin (ed.). *Naturally Occurring Pest Bioregulators*. ACS Symp. Ser. 449, Am. Chem. Soc., Washington, DC, USA, pp. 137-149.

Stephens, S.G. 1957. Sources of resistance of cotton strains to the boll weevil and their possible utilization. *J. econ. Ent.* **50**: 415-418.

Stoner, K.A. 1990. Glossy leaf wax and plant resistance to insects in *Brassica oleracea* under natural infestation. *Envir. Ent.* **19**: 730-739.

Stork, N.E. 1980. Role of waxblooms in preventing attachment to brassicas by the mustard beetle, *Phaedon cochleariae*. *Ent. exp. Appl.* **28**: 99-106.

Tanton, M.T. 1962. The effect of leaf "toughness" on the feeding of larvae of the mustard beetle, *Phaedon cochleariae* Fab. *Ent. exp. Appl.* **5**: 74-78.

Thibout, E., Auger, J. and Lecomte, C. 1982. Host plant chemicals responsible for attraction and oviposition in *Acrolepiopis assectella*. *In*: L.J.H. Visser and A.K. Minks (eds). *Proceedings of the 5th International Symposium of Insect-Plant Relationships*. Wageningen Centre for Agricultural Publishing and Documentation, Wageningen, The Netherlands. 464 pp.

Thompson, K.F. 1963. Resistance to the cabbage aphid (*Brevicoryne brassicae*) in *Brassica* Plants. *Nature* **198**: 209.

Tingey, W.M. 1991. Potato glandular trichomes: defensive activity against insect attack. *In*: P.A. Hedin (ed.). *Naturally Occurring Pest Bioregulators*. ACS Symp. Ser. 449, Am. Chem. Soc., Washington, DC, USA, pp. 126-135.

Tingey, W.M. and Laubengayer, J.E. 1986. Glandular trichomes of a resistant hybrid potato alfter feeding behavior of the potato leafhopper (Homoptera: Cicadellidae). *J. econ. Ent.* **79**: 1230-1234.

Triebe, D.C., Meloan, C.E. and Sorensen, E.L. 1981. The chemical identification of the glandular hair exudate for *Medicago sculettater*.

In: Proceedings of 27th Alfalfa Improvement Conference, ARM-NC-19, U.S. Department of Agriculture, 52 pp.

Tyler, J.M. and Hatchett, J.H. 1983. Temperature influence on expression of resistance to Hessian fly (Diptera: Cecidomyiidae) in wheat derived from *Triticum tauschii. J. econ. Ent.* **76**: 323-326.

van de Klashorst, G. and Tingey, W.M. 1979. Effect of seedling age, environmental temperature, and foliar total glycoalkaloids on resistance of five *Solanum* genotypes of the potato leafhopper. *Envir. Ent.* **8**: 690-693.

van Emden, H.F. 1966. Plant resistance to insects induced by environment. *Sci. Hort.* **18**: 91-102.

van Emden, H.F. (ed.). 1973. *Insect-Plant Relationships.* Royal Entomological Society, London, UK.

Velusamy, R. and Heinrichs, E.A. 1986. Tolerance in crop plants to insect pests. *Insect Sci. Applic.* **7**: 689-696.

Visser, J.H. 1983. Differential sensory perceptions of plant compounds by insects. ACS Symp. Ser. No. 208, Am. Chem. Soc., Washington, DC, USA.

Walker, G.P. 1987. Probing and oviposition behavior of the bayberry whitefly (Homoptera: Aleyrodidae) on young and mature lemon leaves. *Ann. ent. Soc. Am.* **80**: 524-529.

Walker, G.P. 1988. The role of leaf cuticle in leaf age preference by bayberry whitefly (Homoptera: Aleyrodidae) on lemon. *Ann. ent. Soc. Am.* **81**: 365-369.

Walker, G.P. and Aitken, D.C.G. 1985. Oviposition and survival of bayberry whitefly, *Parabemisia myricae* (Homoptera: Aleyrodidae) on age. *Envir. Ent.* **14**: 254-275.

Wallace, L.E., McNeal, H. and Berg, M.A. 1974. Resistance to both *Oulema melanopus* and *Cephus cinctus* in pubescent-leaved and solid stemmed wheat selections. *J. econ. Ent.* **67**: 105.

Walsh, B.D. 1964. On phytophagic varieties and phytophagic species, with remarks on the unity of coloration of insects. *Proc. ent. Soc. Philadelphia* **3**: 403-430.

Walters, D.S., Harman, J., Craig, R. and Mumma, R.O. 1991. Effect of temperature on glandular trichome exudate composition and pest resistance in geraniums. *Ent. exp. Appl.* **60**: 61-69.

Webster, J.A. 1975. *Association of Plant Hairs and Insect Resistance.* An Annotated Bibliography, USDA-ARS Misc. Pub. No. 129.

Wellso, S.G., Howard, D.J., Adams, J.L. and Arnold, J. 1988. Electrophoretic monomorphism in six biotypes of the Hessian fly (Diptera: Cecidomyiidae). *Ann. ent. Soc. Am.* **81**: 50-53.

Whittaker, R.H. 1970. The biological ecology of higher plants. *In*: E. Sondheimer and J.B. Simeone (eds). *Chemical Ecology*. Academic Press, New York, USA, pp. 43-70.

Wiseman, B.R., Leuck, D.B. and McMillan, W.W. 1973. Effects of fertilizers on resistance of Antigua corn to fall armyworm and corn earworm. *Fla. Ent.* **56**: 1-7.

Wood, E.A. Jr. and Starks, K.J. 1972. Effects of temperature and host plant interaction on the biology of three biotypes of greenbug. *Envir. Ent.* **1**: 230-234.

Woodhead, S. 1981. Environmental and biotic factors affecting the phenolic content of different cultivars of sorghum bicolor.-*J. chem. Ecol.* **7**: 1035-1047.

Woodhead, S. and Padgham, D.E. 1988. The effect of plant surface characteristics on resistance of rice to the brown planthopper, *Nilaparvata lugens*. *Ent. exp. Appl.* **47**: 15-22.

Zettler, F.W., Smyly, M.O. and Evans, I.R. 1969. The repellency of mature citrus leaves to probing aphids. *Ann. ent. Soc. Am.* **62**: 399-402.

BOTANICAL PEST CONTROL

R.C. SAXENA
International Centre of Insect Physiology and Ecology
P.O. Box 30772, Nairobi, Kenya

6.1. INTRODUCTION

After about five decades of insecticide-based pest control in agriculture, forestry, household, and in public health, sufficient evidence has accumulated that continued use of broad-spectrum, synthetic insecticides may not only cause irreversible damage to the environment, but might also cause cancer and other unknown human health hazards. Moreover, the increasing incidence of pest resistance to pesticides and pesticide-induced pest resurgence, resulting in a pesticide treadmill situation, is causing serious concern worldwide (Pimentel and Lehman, 1993). It is being recognized that instead of striving for more "green revolutions" with emphasis on miracle seeds, hard-hitting, synthetic and engineered pesticides, and increased use of fertilisers, the future must look to natural ways and processes for augmenting agricultural productivity (Edwards, 1990). Sustainable agricultural systems must be *ecologically sound* for longterm food sufficiency, *equitable* in providing social justice, and *ethical* in respecting both future generations and other species. Technologies which do not assault the nature will, therefore, play key roles in ensuring future food security, improving human health, and rehabilitating and conserving the environment.

As far as pests are concerned, biological control, cultural control, habital management, including deployment of resistant cultivars, and other subtle modes of pest population regulation (discussed in other chapters), formulated as integrated pest management programmes, have been emphasized since the publication of Rachel Carson's *Silent Spring* in 1962. In the past two decades, interest in the use of pesticidal plants has also grown particularly those, that can be harvested, formulated, and easily used.

Botanical pest control is becoming relevant not only to developing countries with limited options for other control measures, but also attracting attention in developed countries which, though food sufficient, are increasingly becoming conscious of food safety. Other motivating factors for renewed interest in pesticidal plants are (i) the likelihood of chance discoveries of sources of new pesticides for commercial extraction, and (ii) that studies of modes of action and elucidation of chemical structures of bioactive principals may provide leads and insights into structural-activity relationships. Novel metabolites thus identified may serve as models for the chemical synthesis of new pesticides with more desirable properties. The greatest spin-off is expected in the field of insect control.

Plant species screened for insecticidal properties exceeded 6,000 by 1971. Of these, nearly 2,400 species belonging to 235 plant families are reported to exhibit measurable to considerable pest control activity (Crosby, 1971; Jacobson, 1975; Grainge and Ahmed, 1988). The plant species possessing pest control potential range from tiny mosses and lichens to giant trees; from grasses and sedges to legumes; from desert cacti and succulents to tropical rain forest trees; and from the vegetation found in the frigid polar regions to that found in torrid equatorial regions. The bioactive principals of pesticidal plants belong to the group of so-called secondary metabolites or allelochemicals, which includes thousands of alkaloids, terpenoids, phenolics and minor secondary chemicals. These substances have little or negligible function in photosynthesis, growth, or other aspects of basic plant physiology; however, their biological activity against pests: insects, nematodes, and pathogens is becoming well documented. Selective pressure exerted by plant pests since about the Devonian period probably led to the evolution of natural protective agents designed to disrupt pest activity. During this evolution, the ineffective chemical barriers were eliminated. However, some have remained effective killers of pestiferous insects. These bioactive compounds may kill directly, or indirectly by inhibiting feeding, thereby causing starvation or adversely affecting insect growth, reproduction, or behaviour. They may by isolated by the use of various extraction methods from either the whole plants or their specific parts such as flowers, seeds, leaves, stems, barks, roots or rhizomes. Plant species containing such chemicals are too numerous to be considered separately, so only those which have found some use in insect control directly or indirectly are dealt with.

6.2. PLANTS WITH INSECTICIDAL PROPERTIES

6.2.1. Pyrethrum

Pyrethrum, derived from the dried flower heads of *Chrysanthemum cinerariaefolium* Vis. (Asteraceae), is an outstanding example of a safe and effective natural insecticide. Though native to former Yugoslavia and Albania, it was introduced to Europe, the USA, Japan, East Africa, and South America in the 19th century. The pyrethrum products for insect control first became available commercially in 1914. The post-World War II surplus of DDT dealt a severe blow to the pyrethrum market. However, as insects developed resistance to DDT and other insecticides, and the hazards of their use were recognized, the pyrethrum demand rose again. Worldwide annual production of pyrethrum now averages 30,000 tons. Kenya is the world's largest producer of high quality pyrethrum. Tanzania, Rwanda, Zaire, Japan, and India also produce pyrethrum. The Persian insect flower, *Chrysanthemum roseum* Adam, is the basis of pyrethrum industry in Papua New Guinea.

Casida (1973) has reviewed in depth the scope of pyrethrum in insect control. The six insecticidal principals or rethrins in pyrethrum extracts are pyrethrins I and II, cinerins I and II, and jasmolins I and II. Pyrethrin I is the most effective. Insect susceptibility to pyrethrum is due to the cuticular permeability and sensitivity of internal tissue "receptors" that control oxidative enzyme systems. Pyrethrins quickly act on the insect central nervous system, causing a 'knockdown' effect. The insecticidal activity of pyrethrins is markedly increased by sesame oil and synergists like piperonyl butoxide. Pyrethrum products are commonly applied as oil-or water-based sprays containing 0.03-0.1 per cent pyrethrins and 5 to 10 times the concentration of synergists. Also, dusts are made by mixing pyrethrum extracts with clay or talc carriers; however, they deteriorate fast. There are five major uses of pyrethrum products, i.e., (i) household insecticides to control flies, mosquitoes, bedbugs, ants, roaches, etc., (ii) livestock or cattle sprays against biting and non-biting flies, (iii) sprays for mills, warehouses, and stored grains, (iv) dusts or sprays against pests of vegetable and fruit crops, and (v) sprays to suppress certain forest defoliators.

Although pyrethrins do not affect certain natural enemies of pests, they are non-persistent and their effects are short-lived. Pyrethrum does not harm bees, birds, fish, and larger animals, including human beings. Occasionally, pyrethrum dust does cause allergic reactions, but symptoms disappear when the allergen is removed. Some resistance to pyrethrins has been reported in cockroaches, bedbugs, houseflies, cattle ticks, body lice, and among a few agricultural pests. These species, however, also have high levels of resistance to organochlorines, organsophosphates, and carbamate insecticides.

The high cost of pyrethrins and their non-persistent character have limited their use in agriculture, but ultra-low volume (ULV) applications of pyrethrum formulations permit rapid treatment of large areas effectively and economically. To overcome the high cost and instability of natural pyrethrins under field conditions, chemists have come up with an array of synthetic analogs called 'pyrethroids' or 'synthetic pyrethroids', such as allethrin, decamethrin, permethrin, cypermethrin, and bioallethrin, which have higher toxicity to insects and remain lethal to insects even after exposure to sunlight (Elliott, 1977).

6.2.2. Rotenone and Rotenoids

For centuries, natives in many tropical countries used plant material from certain legumes, called *toeba* or *tuba, Derris* spp., in the Malay archipelago, and *timbo* or *cube, Lonchocarpus* spp. (all belonging to Fabaceae) in Central and South America, as fish poison or to poison arrows or spearheads. Perhaps the first agricultural use of *Derris* as an insecticide was recorded in Singapore in 1948 when root emulsions were smeared on young nutmeg plants to control insect pests. Then, Chinese farmers started using rotenone emulsions to control pests of vegetable crops. The active ingredient from *Derris chinensis* was called *roten* by natives in Formosa. Because it was a ketone, it was called rotenone. At least 10 chemically related rotenoids have been isolated. The commercially important species are *Derris elliptica* (Roxb.) and *D. malaccensis* which have an average content of 4-5 per cent, and *Lonchocarpus utilis* and *L. urucu* containing 8-10 per cent rotenone in dried roots. A few *Tephrosia* species are good sources of rotenoids. In Eastern and Southern Africa, *T. vogelii* Hook, *T. densiflora* Hook, and *T. candida* (Roxb.) have been cultivated to some extent by peasant farmers for use in crop protection, but have not been promoted to a commercial level.

Low in mammalian toxicity, rotenone is extremely active as a contact and stomach poison against many crop pests such as the Mexican bean beetle, woolly aphids on apple, the European corn borer, the pea aphid, and household pests. It is reasonably safe to honeybees. Rotenone dusts, dispersible powders, and sprays have been used in crop and animal health protection. Crystalline rotenone is used for moth-proofing. The action of rotenone as a contact insecticide and as a stomach poison is slow, often requiring several days to kill. The toxic principals deteriorate rapidly in sunlight and air. Synthetic rotenoids for agricultural purposes have not been used extensively because of the high cost of producing the complex molecules.

6.2.3. Nicotine and Other Alkaloids

Tobacco, *Nicotiana tabacum* L. (Solanaceae) originated in tropical America where it has been cultivated and used in religious ceremonies by the American Indians for at least 1000 years. The insecticidal use of tobacco in Europe began about 300 years ago when tobacco trade flourished between the American colonies and Europe. Although the principal toxic component, nicotine, was not then known, farmers used ground tobacco or its water extract to kill insects.

Nicotine is a fast-acting contact insecticide, which also acts as a fumigant or stomach poison. Nicotine alkaloid occurs in varying amounts in at least 18 *Nicotiana* species, but the commercially important species of *N. tabacum* L. and *N. rustica* L. contain up to 6 and 18 per cent nicotine, respectively. 'Black leaf 40' is one of the commercial formulations. Some unrelated taxa such as *Asclepias syriaca* L., *Atropa belladonna* L., *Equisetum arvense* L., and *Lycopodium clavatum* L. also contain nicotine. On a worldwide basis, about 600,000 kg of nicotine sulfate and 75,000 kg of pure nicotine were being produced annually till recently, mostly in U.K. Although nicotine is highly toxic to insects, it is no longer widely used. It has little residue problem, but the disagreeable smell, the activity limited mostly to minute, soft-bodied insects, such as aphids, and the relatively high production cost limit its use. Related alkaloids, such as nornicotine and anabasine, found in several species of plants, also have limited insecticidal use.

6.2.4. Ryania

Ryania speciosa Vahl. (Flacourtiaceae) is native to the northern part of South America and the Amazon Basin, where the plants have been uscd as a source of arrow poison. Powdered roots, leaves, and stems can be used as an insecticide. The insecticidal material can be extracted using solvents. Ryanodine is the active principal which acts both as a contact and stomach poison to insects. The major use of the insecticide is against the European corn borer, but the high production cost and poor persistence render its use uneconomical. However, its rapid breakdown into non-toxic chemicals shows it could be used safely on fruit and food crops even immediately before harvest.

6.2.5. Quassia

Quassia amara L. (Simarubaceae) is a tropical American tree with a characteristically bitter bark and wood. It occurs from the lower Amazon to Central America and in the West Indies. The yellowish-white wood is the source of 'quassia chips', from which a bitter extract with insect killing activity is made. Quassia chips normally contain <0.2 per cent of the insecticidal constituents, quassin and neoquassin. Despite its rather extensive use earlier, the use of quassia as an insecticide is now limited.

6.2.6. Sabadilla and Hellebore

As early as the 16th century, the natives in South and Central America and Mexico were using an insect-destroying substance made from the powdered seeds of a grassy bulb plant, *Schoenocaulon officinale* Gray (Liliaceae), called 'cebadilja'. The dried and powdered roots of a similar plant, *Veratrum album* L. were used in Europe to produce hellebore powder, which was a powerful insecticide. The insecticidal properties of sabadilla and hellebore are due to 'ceveratrum' alkaloids. Both sabadilla and hellebore lack field persistence and are extremely toxic to honeybees as well as humans. Sabadilla is now used on a limited scale against insect pests of crops, but hellebore is not.

6.2.7. Custard Apple

The dried seeds of *Annona reticulata* L. and *Annona squamosa* L. (Annonaceae) have long been used as an insecticide throughout many tropical countries. The insecticidal property is attributed to the benzylisoquinoline alkaloid, anonaine. *Annona* extracts act both as a contact and a stomach poison.

Besides the above mentioned plant species, Saxena and Dale (1989) have listed more than 100 species of plants with insecticidal properties, some of which have enjoyed temporary or restricted use. However, their insecticidally active principals are chemically obscure.

Resistant varieties of some cultivated crops also possess chemical molecules that are either toxic to insect pests or discourage their feeding. The discovery and identification of chemicals toxic to insect pests in resistant crop cultivars is of much scientific importance and economic significance. These unique chemicals not only serve as markers in breeding programmes, but may also provide basic chemical models for future insecticides with high specificity, low mammalian toxicity, and optimum degradability (Dhaliwal and Dilawari, 1996).

6.3. PLANTS WITH PHOTOTOXIN METABOLITES

Some species of plants produce "photosensitizers" or phototoxins (Downum, 1986; Arnason *et al.*, 1989). The metabolites are unique in that they become toxic to insects in the presence of sunlight, being activated particularly by the UVA region of the spectrum (320-400 nm). Such "solar-powered" defense agents may discourage insect herbivory. They are produced by a variety of biochemical pathways and are, therefore, a structurally diverse group of natural products. Members of Asteraceae and Rutaceae possess the widest range of phototoxic compounds. In field trials conducted in Tanzania, Arnason *et al.* (1989) demonstrated the larvicidal potential of *alpha*-terthienyl (*alpha*-T) against the malaria mosquito, *Anopheles gambiae* Giles. The effectiveness was on par with that of synthetic pyrethroids. However, the potential of phototoxins for insect control has not been much exploited and, so far, only one phototoxin, erythrosin B, has been registered for housefly control.

6.4. PLANTS WITH INSECT HORMONAL ACTIVITY

Hormones regulate the metamorphosis of insects from eggs to the adult stage. The most important insect hormones are the juvenile and molting hormones. If the hormone balance is altered in some way, the insect molts erratically, develops abnormally, and eventually dies. It has been found that many plants contain chemicals with insect hormonal activity, perhaps produced as a defense against insect herbivory.

Forty years ago, the serendipitous discovery of a juvenile hormone (JH) analog in the balsam fir tree, *Abies balsamea* (L.) Mill. (Pinaceae), the source of wood pulp for paper towel industry in North America, created much euphoria for using insect hormones as possible insecticides. Ether extracts of the paper towel induced abnormalities in the bug, *Pyrrhocoris apterus* Linnaeus, thus confirming the hormonal source. The coniferous western red cedar tree, *Thuja plicata* Don. (Cupressaceae), which occurs in East Asia and North America, also possesses insect juvenile hormone activity. The bioactive compound in the plant extract is thujic acid. From sweet basil, *Ocimum basiliocum* L. (Lamiaceae), a shrub common in tropical and subtropical regions, two highly active juvenile hormone analogs have recently been identified. From the common bedding plant, *Ageratum houstoniamum* Mill. (Asteraceae), two compounds, precocene I and precocene II, were identified, which on contact with certain insects induced irregular metamorphosis into small sterile adults. These chemicals were later shown to be anti-juvenile hormones, that prevent hormone production at the corpus allatum gland (Bowers *et al.*, 1976). Other *Ageratum* species also contain anti-juvenile hormones.

Analogs of insect molting hormones or ecdysones, normally produced in the prothoracic glands of immature insects, have also been isolated from plants. Thus, a molting hormone analog, called 'ponasterone A' was isolated from the white pine, *Podocarpus nakaii* Hay (Podocarpaceae). Essential oils obtained from some plants induce sterility in insects. For example, the oil of Indian calamus root, *Acorus calamus* L. (Zingiberaceae), induces sterility in male houseflies, female bean weevils, *khapra* beetles, and red cotton bugs. The active chemical responsible for sterility is β-asarone. Although several plants with hormonal activity have been identified, the natural products or their structural analogs have found little practical use in plant protection (Saxena and Khan, 1987).

6.5. PLANTS AFFECTING INSECT BEHAVIOUR AND PHYSIOLOGY

Some plants are not necessarily toxic but adversely affect the pest behaviour and impair the pest physiology. The most promising and most widely studied among them is neem, *Azadirachta indica* A. Juss. (Meliaceae). This tree is native to the drier parts of the Indian Sub-continent, but is now widely distributed in Africa, through planting during colonial times, and is also growing in Australia, Central and South America, and Caribbean islands, and Middle East, etc. Centuries before insecticides became available, farmers in India protected crops and stored food from insect attacks with natural repellents and feeding deterrents found in neem leaves and seed. In field trials, the pest control value of neem was first demonstrated by Pradhan *et al.* (1962) who during a locust invasion protected the standing crops at the Indian

Agricultural Research Institute, New Delhi, by spraying them with 0.1 per cent neem kernel suspension. Unfortunately, the pest control potential of neem was neglected with increased emphasis on synthetic insecticides during the "green revolution" era. Also, wide publicity given to slogans such as "the only good bug is a dead bug" and identifying traditional uses of neem as backward, gradually weaned Indian farmers away from using neem.

It is only in the past two decades that the pest control potential of neem has been rediscovered. Though subtle, neem's effects such as repellence, feeding and oviposition deterrence, disturbing insect host-endosymbiote equilibrium, growth inhibition, mating disruption, chemosterilization, reduction in egg hatchability, etc. are now considered far more desirable than a quick knock-down effect in integrated pest management programmes as they reduce the risk of exposing pests' natural enemies to poisoned food or starvation (Saxena, 1989; Schmutterer, 1990, 1995; NRC 1992; Ascher, 1993). In spite of high selectivity, neem derivatives affect more than 400 insect species belonging to Blattoda, Caelifera, Coleoptera, Dermaptera, Diptera, Ensifera, Hemiptera, Hymenoptera, Isoptera, Lepidoptera, Phasmida, Phthiraptera, Siphonoptera, and Thysanoptera, one species of ostracod, several species of mites, nematodes, and even noxious snails (Saxena, 1989; Schmutterer, 1995), and fungi, including aflatoxin-producing *Aspergillus flavus* (Zeringue and Bhatnagar, 1990).

The bioactive principals of neem or the "bitters" have been identified as limonoids, comprising more than 100 tetranortriterpenoids, diterpenoids, triterpenoids, pentanortriterpenoids, hexanortriterpenoids, and some nonterpenoids isolated from various parts of the neem tree; still more are being isolated. This formidable array of highly bioactive compounds makes neem a unique plant with potential applications for pest control in agriculture, stored products, forestry, and veterinary and public health. In general, insects feed far less, grow poorly, and lay significantly fewer eggs on plants treated with neem oil, neem cake, or neem extracts. Effects are similar for sucking and chewing insects; however, their predators and parasites are much less affected.

The limonoids in neem belong to nine basic structure groups, such as azadirone (from oil; all four rings of the basic skeleton intact), amoorastatin (from fresh leaves), vepinin (from seed oil), and vilasinin (from green leaves), the seco-systems related to gedunin (from seed oil and bark), nimbin (from leaves and seed), nimbolinin (from kernel), salannin (fron fresh leaves and seed), and the aza group (from neem seed; basically nimbolin type compounds) (Kraus, 1995). The best known molecule azadirachtin (Fig. 6.1) and its analogs have fascinated researchers for the past 30 years because of repellency, antifeedancy, reduced gut motility, growth inhibition, and chemosterlizing effects on insect pests (Ascher, 1993; Mordue (Luntz) and Blackwell 1993; Mordue (Luntz) *et al.*, 1985; Rembold, 1989; Saxena, 1989; Schmutterer, 1990, 1995). The azadirachtin content in neem varies considerably due to edaphic, climatic, or genotypic differences.

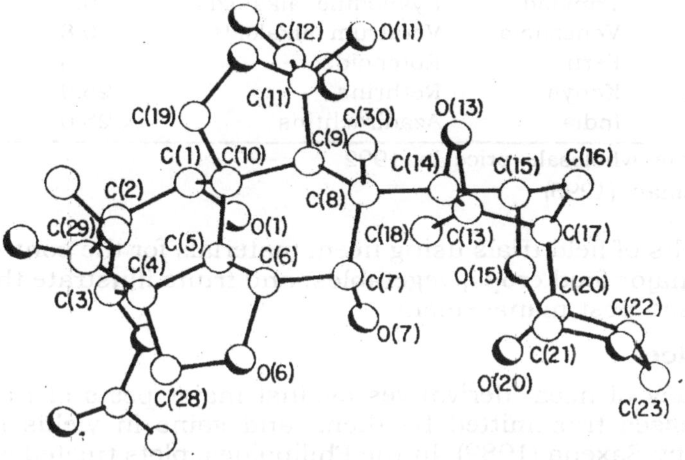

Fig. 6.1. Molecular structure of azadirachtin

Today, several neem-based insecticides are being marketed in many countries. In India, as of 1996, almost 40 neem-based insecticides have been registered provisionally or granted full registration. In North America, several neem-based insecticides have been registered and are now available for use on food, fruit, fodder, and fibre crops. Because pest control with neem is acceptable and safe, the demand for neem materials is growing rapidly. The complex molecular structure of azadirachtin precludes chemical synthesis in the near future. Efforts are underway to synthesize simpler mimetics of azadirachtin (Ley *et al.*, 1993), but the economics of production of "azadirachtoids" and their durability against pests will determine their success.

As neem does not thrive in temperate regions, a sustained supply of the raw material, which has to be shipped from tropical countries, is a limiting factor to increased use of neem in developed countries. This situation has greatly spurred the price of neem seed (from US$ 0.10/kg to US$ 0.20/kg) even in India. In Latin America, neem seed is selling at US$ 0.60 to 1.00/kg. Compared with other botanicals, the wholesale price of technical grade active concentrate of neem is several times higher (Table 6.1) (Isman, 1995). Efforts are underway in some advanced laboratories to produce azadirachtin and other bioactive neem constituents in *A. indica* cell- or callus cultures (Morgan *et al.*, 1996); however, the economic viability of such ventures remains to be ascertained.

Table 6.1. Active ingredient and comparative wholesale price of botanical insecticides (technical grade active concentrate or TGAC) currently registered in USA

Source of insecticide	Country of origin	Active ingredient	Per cent a.i. in TGAC	Price[1] (US$/kg)
Ryania	Trinidad	Ryanodine alkaloids	0.1	3
Sabadilla	Venezuela	Veratrum alkaloids	0.8	3
Rotenone	Peru	Rotenoids	7.3	8
Pyrethrum	Kenya	Rethrins	25.0	75
Neem	India	Azadirachtins	25.0	375

[1]Approximate wholesale price in 1992

Source: Isman (1995)

Results of field trials using neem materials for the control of pests of some major food crops, vegetables, and fruits illustrate the value of neem-based pest management.

6.5.1. Rice

The efficacy of neem derivatives against major pests of rice and the viral diseases transmitted by them, and gains in yields have been reviewed by Saxena (1989). In the Philippines, plots treated with a 2:10 neem cake-urea mixture applied at 120 kg/ha had lower incidence of ragged stunt, grassy stunt, and tungro viruses and yielded significantly more than control plots in both dry and wet seasons. Using enzyme-linked immunosorbent assay it was found that the application of neem cake reduced rice seedling infection by both rice tungro bacilliform and spherical virus particles. Also, in the Philippines, weekly spraying of 50 per cent neem oil-custard-apple oil mixture in 4:1 proportion (vol/vol) at 8 liters/ha from seedling to the maximum tillering stages significantly reduced tungro incidence and increased grain yield (Abdul Kareem *et al.*, 1987). The low input cost of the neem-treatment contributed to a higher net gain compared with the insecticide treatment (Table 6.2). In field trials conducted in India, neem treatments were found effective against populations of the green leafhopper, *Nephotettix virescens* (Distant); the yellow stem borer, *Scripophaga incertulas* (Walker); the rice gall mide, *Orseolia oryzae* (Wood-Mason); the rice leaf folder, *Cnaphalocrocis medinalis* (Guenee) and grasshoppers (Dhaliwal *et al.*, 1996; Nanda *et al.*, 1996).

Table 6.2. Comparative rice tungro virus (RTV) control, grain yield, and net gain in rice fields sprayed with neem oil and custard apple oil (NO:CAO) mixture or an insecticide[1]

Treatment	RTV (%)	Yield (t/ha)	Value of yield ($)	Cost of treatment ($)	Net gain (value of yield less cost of treatment[2])
			Jan-Apr 1984		
NO:CAO	5a	6.1a	1068	44	1024

Table 6.2 contd

1	2	3	4	5	6
BPMC	4a	6.1a	1068	125	943
Control	7a	5.6a	980	12	968
		Jun-Oct 1984			
NO:CAO	4a	5.1a	892	44	848
BPMC	6ab	4.7a	822	125	697
Control	9b	4.6a	805	12	793
		Nov 1984-Mar 1985			
NO:CAO	29a	3.1a	542	44	498
BPMC	56b	2.5b	438	125	313
Control	52b	2.3b	402	12	390

[1]Average of 4 replications per cropping season. Cost of rice = US$ 0.175/kg. NO:CAO mixture and BPMC were applied 8 times during each cropping season. Means followed by a common letter are not significantly different at the 5% level by Ducan's Multiple Range Test

[2]Cost of treatment included labour and materials. Control treatment cost included labour (US$ 10) and 8 pc DC batteries (US$ 2) for applying 1.6% Teepol-water solution (emulsifier) using an ultra-low volume applicator.

Source: Abdul Kareem *et al.* (1987)

6.5.2. Maize, Sorghum and Millet

Recently, in trials conducted at the ICIPE Field Station and in farmers' fields in Kenya, foliar application of powdered neem seed at 3 g/plant or powdered neem cake (deoiled) at 1 g/plant once at 4 weeks after crop emergence (WE) or twice at 4- and 6 WE to maize, which had been infested with eggs of the spotted stem borer, *Chilo partellus* (Swinhoe), significantly reduced foliar damage, stem tunneling, tassel breakage, and populations of the stalk borer larvae. Grain yield in neem-treated maize plots was as high as that obtained with insecticides and significantly higher than that in untreated control plots. Storage of neem cake up to 2 years in the dark did not reduce its effectiveness against pests (Table 6.3) (R.C. Saxena, unpublished). Late infestation of maize by field populations of *Eldana saccharina* Walker was low in neem-treated plots but not in insecticide-treated or in untreated control plots. Similar reduction in pest damage, reduction in pest body size (as measured by the width of larval head capsules), and an increase in yield were recorded when neem cake was applied to the sorghum crop. Because the cost of neem treatment was approximately one-tenth of the cost of the insecticide treatment, the net gain was higher with neem. In trials conducted in Mali, the use of local neem extract resulted in a significant 53 per cent increase in yield of early and main season millet as a result of the control of millet head pests, blister beetle, and head miner (Anonymous, 1996).

Table 6.3. Comparative stem borer infestation, and plant damage by *Chilo partellus* (Swinhoe) and *Eldana saccharina* Walker larvae, and grain yield in plots planted to ICZ5, a maize cultivar moderately resistant to stem borers, and applied with neem cake (NC) or a synthetic insecticide. ICIPE Field Station, Mbita, Kenya. Long-rains Cropping Season, 1994[1]

Treatment	Damage[2]	Plant height (cm)	Tunnel length (cm)		*Chilo* (no.)[3]	*Eldana* (no.)[4]	Tassel breakage (%)[5]	Yield (kg/ha)
	9WE	14WE	7WE	14WE				
Fresh NC	2.4a	180ab	0.6a	15.6a	1.5a	7.8a	8.6a	7458a
1-yr-old NC	2.4a	186a	0.7a	13.0a	1.5a	6.5a	8.1a	7760a
2-yr-old NC	2.4a	185a	0.6a	18.5a	0.3a	9.4a	9.4a	7469a
Dipterex	2.1a	177ab	0.5a	20.3ab	1.5a	17.0a	17.0a	7271a
Control (untreated)	4.6b	167b	7.8b	31.6b	14.3b	25.5b	35.5b	5088b

[1]Within a column, means followed by a common letter are not significantly different at the 5% level by the Least Significant Difference (LSD) Test; averages of 4 replications.

[2]Foliar damage scored visually on 1-9 scale (1=no damage, 9=completely damaged).

[3,4,5]*Chilo* larvae were recorded at 9WE; *Eldana* larvae at 11 WE, and tassel breakage at 11WE.

Source : R.C. Saxena (unpublished)

6.5.3. Grain Legumes and Vegetables

Because of high profitability, especially of vegetables, farmers tend to overuse chemicals which result in hazards to the environment and health of producers and consumers, as well as serious resistance problems. However, neem can provide satisfactory control of insect pests affecting grain legumes and vegetables. In field trials conducted in Kenya, applications of 2 or 3 per cent neem seed extract (NSE) @ 200 1/ha at 38, 47, and 51 days after emergence (DE) of the cowpea crop or ULV spray applications of 5,10 or 20 per cent NSE @ 10 1/ha often significantly reduced the number of larvae of the flower thrips, *Megalurothrips sjostedti* (Trybom), in cowpea flowers recorded 2 days after each treatment (Saxena and Kidiavai, 1997). Also, fewer adults occurred at 51 DE in plots sprayed with 5,10, or 20 per cent NSE. Cowpea grain yield was significantly higher in plots sprayed with 20 per cent NSE than in untreated control plots and was comparable with the grain yield obtained in plots sprayed thrice with cypermethrin. Because of the low cost of NSE treatment, the net gain was more when cowpea was sprayed with NSE than with cypermethrin. Neem derivatives also control a variety of pests on Bengal gram, chickpea, and pigeon pea (Sachan and Katti, 1996; Sinha, 1996).

In cabbage, weekly spray applications of 2.5 or 5 per cent aqueous neem seed kernel extract completely protected the crop against the diamondback moth, *Plutella xylostella* (Linnaeus), in Africa, and was superior to Dipel (Dreyer, 1987). Similar results have been reported from Asia. Other pyralid pests of cabbage and aphids, such as *Lipaphis erysimi* (Kaltenbach) and *Brevicoryne brassicae* (Linnaeus), are also controlled with neem (Ostermann and Dreyer, 1995). Neem treatments in cabbage intercropped with other vegetables in Mauritius controlled *P. xylostella* infestation and the combined effect was more effective than cartap hydrochloride, the recommended insecticide (Facknath, 1996). In tomato, weekly spray applications of 5 per cent aqueous neem seed extract controlled *Helicoverpa armigera* (Hubner) damage and increased the marketable fruit vield (Ostermann and Dreyer, 1995).

Certain azadirachtin-rich formulations have already been commercialized for crop pest control. 'Neemazal-T/S,' an oil-free, stable, emulsifiable concentrated with 1 per cent azadirachtin (a.i.) and other limonoids has been reported to be effective in several countries against major crop pests at 10-20 g a.i./ha (Senrayan, 1996). 'Neemix,' another azadirachtin-based pesticide, has been approved for commercial use on food crops in the USA (Walter, 1996). It has found wide use for controlling pests on peppers, melons, lettuce, pears, citrus celery, and other crops. Because of its inherent safety, 'Neemix' has been exempted from tolerance use at <50 g a.i./ha and has the shortest re-entry time allowed by the U.S. Environmental Protection Agency. Likewise, other neem-based insecticides with different concentrations of azadirachtin have been commercialized for horticultural, agricultural and turf care markets (Immaraju *et al.*, 1996).

6.5.4. Flowers and Ornamentals

Before the use of neem-based insecticides was permitted on food crops in the USA, they were first used successfully for the control of leaf miners on chrysanthemums, aphids on roses, chinch bugs on lawns, thrips on gladiolas, etc. In Holland, a neem oil-based water missible concentrate has been used as a foliar spray and as a soil drench on greenhouse roses and chrysanthemums for the control of whitefly, leafminers, mealy bugs, spider, mites, thrips, and centipedes (Ruiter, 1996). The results are promising. In field trials conducted recently in Kenya, insect control in roses with neem cake has been found to be on par with synthetic insecticides (R.C.Saxena, unpublished).

6.5.5. Fibre Crop

In field trials conducted in India, neem products alone, or with *Bt*, or conventional synthetic insecticides failed to suppress the cotton pest complex; however, the pests were controlled and cotton yield and quality improved when neem products were applied in combination with synthetic pyrethroids (Gupta, 1996). An azadirachtin-rich insecticide, RD-9 Repelin, controlled the bollworm complex on cotton in Punjab (Dhawan and Simwat, 1996) and in Andhra Pradesh, India, on par with quinalphos (Rosaiah and Reddy, 1996).

6.5.6. Forest Trees

Twelve species of lepidopteran and sawfly larvae were found to be highly susceptible to azadirachtin-rich NSE (Helson *et al.*, 1996). In field trials conducted in Canada, mistblower applications of azadirachtin at 50 g/ha during egg hatch significantly reduced the numbers of the pine false webworm larvae, *Acantholyda erythrocephala* (*L.*), infesting red pine and provided excellent control of 1-year old as well as young foliage. ULV spray applications at 10 l/ha (20 g a.i./ha) also gave excellent foliage protection. In field trials, neem treatments also were effective against the introduced pine sawfly, *Diprion similis* (Hartig), and the white pine weevil, *Pissodes strobi* (Peck). In pilot tests, systemic injections of emulsifiable neem concentrates into tree trunks controlled all the three pest species, indicating that this approach also has potential for the management of tree-feeding pests. Foliar and systemic applications of 'Azatin,' an azadirachtin rich insecticide, in a white spruce seed orchard at vegetative budswell and vegetative budflush effectively controlled defoliation by larvae of eastern spruce budworm, *Choristoneura fumiferana* (Clemens), indicating neem's potential in high value forest and urban settings (Wanner *et al.*, 1996).

6.5.7. Stored Products

Worldwide annual losses reach up to 10 per cent of all stored grains, i.e., 13 million metric tons of grain lost due to insects or 100 million tons due to failure to store properly (Wolpert, 1967). Saxena (1995) has reviewed the potential of neem products in reducing postharvest losses at farm level storage and warehouses. Powdered neem seed kernel mixed with paddy (1 to 2 %) significantly reduced infestation in warehouses. Neem leaves (2%) mixed with paddy, bags treated with 2 per cent neem extract, or 20- to 30-cm dried neem leaf barrier between the bags and storage floor, significantly reduced insect infestation and damage to grain during a 3-month storage period; the effectiveness being comparable with methacrifos dust. Likewise, NSE at 7.2 g/90 kg capacity jute bag (100 cm by 60 cm) controlled 80 per cent of the population of major insect pests and checked damage to wheat (Jilani and Amir, 1987). The treatment was effective up to 13 months and provided more than 70 per cent protection as compared with untreated control. The NSE treatment was as effective as that of 0.0005 per cent primiphos methyl mixed with the grain. Using this technology in Sind, Pakistan, high benefit cost-ratios were obtained by small-, medium-, and large-scale farmers.

The effectiveness of neem oil alone or in combination with fumigation was evaluated against five major species of stored grain pests infesting rice and paddy grains in a warehouse trial conducted in the Philippines (Jilani and Saxena, 1988). Rice grain treated with 0.05 to 1 per cent neem oil or treated with neem oil after fumigation with 'Phostoxin,' and stored for 8 months had significantly less *Tribolium castaneum* (Herbst) adults than in untreated control. Both kinds of neem treatments were as effective as the bag treatment with Actellic at 25 $\mu g/cm^2$ or grain treatment with Actellic at 0.0005 per cent and suppressed the pest population by 60 per cent. The population build-up also was reduced when either fumigated or nonfumigated rice was stored in bags treated with neem oil at > 1 mg/cm^2. *Rhyzopertha dominica* (Fabricius), *Sitophilus oryzae* (Linnaeus), *Oryzaephilus surinamensis* (Linnaeus), and *Corcyra cephalonica* (Stainton) were similarly reduced by neem treatments alone or in combination with prior grain fumigation. Fumigation with Phostoxin was effective only for about 2 months against *R. dominica*, and up to 6 months against other pest species, while neem oil treatments were effective up to 8 months. Compared with the pest damage to untreated or fumigated rice, neem oil treatments significantly reduced the damage to rice grain. At 8 months after storage, weevil attacked grains in neem treatments were 50 per cent of those in the fumigated rice and 25 per cent of those in the untreated rice. Neem treatments also reduced the pest populations and the damage in paddy. In studies conducted in Kenya, the growth and development of 1st instars of the maize weevil, *Sitophilus zeamais* Motschulsky, was completely arrested in maize grain treated with neem oil at 0.02 per cent, while the weight loss of treated cobs was less than 1 per cent as compared with a 50 per cent reduction in weight of untreated cobs stored for 6 months (Kega and Saxena, 1996).

While neem cannot completely replace chemical pesticides used in stored product preservation, the amounts of pesticides needed can be reduced, particularly in developing countries, thereby decreasing the pesticide load in food grains. With proper timing and innovative methods of application, their use could well be integrated in stored product pest management.

6.5.8. Blood-sucking Pests

The effects of neem on hematophagous insects affecting humans and livestock have been reviewed (Ascher and Meisner, 1989). Application of a paste made from neem leaves and turmeric in 4:1 proportion to the skin cured 97 per cent of the patients suffering from scabies caused by the mite, *Sarcoptes scabiei* (Linnaeus), in 3-15 days (Charles and Charles, 1992). Monthly sprays of ethanolic extracts of neem or weekly bathing in azadirachtin-rich aqueous 1:20 'Green Gold' controlled the bush tick, *Ixodes holocylus* Neumann, and the cattle tick, *Boophilus microplus* (Canestrini), in Australia, but were less effective against the brown dog tick, *Rhipicephalus sanguineus* (Latreille) (Rice, 1993). In Jamaica, neem kernel extract controlled ticks on cattle and dogs (Wilson, 1994). Neem products repel and affect the development of mosquitoes (Sharma, 1996). Two per cent neem oil mixed in coconut oil, when applied to exposed body parts of human volunteers, provided complete protection for 12 h from bites of all anophelines. Kerosene lamps containing 0.01-1 per cent neem oil, lighted in rooms containing human volunteers, reduced mosquito biting activity as well as catches of mosquitoes resting on walls in the rooms; protection was greater against *Anopheles* than against *Culex*. Effectiveness of mats with neem oil against mosquitoes has also been demonstrated. The vaporizing oil repelled mosquitoes for 5-7 h at almost negligible cost. The sandfly, *Phlebotomus argentipes* Annandale & Brunetti, also was totally repelled by neem oil, mixed with coconut or mustard oil, throughout the night under field conditions in India. Application of neem cake @ 500 kg/ha, either alone or mixed with urea, in paddy fields in southern India reduced the number of pupae of *Culex tritaeniorhynchus* Giles, the vector of Japanese encephalitis, and also resulted in higher grain yield (Rao *et al.*, 1992).

The pest control potential of Thai neem, *Azadirachta siamensis* (Val.) (formerly *A. indica* var. *siamensis* Val.), the marrango tree, *Azadirachta excelsa* (Jack) (formerly *Melia excelsa* Jack), the Persian lilac or Chinaberry tree, *Melia azedarach* L., *Melia toosendan* Sieb. & Zucc., and *Melia volkensii* Gurke, has also been investigated (Schmutterer, 1995). Also, other Meliaceous plants, such as several species of *Aglaia* and *Trichilia*, have been identified possessing significant biological activity (Isman, 1995). Although the biological efficacy of *A. siamensis* seed kernel extracts approximates that of *A. indica* (Somabtsiri and Tigvattanont, 1984), the overall azadirachtin content and oil is less (Chiratharmjaree and Ermel, 1996). In field trials conducted in Thailand, spray applications of Thai neem extracts effectively controlled the insects attacking Chinese kale, asparagus, leech lime or porcupine orange, cotton, and soybean (Sombatsiri *et al.*, 1995). A few Thai neem-based insecticides are already being marketed locally. Thai neem with reddish young leaves is considered more active in pest control that with green leaves.

The marrango tree, *A excelsa,* occurs in Southeast Asia. Its bioactive constituent, marangin, has been found to be considerably more effective than azadirachtin in growth disturbing properties (Schmutterer and Ermel, 1995). Unfortunately, due to extensive felling, this tree has become very rare in some regions. Large stands of marrango will have to be re-established in order to produce enough raw material for insecticides.

Compared with neem, *Melia* spp. are adapted to low temperature. *M. azedarach* and *M. toosendan* are widely distributed in China in regions south of the Yellow River, while *M. volkensii* grows in the wild in semi-arid areas of Ethiopia, Somalia, Kenya, and Tanzania. Their pesticidal and chemical aspects have been reviewed in depth (Chiu 1989; Kraus *et al.*, 1993; Rembold and Mwangi, 1995). Toosendadnin is the major bioactive material in the bark of *M. azedarach* and *M. toosendan,* while meliantriol and melianone occur in fruits of *M. azedarach.* Volkensin occurs in the fruit extract of *M. volkensii.* These and other *Melia* constituents possess strong antifeedant properties and also inhibit insect growth and development. An insecticide with toosendanin as the main component (0.5 EC) has been registered as a commercial insecticide in China (Chiu, 1995). Field trials with the material at a 800 to 1000 times dilution have shown promise against the cabbage worm, *Pieris rapae* (Linnaeus) and other vegetable and fruit insects. In Cuba, ground fruits of *M. azedarach* are used in the production of insecticidal extracts (H. Schmutterer, personal communication). In field trials, oil-formulated *M. volkensii* materials have also shown excellent potential for control of the desert locust, but production and commercialization of volkensin-based insecticides have not yet taken place. The presence of toxic meliacins in *Melia* fruits limits their use as insecticides. *Citrus* (Rutaceae) limonoids, limonin, obacunone, and nomilin, possessing strong antifeedant activity and toxicity against some insects (Alford *et al.*, 1987; Mendel *et al.*, 1991) have been considered potential candidates for synthesis but little progress has been made so far.

6.6. BOTANICAL PEST CONTROL AND INSECT RESISTANCE

A predictable response to insecticide use, whether synthetic or botanical, is the likelihood of evolution of resistant populations of insect pests. Taylor (1984) indicated that insect pests may possibly adapt to limonoids rather quickly. However, Vollinger (1987) demonstrated that two genetically different strains of *P. xylostella* treated with a neem seed extract showed no sign of resistance in feeding and fecundity tests up to 35 generations. In contrast, two deltamethrin-treated lines developed resistance factors of 20 in one and 35 in the other. Also, there was no cross resistance between deltamethrin and neem seed extract in the deltamethrin-resistant lines. The activity of detoxification enzymes, esterases and mixed function oxidases did not change during the 35 generations. The diversity of bioactive ingredients and their combined behavioral and physiological effects on insect pests seem to confer a built-in resistance prevention or delay mechanism in crude or unrefined neem materials. Wisdom demands that pesticide industries should refrain from producing pure azadirachtin-rich insecticides and, instead, market standardised "neem-rich" extracts or materials containing multiple active ingredients. Also, the users should be encouraged to avoid exclusive and extended applications of botanical insecticides based on single active ingredients. For durability sake, even novel neem-rich insecticides must be applied within the framework of integrated pest management programmes.

6.7. CONCLUSIONS

The use of botanicals for pest control originated as a need and not as a science in many parts of the world in ancient times. With the development of better techniques and analytical tools for detection, isolation, and identification of bioactive constituents, the scientific basis of botanical pest control has developed on a firm footing in the past 50 years. Although the pesticide industry, in general, is still hooked on to the idea of using natural molecules as models for chemical synthesis or improvisation into more potent pest control materials for eventual patents and corporate gains, there is a growing acceptance of the use of standardized natural plant products for pest control both by the industry as well as farmers and other pest control practitioners.

Of nearly 2400 plant species with insect pest control potential, only pyrethrum, nicotine, rotenone, ryania, sabadilla, and neem have been registered for significant use as insecticides. To be commercially viable, botanical insecticides must meet the following criteria: (i) Effective against pests but relatively safe to natural enemies and other non-target organisms, including man, (ii) biodegradable, (iii) non-phytotoxic, (iv) easy to harvest, (v) sustained availability of the raw material, and (vi) cost effectiveness. Judged by these criteria, only neem and pyrethrum emerge as the most desirable botanicals. This situation augurs well for developing countries as both pyrethrum and neem at present are largely grown and produced in those developing countries, which would gain not only from their use but also from increased export of the raw or processed material. A case in point is the neem industry in India which, from near scratch, has proliferated and now exports large quantities of enriched neem materials or concentrates for the manufacture of insecticides in North America. It is expected that in the coming decades, neem-based insecticides will occupy at least a 10 per cent of the global insecticide market. While the issue of patent rights will continue to erupt now and then, it should be resolved for the overall societal benefit and to promote development of biological resources in a sustainable way.

ACKNOWLEDGEMENT

Some of the information on neem as a biopesticide, reported above, has been generated in the Neem Awareness Project supported from funds given by the Government of Finland and the United Nations Environment Programme.

REFERENCES

Abdul Kareem, A., Saxena, R.C. and Justo, Jr., H.D. 1987. Cost comparison of neem oil and insecticide in rice tungro virus (RTV). *Int. Rice Res. Newsl.* **12** (4): 28-29.

Alford, A.R., Cullen, J.A., Storch, R.H. and Bentley, M.D. 1987. Antifeedant activity of limonin against the Colorado potato beetle. *J. econ. Ent.* **80:** 575-578.

Anonymous 1996. The program director's view. *IPM Collab. Res. Support Prog. Update* **2**(1): 1.

Arnason, J.T., Philogene, B.J.R., Morand, P., Imrie, K., Iyenger, S., Duval, F., Soucy-Breau, C., Scaiano J.C., Werstiuk, N.H., Hasspieler, B. and Downe, A.E.R. 1989. Naturally occurring and synthetic thiophenes as photoactivated insecticides. *In:* J.T. Arnason, B.J.R. Philogene and P. Morand (eds). *Insecticides of Plant Origin.* ACS Symp. Ser. 387, Am. Chem. Soc., Washington, DC, USA, pp. 164-172.

Ascher, K.R.S. 1993. Nonconventional insecticidal effects of pesticides available from the neem tree, *Azadirachta indica. Arch Insect Biochem. Physiol.* **22:** 443-449.

Ascher, K.R.S. and Meisner, J. 1989. The effects of neem on insects affecting man and animal. *In:* M. Jacobson (ed.). *Focus on Phytochemical Pesticides. Vol. 1. The Neem Tree.* CRC Press, Boca Raton, pp. 113-131.

Bowers, W.S., Ohta, T., Cleere, J.S. and Marsella, P.A. 1976. Discovery of insect anti-juvenile hormones in plants. *Science* **193:** 542.

Carson, R. 1962. *Silent Spring.* Houghton Mifflen, Boston.

Casida, J.E. (ed.). *Pyrethrum: The Natural Insecticide.* Int. Symp. on Recent Advances with Pyrethrum, the Natural Insecticide, University of Manitoba, 1972. Academic Press, London, UK. 329 pp.

Charles, V. and Charles, J.X. 1992. The use and efficacy of *Azadirachta indica* ADR ('neem') and *Curcuma longa* ('turmeric') in scabies. *Trop. Geogr. Med.* **44:** 178-181.

Chirathamjaree, C. and Ermel, K. 1996. Azadirachtin content of neem seed kernels from selected locations in Thailand. Paper presented at *Int. Symposium on Toxicity, Safety, and Proper Use of Biopesticides,* Phitsanulok, Thailand, Oct. 1996 (Abstract).

Chiu, Shin-Foon 1989. Recent advances in research on botanical insecticides in China. *In:* J.T. Arnason, B.J.R. Philogene and P. Morand (eds). *Insecticides of Plant Origin.* ACS Symp. Ser. 387, Am. Chem. Soc., Washington, DC, USA, pp. 69-77.

Chiu, Shin-Foon 1995. *Melia toosendan* Sieb. & Zucc. *In:* H. Schmutterer (ed.). *The Neem Tree, Azadirachta indica A. Juss., and Other Meliaceous Plants: Source of Unique Natural Products for Integrated Pest Management, Medicine, Industry and Other Purposes.* VCH, Weinheim, pp. 642-646.

Crosby, D.G. 1971. Minor insecticides of plant origin. *In:* M. Jacobson and D.G. Crosby (eds). *Naturally Occurring Insecticides.* Marcel Dekker, Inc., New York, USA, pp. 177-239.

Dhaliwal, G.S. and Dilawari, V.K. 1996. Host plant resistance in integrated pest management. *In:* R.K. Upadhyay, K.G. Mukerji and R.L. Rajak (eds). *IPM System in Agriculture. Vol. I. Principles and Perspectives.* Aditya Books Private Ltd., New Delhi, India, pp. 264-315.

Dhaliwal, G.S., Singh, J. and Dilawari, V.K. 1996. Potential of neem in insect pest management in rice. *In:* R.P. Singh, M.S. Chari, A.K. Raheja and W. Kraus (eds). *Neem and Environment. Vol. I.* Oxford & IBH Publishing Co. Pvt. Ltd., New Delhi, India, pp. 425-431.

Dhawan, A.K. and Simwat, G.S. 1996. Field evaluation of some botanical insecticides alone or in combination with other insecticides for management of bollworm complex on cotton. *In:* R.P. Singh, M.S. Chari, A.K. Raheja and W. Kraus (eds). *Neem and Environment. Vol. I.* Oxford & IBH Publishing Co. Pvt. Ltd., New Delhi, India, pp. 485-491.

Downum, K.R. 1986. Photoactivated biocides from higher plants. *In:* M.B. Green and P.A. Hedin (eds). *Natural Resistance of Plants to Pests: Roles of Allelochemicals.* ACS Symp. Ser. 296, Am. Chem. Soc., Washington, DC, USA, pp. 197-205.

Dreyer, M. 1987. Field and laboratory trials with simple neem products as protectants against pests of vegetables and field crops in Togo. *In:* H. Schmutterer and K.R.S. Ascher (eds). *Natural Pesticides from the Neem Tree (Azadirachta indica A. Juss) and Other Tropical Plants.* Proc. 3rd Int. Neem Conf., Nairobi, Kenya, Jul 1986. GTZ, Eschborn, Germany, pp. 431-447.

Edwards, C.A. 1990. The importance of integration in lower input agricultural systems. *Agric. Ecosys. Envir.* **27:** 25-35.

Elliott, M. (ed.). 1977. *Synthetic Pyrethroids.* ACS Symp. Ser. 42, Am. Chem. Soc., Washington DC, 229 pp.

Facknath, S. 1996. Application of neem and intercropping for control of important cabbage pests. Paper presented at *Int. Neem Conf.,* Gatton College, Australia, Feb. 1996 (Abstract).

Grainge, M. and Ahmed, S. 1988. *Handbook of Plants with Pest Control Properties.* John Wiley & Sons, New York, USA, 469 pp.

Gupta, G.P. 1996. Use of neem in management of pest complex in cotton. *In:* R.P. Singh, M.S. Chari, A.K. Raheja and W. Kraus (eds). *Neem and Environment, Vol. I.* Oxford & IBH Publishing Co. Pvt. Ltd., New Delhi, India, pp. 469-483.

Helson, B., Lyons, B. and Groot, P. de 1996. Development of neem seed extract for forest insect pest management in Canada. Paper presented at *Int. Neem Conf.*, Gatton College, Australia (Abstract).

Immaraju, J., Wood, T., Nelson, R., Ruggero, W. and Wells, S. 1996. Efficacy profile of a commercial neem insecticide. *In:* R.P. Singh, M.S. Chari, A.K. Raheja and W. Kraus (eds). *Neem and Environment. Vol. 2.* Oxford & IBH Publishing Co. Pvt. Ltd., New Delhi, India, pp. 897-906.

Isman, M.B. 1995. Leads and prospects for the development of new botanical insecticides. *Rev. Pestic. Toxicol.* **3:** 1-20.

Jacobson, M. 1975. Insecticides from plants, a review of literature, 1954-1971. *Agric. Handbook 461,* Agric. Res. Serv., U.S. Dept. Agric.

Jilani, G. and Amir, P. 1987. *Economics of Neem in Reducing Wheat Storage Losses: Policy Implications.* Tech. Bull. No. 2, SEARCA, Philippines.

Jilani, G. and Saxena, R.C. 1988. Evaluation of neem oil, combination of neem oil and fumigation, and Actellic as paddy/rice protectants against storage insects. *Proc. Final Workshop on Botanical Pest Control in Rice-based Cropping Systems.* Int. Rice Res. Inst., Manila, Philippines.

Kega, K.M. and Saxena, R.C. 1996. Neem derivatives for management of *Sitophilus zeamais* Motschulsky [Coleoptera: Curculionidae]. Paper presented at *Int. Neem Conf.*, Gatton College, Australia, Feb. 1996 (Abstract).

Kraus, W. 1995. Biologically active ingredients. *In:* H. Schmutterer (ed.). *The Neem Tree, Azadirachta indica A. Juss., and Other Meliaceous Plants: Source of Unique Natural Products for Integrated Pest Management, Medicine, Industry and Other Purposes.* VCH, Weinheim, Germany, pp. 34-88.

Kraus, W., Bokel, M., Schwinger, M., Vogler, B., Soellner, R., Wendisch, D., Steffens, R. and Wachendorff, U. 1993. The chemistry of azadirachtin and other insecticidal constituents of Meliaceae. *In:* T.A. Van Beek and H. Breteller (eds). *Phytochemistry and Agriculture.* Proc. Phytochem. Soc. Europe, Oxford University Press, Oxford, UK, pp. 18-39.

Ley, S.V., Denholm, A.A. and Wood, A. 1993. The chemistry of azadirachtin. *Nat. Prod. Reports* **10:** 109-157.

Mendel, M.J., Alford, A.R., Rajab, M.S. and Bentley, M.D. 1991. Antifeedant effects of citrus limonoids differing in A-ring structure on Colorado potato beetle (Coleoptera: Chrysomelidae) larvae. *J. econ. Ent.* **84:** 1158-1162.

Mordue (Luntz), A.J. and Blackwell, A. 1993. Azadirachtin: an update. *J. Insect Physiol.* **39:** 903-924.

Mordue (Luntz), A.J., Cottee, P.K. and Evants, K.A., 1985. Azadirachtin: its effect on gut motility, growth and moulting in *Locusta. Physiol. Ent.* **10:** 431-437.

Morgan, E.D., van der Esch, S.A., Jarvis, A.P., Maccioni, O., Giagnacovoi, G. and Vitale, F. 1996. Production of natural insecticides from *Azadirachta* species by tissue culture. Paper presented at *Int. Neem Conf.*, Gatton College, Australia, Feb 1996 (Abstract).

Nanda, U.K., Parija, B., Pradhan, N.C., Nanda, B. and Dash, D.D. 1996. Bioefficacy of neem derivatives against insect pest complex of rice. *In:* R.P. Singh, M.S. Chari, A.K. Raheja and W. Kraus (eds). *Neem and Environment. Vol. I.* Oxford & IBH Publishing Co. Pvt. Ltd., New Delhi, India, pp. 517-527.

NRC (National Research Council). 1992. *Neem—A Tree for Solving Global Problems.* Report of an Adhoc Panel of the Board on Science and Technology for International Development, National Academy Press, Washington, DC, USA, 141 pp.

Ostermann, H. and Dreyer, M. 1995. Vegetables and grain legumes. *In:* H. Schmutterer (ed.). *The Neem Tree, Azadirachta indica A. Juss., and Other Meliaceous Plants: Source of Unique Natural Products for Integrated Pest Management, Medicine, Industry and Other Purposes.* VCH, Weinheim, Germany, pp. 392-403.

Pimentel, D. and Lehman, H. 1993. *The Pesticide Question: Economics, and Ethics.* Chapman & Hall, New York, USA, 441 pp.

Pradhan, S., Jotwani M.G. and Rai, B.K. 1962. The neem seed deterrent to locusts. *Indian Fmg.* **12**(8): 7-11.

Rao, D.R., Reuben, R., Venugopal, M.S., Nagasampagi, B.A. and Schmutterer, H. 1992. Evaluation of neem, *Azadirachta indica,* with and without water management for the control of culicine mosquito larvae in rice fields. *Med. Vet. Ent.* **6:**318-324.

Rembold, H. 1989. Azadirachtins: their structure and mode of action. *In:* J.T. Arnason, B.J.R. Philogene and P. Morand (eds). *Insecticides of Plant Origin.* ACS Symp. Ser. 387, Am. Chem. Soc., Washington, DC, USA, pp. 150-163.

Rembold, H. and Mwangi, R.W. 1995. *Melia volkensili* Gurke. *In:* H. Schmutterer (ed.). *The Neem Tree, Azadirachta indica A. Juss. and Other Meliaceous Plants: Source of Natural Products for Integrated

Pest Management, Medicine, Industry and Other Purposes. VCH, Weinheim, Germany, pp. 647-652.

Rice, M. 1993. Development of neem research and industry in Australia *In: Souven. World Neem Conf.,* Bangalore, India, pp. 8-24.

Rosaiah, B. and Reddy, A.S. 1996. Utility of botanicals and biocides as additives to insecticides in *Helicoverpa armigera* Hubner management on cotton. *In:* R.P. Singh, M.S. Chari, A.K. Raheja and W. Kraus (eds). *Neem and Environment. Vol. 2.* Oxford & IBH Publishing Co. Pvt. Ltd., New Delhi, India, pp. 1101-1107.

Ruiter, A.A. 1996. Neem in Holland. *In:* R.P. Singh, M.S. Chari, A.K. Raheja and W. Kraus (eds). *Neem and Environment. Vol. I.* Oxford & IBH Publishing Co. Pvt. Ltd., New Delhi, India, pp. 48-50.

Sachan, J.N. and Katti, Gururaj 1996. Neem in *Helicoverpa (Heliothis) armigera* management: pulses ecosystem. *In:* R.P. Singh, M.S. Chari, A.K. Raheja and W. Kraus (eds). *Neem and Environment. Vol. I.* Oxford & IBH Publishing Co. Pvt. Ltd., New Delhi, India, pp. 459-467.

Saxena, R.C. 1989. Insecticides from neem. *In:* J.T. Arnason, B.J.R. Philogene and P. Morand (eds). *Insecticides of Plant Origin.* ACS Symp. Ser. 387, Am. Chem Soc., Washington, DC, USA, pp. 110-135.

Saxena, R.C. 1995. Pests of stored products. *In:* H. Schmutterer (ed.). *The Neem Tree, Azadirachta indica A. Juss., and Other Meliaceous Plants: Source of Unique Natural Products for Integrated Pest Management, Medicine, Industry and Other Purposes.* VCH, Weinheim, Germany, pp. 418-432.

Saxena, R.C. and Dale, D. 1989. Plant defence mechanisms against insects. *In:* M.S. Swaminathan and S.L. Kochar (eds). *Plants and Society.* Macmillan, London, UK, pp. 351-366.

Saxena, R.C., and Khan, Z.R. 1987. New bioactive products: growth regulators, antifeedants, pheromones and other attractants *In:* G.B. Marini-Bettolo (ed.). *Towards a Second Green Revolution: From Chemical to New Biological Technologies in Agriculture in the Tropics.* Academia Nazionale Delle Scienze Detta Dei XL, Roma, pp. 304-317.

Saxena, R.C. and Kidiavai, E.L. 1997. Neem seed extract spray applications as low-cost inputs for management of flower thrips in cowpea. *Phytoparasitica* (In press).

Schmutterer, H. 1990. Properties and potential of natural pesticides from the neem tree, *Azadirachta indica. A. Rev. Ent.* **35:** 271-297.

Schmutterer, H. (ed.). 1995. *The Neem Tree, Azadirachta indica A. Juss., and Other Meliaceous Plants: Source of Unique Natural Products for Integrated Pest Management, Medicine, Industry and Other Purposes.* VCH, Weinheim, Germany.

Schmutterer, H. and Ermel, K. 1995. The sentang or marrango tree: *Azadirachta excelsa* (Jack). *In:* H. Schmutterer (ed.). *The Neem Tree, Azadirachta indica A. Juss., and Other Meliaceous Plants: Source of Unique Natural Products for Integrated Pest Management, Medicine, Industry and Other Purposes.* VCH, Weinheim, Germany, pp. 598-604.

Senrayan, R. 1996. Efficacy profile and biosafety of a neem formulation Neemazal-T/S. Paper presented at *Int. Neem Conf.*, Gatton College, Australia, Feb. 1996 (Abstract).

Sharma, V.P. 1996. Repellent action of neem oil on insects of public health importance. *In:* R.P. Singh, M.S. Chari, A.K. Raheja and W. Kraus (eds). *Neem and Environment. Vol. I.* Oxford & IBH Publishing Co. Pvt. Ltd., New Delhi, India, pp. 19-26.

Sinha, S.N. 1996. Neem in integrated management of *Helicoverpa armigera* Hubn. infesting chickpea. *In:* R.P. Singh, M.S. Chari, A.K. Raheja and W. Kraus (eds). *Neem and Environment. Vol. I.* Oxford & IBH Publishing Co. Pvt. Ltd., New Delhi, India, pp. 439-446.

Sombatsiri, K. and Tigvattanont, S. 1984. Effects of neem extracts on some insect pests of economic importance in Thailand. *In:* H. Schmutterer and K.R.S. Ascher (eds). *Natural Pesticides from the Neem Tree and Other Tropical Plants.* Proc. 2nd Int. Neem Conf., Rauischholzhausen, Germany, May 1983. GTZ, Eschborn, pp. 95-100.

Sombatsiri, K., Ermel, E. and Schmutterer, H. 1995. The Thai neem tree: *Azadirachta siamensis* (Val.). *In:* H. Schmutterer (ed.). *The Neem Tree, Azadirachta indica A. Juss., and Other Meliaceous Plants: Source of Unique Natural Products for Integrated Pest Management, Medicine, Industry and Other Purposes.* VCH, Weinheim, Germany, pp. 585-597.

Taylor, D.A.H. 1984. The chemistry of limonoids from Meliaceae. *In:* W. Herz, H. Grisebach and G.W. Kirby (eds). *Progress in Chemistry of Organic Natural Products.* Vol. 45, Springer-Verlag, Vienna, pp. 1-102.

Vollinger, M. 1987. The possible development of resistance against neem seed kernel extract and deltamethrin in *Plutella xylostella. In:* H. Schmutterer and K.R.S. Ascher (eds.). *Natural Pesticides from the Neem Tree (Azadirachta indica A. Juss) and Other Tropical Plants.* Proc. 3rd Int. Neem Conf., Nairobi, Kenya, Jul. 1986, GTZ, Eschborn, pp. 543-554.

Walter, J.F. 1996. Neemix and Neemix 4.5 to control fruit and vegetable pests on a commercial scale. Paper presented at *Int. Neem Conf.*, Gatton College, Australia, Feb. 1996 (Abstract).

Wanner, K.W., Kostyk, B.C. and Helson, B.V. 1996. Efficacy of neem seed extract for control of eastern spruce budworm defoliation damage to white spruce seed orchards in Ontario, Canada. Paper presented at *Int. Neem Conf.*, Gatton College, Australia, Feb. 1996. (Abstract).

Wewetzer, A. 1996. Production of azadirachtin by callus cultures from *Azadirachta indica* A. Juss. Paper presented at *Int. Neem Conf.*, Gatton College, Australia, Feb. 1996 (Abstract).

Wilson, A. 1994. Pesticidal properties of some selected plant extracts and their formulations. M.Phil. Thesis, Univ. of West Indies, Jamaica.

Wolpert, V. 1967. Needless losses. *Far Eastern Econ. Rev.* **55:** 411-412.

Zeringue, Jr., H.J. and Bhatnagar, D. 1990. Inhibition of aflatoxin production in *Aspergillus flavus* infected cotton bolls after treatment with neem (*Azadirachta indica*) leaf extracts. *J. Am. Oil Chem. Soc.* **67:** 215-216.

7

PESTICIDE MANAGEMENT : ISSUES AND STRATEGIES

B.S. PARMAR AND P. DUREJA

Division of Agricultural Chemicals

Indian Agricultural Research Institute, New Delhi-110 012, India

7.1. INTRODUCTION

Synthetic as well as natural pesticides had a glorious past and did yeoman's service to mankind in increasing agricultural production and improving food and health quality. However, lately, there has been ever-mounting criticism against their use. In spite of this, pesticides will continue to be a powerful tool to combat pest problems in the 21st century. There is, however, no denying the fact that the harmful effects attributed to their use will have to be mitigated.

Past experience reveals that the non-judicious use of pesticides (improper selection, repeated application, sub-standard product, excessive dosage, etc.) has been responsible for various ills. Most of these problems can be overcome through a proper pesticide selection, application, post-application care and monitoring to identify undesirable effects, corrective measures to overcome the defects, etc., the aspects which constitute the term pesticide management.

7.2. PESTICIDES IN RETROSPECT

Cultural and mechanical practices were the dominant pest control methods before the second world war. In some cases, botanical insecticides like nicotine sulphate, pyrethrum, ryania and mineral insecticides like lead arsenate, sodium fluorosilicate, etc. were also used. These chemicals are now conveniently described as "first generation pesticides". The discovery of insecticidal properties of DDT heralded the era of "second generation chemicals". DDT was followed by numerous other pesticides belonging to organochlorine, carbamate, organophosphorus and synthetic pyrethroid groups.

The era of "third generation chemicals" witnessed the introduction of juvenile hormones, JH mimics, anti-JH, moulting hormones and insect growth disruptors, which acted primarily by disrupting growth or metamorphosis in insects. Insect behaviour modifying chemicals such as pheromones, antifeedants, etc. are now being developed as "fourth generation pesticides". The salient landmarks in pesticides development are listed in Table 7.1.

Table 7.1. Salient landmarks in pesticide development

Period	Landmark	Country/ Region
500 BC	Sulphur as an acaricide and insecticide	Sumaria
1200 BC	Chalk and wood ash	China
1670	Whale oil against paddy pests	Japan
1763	Crude tabacco extract	-
1800	Chrysanthemum flower	Dalmatia
1848	Rotenoids	Asia
1867	Paris green against Colorado beetle	USA
1885	Bordeaux mixture against downey mildew of vine	USA
1888	Biological control of Vedalia ladybird beetle	USA
1899	HCN fumigant	USA
1913	Organomercurials as fungicide	-
1915	*Bacillus thuringiensis (Bt)*	-
1930	Alkyl thiocyanate insecticide	-
1931	Shirlan as fungicide	-
1934	Dithiocarbamate fungicides	-
1938	Chloranil fungicides	-
1938	Tetra-ethyl pyrophosphate (TEPP)	-
1939	DDT against Colorado beetle	Switzerland
1942	Hexachlorocyclohexane (HCH)	-
1943	2,4-Dichlorophenoxy acetic acid	England
1945	Cyclodiene insecticides	-
1947	Carbamate (Sevin)	Switzerland
1950	Organophosphorus insecticides	-
1951	Captan	USA
1952	Triazine herbicides (Simazine, atrazine, etc.)	-
1954	Allethrin	-
1955	Sterile insect release methods/DEET as repellent	USA
1958	Bipyridylium herbicides (Diquat and Paraquat)	England
1960	Azole fungicides (Triadimefon)	-
1961	Neem, *Azadirachta indica,* as antifeedant	India
1967	Benzimidazole herbicides, Farnesole	USA
1970	Dispalure	-
1973	Carbendazim	-
1974	Resmethrin	England
1975-90	Photostable pyrethroids	England/ Japan
1976	Precocenes I & II	USA
1982	Sulfonylurea herbicides (i.e., Chlorsulfuron)	-
1983	Imidazoline fungicide (i.e., Imazaquin)	-
1987	Flupropadine (rodenticide)	-

- = Not known

During the year 1994-95, the global pesticide market stood at 27.8 million U.S. Dollars. It showed an increase of 10.1 per cent over the previous year (Annual Report 1994-95, British Agrochemical Association). Pesticide sales across Western Europe revealed a rise of 12 per cent, being the largest in France, Italy, Germany and UK. Other areas of expansion were North America, Latin America (particularly Brazil) and Argentina, Japan and several other Asian markets notably Thailand, Indonesia and India. The consumption of pesticides in some selected regions of the world, a category-wise split of pesticide use and the per hectare pesticide consumption are reported in Tables 7.2-7.4.

Table 7.2. Worldwide consumption of pesticides during 1995

Region	Pesticide use (% of total)
North America	26
Latin America	11
Eastern Europe	3
Western Europe	24
Africa	3
Japan	16
Asia (excluding Japan)	17

Source: Anonymous (1996)

Table 7.3. World pesticide market by category in 1995

Category	Use (%)
Insecticides	28.9
Fungicides	19.3
Herbicides	47.2
Others	4.6

Source: Anonymous (1996)

Table 7.4. Per hectare pesticide consumption in some countries/regions of the world

Country	Pesticide use (kg ha^{-1})
Japan	10.80
Korea	6.60
Europe	1.90
USA	1.50
Thailand	1.40
India	0.60 (0.28*)
Latin America	0.30
Oceania	0.20
Africa	0.13

Source: Mehrotra (1994); *Tomar and Parmar (1993)

The ever increasing use of different types of pesticides the world over has been associated with serious environmental problems. In 1962, Madam Rachel Carson's book *Silent Spring* highlighted some of these. However, the scientific evidence to suggest that the judicious use of pesticides has caused any significant damage to the human and animal population or their environment is still wanting.

Theoretical life long consumption of large scale 'safe' quantities of food commodities each day may not cause any adverse effect on human health (Bates, 1990). Thus the magnitude of health risk associated with pesticide use in agriculture and public health may be ill-found. In fact, some of the non-pesticidal natural and synthetic products are reportedly equally or far more toxic (Table 7.5).

Still the fact remains that pesticides are basically poisons. It is also to be acknowledged that the global poverty and illiteracy levels are not likely to improve in the near future, thus ruling out the possibility of a judicious pesticide usage in the foreseeable future. The harmful effects of pesticides may thus be inevitable.

The salient harmful effects of pesticides include the contamination of the environment and food, feed and fibre, disruption of non-target organisms, pest resistance, pest resurgence, and so on. These defects have to be overcome while planning pesticide management strategies.

7.3. PESTICIDE MANAGEMENT ISSUES

The key issues necessitating pesticide management are as follows:

7.3.1. Inherent Toxicity

Pesticides are basically poisonous chemicals, thus inherently toxic. Both acute and chronic toxicities are of concern. Even the degradation/ transformation products of pesticides during storage are sometimes more toxic than the parent compound (e.g., isomalathion, isofenitrothion, etc.). Proper safety precautions have, therefore, to be taken at the manufacture, storage, transportation, handling, application and other levels of their use. Application at correct dosages is a prelude to an economic, efficient and safe pest control. Most of the accidents attributed to pesticides are because of mismanagement at one or the other steps in the hierarchy of their processing and use. Most of the non-pesticidal chemicals are also toxic materials if not handled carefully (Table 7.5). A careful handling and proper use can render most pesticides safer.

7.3.2. Pest Resistance

"If Darwin were alive today, the world would delight and astound him with its impressive verification of his theories on the survival of the fittest. Under the stress of intensive chemical spraying, the weaker

Table 7.5. Toxicological information on some common materials

Material	Natural toxin	LD_{50} (mg kg^{-1} rat)	Effect(s)
1	2	3	4
Common salt	Sodium chloride	3,300	Blood pressure, stomach cancer
Groundnut	Aflatoxins	7.2	Carcinogenic, liver and kidney cancer
Carrot	Cartoxines and cicutoxines	-	Nerve poison
Pepper	Piperin	-	Mutagen
Parsley/Celery	Psoralin	-	DNA damage, cancer promotion
Onion/Raddish	Sulfhydrilla	-	Goitrogen
Mushroom	Agaritin	-	Stomach tumor
Mustard	Allylisothiocyanate	-	Carcinogenic
Honey and herbal medicines	Pyrrolizidine	-	Carcinogenic, teratogenic
Coffee	Catechole	-	DNA damage
Cheese	Amines	-	Blood pressure
Apples	Phloricine	-	Glucosuria
Potato	Solanin	-	Death
Cabbage	Glucosinolates	-	Goitre
Sucrose	-	29,700	-
Ethyl alcohol	-	13,700	-
Aspirin	-	1,750	-
Ascorbic acid	-	518	-
Tetrachloro dibenzodioxin (TCDD)	-	0.0006 (Guinea pig)	-
Organically grown grain	Fungus	-	Glucosuria
DDT	-	118	-
Methyl parathion	-	2	-
Malathion	-	2,500	-
HCH	-	80-270	-
Monocrotophos	-	21	-
Isoproturon	-	1,875	-
Fenvalerate	-	451	-

- = Not available

Source: Faust (1990)

members of the insect populations are being weeded out" (Carson, 1962).

Evolution of resistance to pesticides by pests not only confirms Darwin's theories, but also poses a serious threat to agriculture and public health throughout the world (NAS, 1986; Roush and Tabashnik, 1990; Denholm and Rowland, 1992). Insecticide resistance occurs as a result of inappropriate and large scale use of pesticides particularly at sub-lethal doses, repeated application of the same pesticide or similar group of pesticides over a period of time as well as underdosing due to substandard pesticidal formulations. The situation, besides bringing drastic changes in the pest complex and crop environment, has culminated in outbreaks of several secondary pests. Even some minor ones have assumed the status of major pests.

Resistance of San Jose Scale, *Quadraspidiotus perniciosus* (Comstock), to lime sulphur in the State of Washington, USA was the first case of insect resistance to an insecticide (Melander, 1914). By 1945, about a dozen cases of resistance to pesticides were recorded. The number rocketed with the wide spread use of DDT and other synthetic organic insecticides. More than 4,000 examples of resistance in arthropods have been recorded (Georghiou and Lagunes-Jejida, 1991).

Resistant field populations have been reported in more than 500 species including mites, spiders and 14 orders of insects (Georghiou and Lagunes-Jejida, 1991). The distribution in the key orders was Diptera (177 species), Lepidoptera (74 species), Coleoptera (72 species), and Homoptera (51 species) and mites (71 species). Agricultural pests accounted for 56 per cent of resistant species, the medical and veterinary pests for 39 per cent and beneficial organisms such as predators and parasites for about 5 per cent. Pests have shown an ability to resist all the categories of insecticides including inorganics, chlorinated hydrocarbons, organophosphates, carbamates, pyrethroids, juvenile hormone analogues, chitin inhibitors, avermectins and microbials.

Since the first report of insecticide resistance in India against Singhara beetle, *Galerucella birmanica* (Jacoby) in 1963 (Pradhan *et al.*, 1964), many pests have developed resistance to insecticides (Bhatia, 1986). A recent noteworthy example is that of *Helicoverpa armigera* (Hubner) on cotton which has become 150 to 300 fold resistant to cypermethrin in parts of Andhra Pradesh (Mehrotra, 1989).

Among the public health pests, *Culex* sp. has been reported to have developed resistance to pesticides (Raghavan *et al.*, 1967). The bed bug, *Cimex lectularius* Linnaeus, is resistant to both DDT and HCH (Kalra and Krishnamurthi, 1965; Shetty *et al.*, 1975).

7.3.3. Pest Resurgence and Outbreaks of Secondary Pests

With the introduction of DDT after World War II and the subsequent increased use of various classes of pesticides, there has been a spurt in reports on pest resurgence, apparently resulting from destruction of natural enemies by pesticide use (Bartlett and Ewart, 1951; McClure, 1977; Heinrichs *et al.*, 1982; Chelliah and Heinrichs, 1984). By 1956, Ripper had listed more than 50 herbivore species that increased in abundance.

A statistically significant increase in the pest population as a consequence of pesticide use, in spite of a good initial kill at the time of treatment, was called as "resurgence" or flareback. McClure (1977) attributed the deleterious effect of pesticides on natural enemies and competitive pest species and the improved nutritive quality and growth of host plants as the major causes of pest resurgence. Evidences were also forthcoming that insecticide residues in host plants and insecticides applied at sublethal doses stimulated the reproduction and survival of phytophagus insects and mites, leading to pest resurgence (Dittrich *et al.*, 1974). Insecticide - induced pest outbreaks have been documented in many crops (Bartlett and Ewart, 1951; Bottrell and Rummel, 1978).

In Asia, the resurgence of brown planthopper following insecticide application was reported in rice (Nagata and Moringa, 1974; IRRI, 1974; Alam and Karim, 1977; AICRIP, 1978; Oka, 1978; Stapley *et al.*, 1979). In cotton, the use of synthetic pyrethroids to control bollworm enhanced the population of whitefly, *Besimia tabaci* (Gennadius) (Gerling and Horowitz, 1984; Reddy *et al.*, 1986). The key examples of pest resurgence in India are listed in Table 7.6.

7.3.4. Destruction of Natural Enemies

Parasites and predators attack a great variety of pests and in some cases provide the primary means of their control (DeBach, 1974; Huffaker, 1980). When insecticides or other pesticides are employed, the poison not only destroys the pest, but also has a serious impact on its natural enemies. In some cases, the destroyed natural enemies are important in controlling certain other pests also. When these natural enemies are eliminated, it may result in outbreak of pests that were previously not a problem in the target crop (Pimentel and Edwards, 1982).

Table 7.6. Insecticide-induced resurgence of insects and mites in India

Crop	Resurged pest(s)
Rice	Brown planthopper, green leafhopper, whitebacked planthopper, leaf folder
Sorghum	Sorghum aphid
Sugarcane	Scale insect
Cotton	Aphid, whitefly, spider mite, mealy bug, jassid
Vegetables	Aphid, whitefly, spider mite, mealy bug, yellow mite, leafhopper
Mustard	Aphid

When pesticide use has destroyed natural enemies and resulted in secondary pest outbreaks in a crop, additional insecticide treatments are usually made to deal with these particular pests. In cotton, at least 4 to 5 additional treatments have been used to control the bollworm and tobacco budworm due to the destruction of the natural enemies of these two pests, when pesticides were applied to control the boll weevil (Pimentel *et al.*, 1977).

7.3.5. Pesticide Residues

Wide spread use of pesticides in agriculture and public health has led to a large scale environmental exposure to these chemicals. It has resulted in residue problems of large magnitude, even in areas far away from the sites of application (Dhaliwal and Singh, 1993). Among the various contaminants reported, the organochlorine pesticide residues have been found dominating (Lal *et al.*, 1989). Pesticides occur in agricultural commodities (Anonymous 1984, 1988) as a result of the intentional use of pesticides for protection of growing crops, and accumulation in mammals from the ingestion of feeds containing pesticide residues.

In the United States, it is estimated that at least 50 per cent of the food samples have detectable levels of pesticides (Johnson and Manske, 1977; Johnson *et al.*, 1981). On the Indian scene, the overall assessment of available data today confirms that a large variety of food and feeds are contaminated with the persistent organochlorine group of pesticides, viz., DDT, HCH, chlordane, etc. Similarly, bioaccumulation of these chemicals in milk, butter, meat and even human fat has been observed (Agnihotri, 1983; Kalra and Chawla, 1985; Awasthi, 1987; Singh and Dhaliwal, 1993).

Soil serves as a major sink for the bulk of pesticide used in agriculture and public health programmes. Studies conducted on persistence and dissipation of organochlorine pesticides in different soils revealed that 86 per cent of the samples collected from agricultural lands of Punjab were contaminated with DDT, HCH, endrin and lindane (Kalra and Chawla, 1983; Agnihotri *et al.*, 1977). Pesticides are taken up by crops from soil and they may also contaminate the ground water. There are several reports of water contamination with pesticide residues. For example, residues of DDT have been detected in Yamuna river in New Delhi (Agarwal *et al.*, 1986; Pillai and Agarwal, 1979) and Ganga river around Farrukhabad region of Uttar Pradesh (Agnihotri *et al.*, 1995). Kulshrestha (1989) reported the occurrence of DDT, aldrin and HCH residues in water samples collected from selected environmental locations of Madhya Pradesh and Rajasthan.

7.3.6. Effects on Non-target Organisms

Although pesticides provide benefit in our battle to control pests that destroy more than a third of all food, their use also results in hazards to biota exposed to it, either directly or through the food chain in the ecosystem.

Humans. Human pesticide poisoning is a major offshoot of pesticide use in agriculture. Globally, the number of deaths from pesticide poisoning has been estimated at about 80,000 per annum, using a case fatality rate of 0.5 per cent for the developing countries and 0.25 per cent for the developed ones (Levine, 1986). An estimated 200 fatalities occurred annually due to pesticides with about 50 as accidental poisonings in USA (EPA, 1974).

Humans contact pesticides through various means. For example, workers contact pesticides, during their production and handling (including application, drift, etc.) and so on. Other sources of pesticide contamination of humans include food, water, air, soil and other commodities of human consumption.

Honeybees. Honeybees and wild bees are well known for their role in the pollination of the flowers of fruits, vegetables, forage crops and other plants. If pesticides destroy or reduce their population, it may have a significant impact on the agroecosystem as well as the natural vegetation. Between 1981 and 1991, 1168 cases of suspected honeybee poisoning in England, Wales and Scotland were submitted for chemical analysis (Greig-Smith *et al.,* 1994). Of these, 667 (57%) contained residues that suggested the cause of death to be pesticide poisoning and it was attributed to 545 separate spraying incidents. Heavy use of pesticides has reduced the bee population and, therefore, reduced crop pollination.

Wildlife. Bird and mammal populations can suffer from exposure to pesticides. The deleterious effects include death from direct exposure to high pesticide dosages, reduced survival, growth and reproduction from exposure to sub-lethal dosages and habitat reduction through the elimination of various food resources.

There are reports suggesting that the pesticide contamination of birds can cause egg-shell thinning, increasing the chance of egg breakage, higher mortality at hatching and congenital abnormalities of the chick leading to an extirpation of the species in the wild life ecosystem (Regupathy and Kuttalam, 1992).

Aquatic organisms. Aquatic ecosystems are frequently under stress from pesticides and a significant loss to fish is caused. High pesticide concentrations in water kill fish directly and low level dosages may kill highly susceptible fish or eliminate essential fish foods like insects or other invertebrates (Pimentel and Edwards, 1982).

Soil flora and fauna. The soil is the major sink for pesticides. The pesticides applied to plant, water or in any other aerial application, ultimately reach the soil and cause adverse effects on the flora and fauna inhabiting it. Earthworms are adversely affected.

Pesticides can also alter the chemical make-up of plants. These changes are often plant or pesticide specific. For example, certain organochlorine insecticides have been found to increase or reduce the amount of some macro- and micro-element constituents of corn and beans (Cole *et al.*, 1968). Heptachlor in soil caused significant changes in the elements N, P, K, Ca, Mg, Fe, Cu and Zn measured in the corn and bean plants. Changes in plant constituents caused by pesticides can influence the insects that feed on them. For example, when corn plants were exposed to recommended dosages of 2, 4-D in the field, the nitrogen content of the corn increased, and this in turn, resulted in nearly a three fold increase in corn leaf aphid population (Oka and Pimentel, 1974). The corn was also 26 per cent more susceptible to corn borer attack.

7.4. PESTICIDE MANAGEMENT STRATEGIES

The pesticide management strategies have to ensure that the undesirable traits of pesticides are overcome. These have to blend the best of the past and the present technologies and ensure their integration with the future needs and developments. Ruthless identification and culling out of the culprit chemicals and a farsighted planning of the newer approaches to combat pests to achieve an economic, efficient and safer pest control, have to be the hallmarks of this strategy. The key components of this strategy are as follows:

7.4.1. Integration of Man-made and Natural Pest Control Agents

A majority of the conventional synthetic pesticides are not compatible with the natural pest control agents. This has been a serious bottleneck in their combined use. Till recently (Singh, 1995a, b), the information on compatibility of the synthetic and botanical chemicals with the biological agents has been lacking. Even the available information has not been generated under organised compatibility studies.

Most biocontrol agents suffer from a narrow target range, making the control of two or more target species at the same time difficult. Properly selected synthetic or botanical pesticides will come handy in such situations by controlling the pests not susceptible to the biological organisms.

Even the concept of resistance to synthetic pesticides can be fruitfully employed in such combinations by including resistance to the pesticides in the bio-organisms.

7.4.2. Use of Selective and High Potency Chemicals

New chemicals. Structural activity relationship studies on compounds such as natural pyrethrins have resulted in the discovery of a large number of new synthetic pyrethroid molecules such as permethrin, allethrin, cypermethrin, decamethrin, fenvalerate, flucythrinate, etc. Being photostable and reasonably biodegradable, these products are suitable in agricultural applications. They are extremely potent, often used at 10-100 g ha^{-1}. Novel non-ester pyrethroids such as MTI-800 and NRDC-200 having no structural resemblance to natural rethrins have been developed. These are safer to fish and possess a broad spectrum of bioactivity. Lamda cyhalothrin promises to be a substitute for DDT in malaria control. It is already in use in Nepal, Brazil and Indonesia.

Similar developments have taken place in the fields of fungicides and herbicides. In 1992, three new fungicides of the azole group namely tetraconazole, bromoconazole and fenbuconazole have shown promise. Herbicides such as nicosulfuron (40-50 g ha^{-1}), rimsulfuron (12.5 g ha^{-1}) and amidosulfuron (40 g ha^{-1}) have been reported.

It is certain that pesticides will continue to be used for a considerable period of time in the future but the notorius chemicals will need to be culled out. The future chemicals will be high potency chemicals requiring a small dosage per unit area of effectiveness. This trend is already apparent from the newer chemicals reported in the literature (Table 7.7). The newer pesticides will also need to be highly selective, the selectivity between mammals and fish on one hand and the pest on the other being highly desirable. This has been achieved to a considerable extent (Table 7.8). The more recently developed chemicals are less toxic to mammals.

Newer concepts. Keeping the selectivity, high potency and safety needs in view, newer concepts to monitor these attributes have been proposed. Some of these are:

(a) *Mammalian safety ratio (MSR).* It expresses the qualitative and quantitative strength of a pesticide based on its bioactivity, safety and toxicology. The mammalian toxicity (LD_{50} rat) and insect toxicity (LC_{50} insect) have been utilized to derive the mammalian safety ratio (MSR), reflecting the safety potential of a given pesticide to mammals. It is endeavoured to develop products of high MSR (Table 7.9).

(b) *Pesticide suitability rating (PSR).* Reported by Dev *et al.* (1988), it is derived by taking into account the toxicity to the pest (LD_{50} housefly), safety to mammals (LD_{50} rat), safety to beneficial insects, mammalian safety ratio and cost effectiveness. It offers a rationale in the use of pesticides under a given set of conditions and underlines that only those pesticides, natural or synthetic that are bioactive, biodegradable and environment friendly, should be included in pest control strategies. These chemicals should not have undesirable traits such as high persistence, photolability, high mammalian toxicity and phytotoxicity and toxicity to beneficial insects and non-target organisms including man and animals.

Table 7.7. Application doses of some pesticides

Pesticide	Dosage (g a.i. ha^{-1})
Aldicarb	1000-1500
Bensulfuron-methyl	5.0-7.5
Carbofuran	500-1500
Cartap	160
Chlorosulfuron	0.16-0.20
Carbaryl	1000
Deltamethrin	5-20
Fenitrothion	360
Fenvalerate	20
Flucythrinate	25
Fluvalinate	25
Metasystox	750
Methamidophos	500
Malathion	500-600
Paraquat	120
Quinalphos	300-500
Triazole fungicide	100
Thiobencarb	500
Teflubenzuron	1.65

Source: Shishido (1990)

Table 7.8. Relative toxicity of some pesticides

Pesticide	LD_{50} (mg kg^{-1})		Selective toxicity ratio (A/B)
	Rat (oral toxicity) A	Housefly (topical) B	
Parathion	3.6	0.9	4.0
DDT	118	2	59
Fenitrothion	570	2.3	248
Chlorofurazuron	>8,500	0.24	>35,416
Methoprene	>34,600	0.020	>1,730,000

Source : Shishido (1990)

Table 7.9. Relative mammalian safety ratio (MSR) of some pesticides

Chemical	MSR (LD_{50} rat/LD_{50} house fly)
Organochlorines	93
Organophosphates	11
Parathion	2
DDT	11.3
Carbaryl	17
Fenvalerate	900
Permethrin	1400
Cypermethrin	4200
Decamethrin	>6000

Source: Mukherjee (1986)

(c) Newer targets. Most of the synthetic insecticides in current use are neurotoxicants and fast acting and have a broad activity spectrum. For quite some time insecticide chemists have been aware of the other targets in insects, which can be explored in developing toxicants. Receptors and metabolizing enzymes for glutamic acid (GA) and *gamma*-aminobutyric acid (GABA) are important targets. GA and GABA are chemical transmitters at excitatory and inhibitory synapses, respectively and express activity at the neuro-muscular junction of invertebrates in contrast to vertebrates in which they only function in the central nervous system.

Receptors and metabolising enzymes for GABA are important targets for neurotoxicants, which are still important insecticides because of their fast acting properties. Amidines of which chlordimeform and amitraz are well known examples act by inhibiting the enzyme monoamine oxidase, specially interfering with octopamine receptors. These have a broad spectrum of activity, insecticidal as well as acaricidal (Dittrich, 1966).

Development of anti-GABA agents is another promising approach for developing new pesticides. Certain caged bicyclic phosphates (BPs) are known as anti-GABA agents, binding to the GABA receptor ionophore complex. The toxicity of BP is greatly affected by the substitute at the 4-position. For example, *tert*-butyl BP is more toxic to mammals, binding well with the brain synaptic membrane, whereas the less toxic propyl BP is more insecticidal binding tightly with the insect synaptic membrane. This suggests the possibility of developing anti-GABA agents as selective insecticides. Recently, bicyclic *ortho*-benzoates have been discovered as anti-GABA insecticides.

7.4.3. Biocides to Bioregulators

Agrochemical research attempts to develop products that are safe to non-target organisms. In this regard, the replacement of biocides by bioregulators assumes importance.

A knowledge of bioregulatory mechanisms of natural bioactive compounds, their mode of action in different organisms and the molecular basis of ecological interactions among various biological species is highly desirable (e.g., host selectivity of pests, exhibition of various symptoms caused by plant pathogens, resistance in crops against attack by pests, etc.). New compounds need to be developed as 'soft' pesticides which have a minimal impact on the environment. Differences in the molecular mechanisms of physiology and ecology between mammals and other species as well as between crop plant and pest bio-organism could be utilized to design selective toxic agents. Living organisms are endowed with the ability to regulate their own growth and differentiation of behaviour. Metamorphosis in insects is such a typical phenomenon. Some juvenoids have been developed for the control of hygienic insect pests. There is a possibility that they can be used for agricultural insect pests because of their ability to suppress insect reproduction. Chemicals such as anti-juvenile hormone (JH) agents and juvenoids are important in plant protection for the control of larval pests. Anti-JHs such as precocene and its active metabolites may be classified into JH receptors, inhibitors of JH biosynthesis and cytotoxicants for the corpora allata. Some juvenoids compete with JH and also inhibit JH biosynthesis in the manner of feed back. Compactin inhibits HMGCoA reductase, a key enzyme for JH biosynthesis and fluoromevalonolactone inhibits mevalonate kinase.

It is known that insects can not synthesise sterol skeleton and utilize phytosterols for their need of sterol including ecdysone. For conversion of phytosterols into ecdysone, several metabolic steps, particularly oxygenation processes, are required. Inhibitors for these oxygenases can be developed as insect growth regulators.

Thus, with a deeper understanding of insect behaviour and physiology more candidate ecochemicals which decrease the population and reduce insect-caused damage can be developed. Among crops, weeds, insects, natural enemies, micro-organisms, etc., a chemical communication system has evolved in nature. By manipulating such communication systems with ecochemicals, the prospects of controlling harmful organisms with minimum consequences on the ecosystem seem realistic.

These trends towards bioregulators are also seen in the fungicide and herbicide field. In the field of fungicides, regulating the life cycle of pathogenic fungi by chemicals will be the main target. For herbicides, bioregulators selectively interfere with a particular function such as photosynthesis, cell division, etc. in plants.

Plant growth regulators constitute another area of interest. These influence several plant physiological functions. Five plant hormones namely auxin (IAA), gibberelin, cytokinin, abscisic acid and ethylene are known. Isolation and characterisation of new types of plant growth regulators such as flower inducing substances, cell division factors, etc. are expected. Dynamic changes in endogenous plant hormones and other plant growth regulators at various growth stages of the life cycle and their molecular mechanisms in relation to biological activity must be investigated.

Investigation of the effects of environmental factors on the proceedings of each physiological step of the life cycle and on gene expression at the cell, organ and individual plant level is also very important and needs to be clearly elaborated. Bioregulators may stand out as the chemicals of the future.

7.4.4. Exploring Leads from Nature

Nature has always provided leads for the development of new chemicals including pesticides. A large number of natural products having biocidal activity have been discovered from plant, animal and microbial sources, but have been exploited only to a limited extent. The synthetic pyrethroids, carbamates, the JH mimics and plant growth regulators are the result of follow up of such leads. A very attractive field for future development is that of phagorepellents which will play an important role to prevent the spread of viral diseases of plants and for repelling their vectors. The crude extract of neem, containing azadirachtin as the repelling factor, is now being increasingly used in agriculture but a large number of other structures such as azugarhin and glucosides like xylomonin are known phagorepellents. These natural compounds serve as models for future synthetic analogues. Structural activity relationship (SAR) studies with such synthetic models may create new breakthroughs as in the case of synthetic pyrethroids. One of the most important approaches to improve the possibility of finding a lead compound may be made in the search of bioactive compounds playing a role in the interaction between crop and pest organisms. Investigations of bioactive chemicals that are responsible for the host selectivity of pests, the exhibition of various pathogenic symptoms caused by plant pathogens and the resistance in crops against the attack of pests must make possible the development of new plant protection technologies.

Botanical pesticides. Natural products of botanical origin are a vast source of bioactive substances that have been exploited to a limited extent only. In spite of the fact that a variety of plant products afford plant protection against pests, relatively only a few of these, namely nicotine, rotenone and pyrethrum, have been used in practice and have survived as commercial products. Natural products such as physostigmine from *Physostigma venenosum* and rethrins isolated from *Chrysanthemum cinerariaefolium* served as prototypes for the development of a large number of carbamates and synthetic pyrethroid pesticides, respectively. Typical mode of action of juvenile and moulting hormones offers further opportunities to chemists and entomologists seeking a biorational approach to insect control. Based on the JH model, a number of products such as Stauffer 20458, CG 34301, Roche 135223 and mimics have been developed.

Over the last 50 years, more than 2,000 plant species belonging to different families and genera have been reported to contain toxic principles that are effective against pests. Among the well reputed plants which still enjoy popularity as insecticides are *Azadirachta indica* (neem), *Annona squamosa, Pongamia glabra, Acorus calamus, Ocimum sanctum, Anethum sowa, Derris., Ryania speciosa,* sorghum, marigold and bougainvillaea. Some candidate products for future use include *Asimina, Citrus paradisi, Melia*spp., *Tagetes* spp., and *Piper* spp. (Isman, 1994).

The well known neem, *Azadirachta indica,* has shown selective activity against a wide array of insects (Mordue and Blackwell. 1993). Besides the antifeedant action of neem seed oil (Saxena, 1986), azadirachtin and neem seed products exhibit pronounced behavioural and physiological effects. Neem is also effective against non-insects like nematodes, snails and crustaceans. It also has antifungal activity and neem oil acts as a seed protectant acting against serious fungal diseases. Roots of marigold (*Tagetes patula*) accumulate thiophenesheterocyclic sulfurous compounds with strong biological activity.

Keeping in view the versatile nature of plant derived products, it is certain that these materials will offer environmentally safe management of insect pests in the future. Moreover, the current global thrust on botanical pesticides may yield new leads and prototypes to guide future research programmes in plant protection.

Chemicals of microbial origin. Among the environmentally safe alternatives, biopesticides or pesticides of microbial origin have also gained prominence due to increased safety and ease of production. There are presently more than 1500 naturally occurring microorganisms or microbial metabolites, which are known to possess insecticidal properties. Fungi, viruses, bacteria, nematodes, etc. are used as microbial pesticides with commercial potential (Table 7.10). With the help of natural selection and genetic engineering, novel toxin producing strains are being developed for future exploitation.

(a) Bacterial pathogens. The most prominent bacterial biopesticide, *Bacillus thuringiensis* (*Bt*), a soil dwelling, gram positive, spore forming,

Table 7.10. US market for biopesticides and pyrethroids

Pest control agent	Sales (Million US $)			Average annual growth rate(%)
	1992	1997	2002	
Pyrethroids	410	487	560	3.2
Bacterial	68	137	275	15.0
Viral and fungal	26	44	69	10.3
Pheromones	42	56	75	6.0
Others	13	16	23	5.6
Total	559	740	1002	5.6

Source: Krishnakumari and Venkataraman (1995)

rod shaped bacterium, was developed for pest control during the 1950s. Since then, over 30 strains of *Bt* have been discovered which are effective against Diptera, Coleoptera, nematodes, animal ectoparasites (mites) and endoparasitic protozoans (Lisansky and Coombs, 1992; Federici, 1991). Several commercial products based on *Bt* are formulated and used (Table 7.11).

Several *Bt* preparations are marketed in developed and developing countries of the world. The toxicity of *Bt* has been attributed to a glycoprotein often referred to as *delta*-endotoxin, which is produced during sporulation/fermentation. When *Bt* spores are ingested by the larvae, these are activated within the gut to produce a toxin, that disrupts the cells which become distorted and eventually burst. *Bt* spores are resistant to desiccation, high temperature, UV irradiation and biodegradation.

(b) *Viral agents (Baculoviruses).* Among the various entomopathogens, the baculoviruses are the most important and are highly safe to non-target organisms including man and the environment. Nearly ten types of such viruses have been isolated and studied for their potential in pest control (Mathews, 1982). Baculoviruses are generally of two types, i.e., granulosis virus (GV) and nuclear polyhedrosis virus (NPV).

Granulosis viruses are rod shaped. Each virus particle is enclosed in a protein capsule. The efficacy of GV as a microbial pesticide was investigated as early as 1973. Aqueous suspension of the virus gives an effective control against infestation by the Indian meal moth, *Plodia interpunctella* (Hubner), infesting peanuts, almonds and walnuts.

NPVs are cross infective among lepidopteran species. The successful mass production and field use of NPV of *Helicoverpa armigera* (Hubner), *Spodoptera litura* (Fabricus) and sugarcane shoot borer are good examples. It is estimated that these viruses can be used against nearly 30 per cent of the major pests of food and fibres (Table 7.12).

In the USA, NPV has been registered for use against cotton bollworm and tobacco budworm at application rates of 140-250 g ha^{-1}. The wettable powder of NPV developed by M/s Sandoz contains at least 4 billion polyhedral inclusion bodies (PIB) of *Helicoverpa* NVP g^{-1} of the product.

Table 7.11. *Bacillus thuringiensis*-based commercial products

Bacillus thuringiensis (Bt) varieties	Product[(R)]	Company
Kurstaki (Btk)	Bactospeine	Duphar
	Biobit	Novo Labs
	Condor^	Ecogen
	Cutlass*	Ecogen
	Dipcl	Abbott Labs
	Javelin	Sandoz, Inc.
	Larvo Bt	Knoll Labs
	Thuricide	Sandoz, Inc.
	Bactucide	CRC
	Delfin	Sandoz, Inc.
	Sok	Nor-American
Encapsulated *kurstaki* endotoxin	MVP**	Mycogen
israelensis	Skeetal	Novo Labs
	Teknar	Sandoz, Inc.
	Vectobac	Abbott Labs
	Bactimos	Philips Duphar
	Thurimos	Novo Labs
	Bactics	CRC
san diego	M-One	Mycogen
tenebrionis	Trident	Sandoz, Inc.
tenebrionis/kurstaki	Foil***	Ecogen
Encapsulated *san diego* endotoxin	M-One Plus***	Mycogen
aizawall	Certan	Sandoz, Inc.
Bt exotoxin	Dibeta	Abbott Labs
	Exobac	CRC
Bt serotype I	Muscabac	Farmos
san diego/tenebrionis	DiTerra	Abbott Labs.

*Genetically manipulated transconjugant; **Genetically engineered;
***Genetically manipulated transconjugant with beetle and caterpillar activity
Source: Lisansky (1989); McDonald (1991)

In India, in Andhra Pradesh, Gujarat, Maharashtra and Tamil Nadu, NPV of *S. litura, S. obliqua* and *H. armigera* based on extensive field testing have been found effective pesticidal agents (Jayaraj, 1985;

Pawar and Thombre, 1992). The potential of baculoviruses against lepidopteran pests in Punjab has also been reported (Battu, 1992).

(c) Fungal pathogens. About 750 fungal species (56 genera) attack terrestial and aquatic arthropods. The most commonly encountered fungi pathogenic to insects belong to three sub-divisions of the Eumycotina. Entomopathogenic fungi are known to produce toxin and nearly 33 toxins have been listed (Burges, 1981) of which some might come as novel insecticidal compounds in the near future. At present there is no fungal toxin commercially available as an insecticide. The product 'Bio-1020' developed by Bayer is based on *Metarrhizium anisopliae* for the control of coleopterous pests of various ornamental crops and it leaves no toxic risk or hazard (McDonald, 1991). 'Mycar' is the product produced by Abbot Laboratories, USA, using *Hirsutella thompsoni* for the control of citrus mite, *Phyllocoptruta oleivora* (Ashmead). Table 7.13 gives a list of entomopathogenic fungi in commercial and experimental production.

(d) Microbial antibiotics. Microbial antibiotics such as streptomycin, tetracycline, chlorophanicol, etc. were initially to protect crops against pathogenic bacteria. They were also used as plant growth regulators, e.g., cycloheximide and griseofulvin. Since many of the microbial products were phytotoxic and biodegradable, a search for more potent products continued. With the discovery of Blasticidin-S as fungicide to control rice blast, many useful products such as Kasugamycin, mildiomycin and polyoxins were also discovered (Fischer and Bellus, 1983).

Among the pesticides listed in Table 7.14, validamycins are exceptional. They are not specific but interfere only with the normal mycelial growth of the pathogen on plants. Probably because of such moderate activity, no occurrence of resistant strains has been reported despite having the largest consumption of all pesticides of microbial origin.

Recently, components of avermectin, were successfully transformed into hundreds of times more active derivatives by chemical modification of the original structure (Dybas *et al.*, 1989).

7.4.5. Safer Formulations

Development of safer formulations by incorporating features such as safety to environment, improved effectiveness, better cost-benefit ratio and convenient and safe application is another promising means of pesticide management. Conventionally, wettable powder (WP), dust (D), granules (G), emulsifiable concentrate (EC), etc. formulations have been used. Of late, a number of newer formulations such as dry and wet flowables and controlled release products have been produced, which are more effective and environment benign than the conventional products.

Table 7.12. Key pests controlled by baculoviruses

Virus	Pest(s)
Nuclear polyhydrosis virus	*Helicoverpa armigera* (Olivier) *Spodoptera litura* (Fabricius) *Amsacta albistriga* (Walker) *Plutella xylostella* (Linnaeus) *Spilosoma obliqua* Walker *Hyblea puera* (Cramer)
Granulosis virus	*Chilo infuscatellus* Snellen *Sacchariphagus indicus* *Agrotis segetum* (Denis & Schiffermuller) *H. armigera* *Cydia pomonella* (Linnaeus) *Adoxophyes orana* (Fisher von Roeslerstamm)
Non-occlusion virus	*Oryctes rhinoceros* (Linnaeus)

The dry and wet flowables are characterised by either their dilution in water before application or use of water as a carrier as well as

Table 7.13. Entomopathogenic fungi in commercial and experimental production

Fungus	Target	Product	Manufacturing country
Beauveria bassiana	Colorado potato beetle, coding moth, European corn borer, pine catterpillar	Boverin	USA
Culicinomyces clavisporus	Mosquito larvae	EAO	USA
Hirsutella thompsonii	Citrus mite	Mycar	USA
Metarrhizium anisopliae	Spittle bug, sugarcane froghopper	Metaquino	Brazil
Nomuraea rileyi	Lepidopteran larvae	EAO	USA
Verticillium lecanii	Aphid, coffee green bug	Vertalec	UK
	Whitefly, thrips	Mycotal	UK
Colletotrichum gloeosporioides	Northern joint vetch	—	USA
Cercospora rodmanii	Water hyacinth	EAO	USA
Peniophora gigantea	*H. annosum* of conifer tree roots		UK
Trichoderma viridae	Apple canker diseases		

- = Not available

Source: McDonald (1991)

diluent. The dry flowables are exemplified by water dispersible granules (WDG), floating granules (FG), etc. and the potential aqueous flowables (AF) are emulsion in water (EW), suspension concentrate (SC), etc. products. The pesticides which can be formulated in either one or the other of the above forms, will carry advantages of environmental and mammalian safety as compared to conventional products. The controlled release products that regulate the release of the active ingredient into the environment, prevent contamination of the environment, provide an effective and efficient pest control at the same time. Such innovative pesticide formulation advances will go a long way in planning effective pesticide management strategies in the future. Besides, the improvement of the shelf life of formulated products would require a serious research input if pesticide management has to make a worthwhile impact in future. Many of the problems of the formulated products can be attributed to their poor shelf life due to which the active ingredients undergo several undesirable changes. Finalisation of an optimum recipe for any product would require information on toxicant-carrier-auxiliary interaction and other concomitant studies.

Table 7.14. Microbial products used in agriculture

Substance	Application/use
Antifungal	
Blasticidin S	Rice blast
Kasugamycin	Rice blast
Polyoxins	Rice sheath blight
Validamycins	Rice sheath blight
Mildiomycin	Powdery mildew
Antibacterial	
Streptomycin	Bacterial diseases of fruits and
Oxytetracyclin	vegetables
Novobiocin	Bacterial canker of tomato
Insecticidal	
Tetranactin	Mites of fruits and trees
Herbicidal	
Bialophos	Weeds in mulberry fields
Nematicidal	
Avermectin	Veterinary application

Source: Fischer and Bellus (1983)

7.4.6. Miscellaneous Strategies

Pesticide rotation. One of the ways to prevent or delay the onset of certain ills of pesticide use is to introduce their rotational use. Depending on the pest situation the rotation may be amongst chemicals of a group or different groups of synthetic or synthetic and botanical or biopesticides. Such a rotation improves pest control efficiency, prevents or delays the development of resistance and will even help in mitigating pesticide residues.

Pesticide mixtures. Another approach gaining momentum to overcome the harmful effects of certain pesticides is the use of a mixture of two or more pesticides. This approach helps in simultaneous control of two or more pests by broadening the spectrum of bioactivity of the product, and overcomes or delays the development of insecticide resistance in insect pests (Tsuboi *et al.*, 1981). In Japan, the use of pesticide mixtures for effective insect pest and disease management, particularly in rice, has been widely adopted (Ozaki, 1989).

Use of synergists. Insecticide synergists increase or extend the toxicity of insecticides by inhibiting the action of enzymes that normally metabolise them. These provide an opportunity to tackle the problem of insecticide resistance and also reduce the dosage of pesticides, thereby enabling an economical, efficient and safe pest control.

Natural pyrethrins have been extensively synergised in the past, particularly for public health pest control applications. Synthetic pyrethroid formulations synergised with suitable synergists offer a solution to combat the problem of insecticide resistance. Some of the established synergists are piperonyl butoxide, MGK 264, Bucarpolate, Tropital, dillapiole, Sesamex, WARF-antiresistant, triphenyl phosphate, etc.

Propesticides. Another field of possible future activity is the modification of certain well known chemicals, that are good insecticides but whose insecticidal activity can not be utilised because of high mammalian toxicity. It these can be made selective, their utilization will be increased. Examples are acephate, ONCOL, carbosulfan, tralomethrin, cartap, larvin, etc.

Management of pesticidal wastes and toxic residues. Municipal and industrial treatment plants, discharge from factories, accidental spillage during storage or transportation, deliberate dumping of pesticide wastes, expired formulations and used containers as well as leakage from waste disposal sites are key sources contributing significantly to environmental pollution. From the environmental safety point of view, serious attention is being given the world over to develop environmentally compatible disposal and decontamination techniques. Some of the successful operations include construction of disposal pits and evaporation beds, filtration systems, incineration, photodecomposition, ozonation and microbiological processes. Haal (1984) and Winterlin *et al.* (1989) designed disposal pits and evaporation beds with hydrated lime which are able to dispose of a large number of pesticides through a combination of chemical and microbial processes. Recent advances in the field of microbial enzymes, co-metabolism, plasmid technology and cloning techniques have opened up exciting possibilities to clean up, and for bio-restoration of pesticide contaminated sites (Walia and Parmar, 1993).

7.5. CONCLUSION

From the foregoing account it is clear that there is a vast array of possibilities and opportunities in pesticide management to combat the various ills associated with pesticide use. Human knowledge and experience about the survival and multiplication capacities of pests indicated that banning all pesticides is not a desirable action but the selection of the right chemicals is the wisest option. The future, thus belongs to a judicious pesticide management strategy.

REFERENCES

Aggarwal, H.C., Mittair, P.K., Menon, K.G. and Pillai, M.K.K. 1966. DDT residues in the river Jamuna in Delhi, India. *Water Air Soil Pollu.* **288**: 28-89.

Agnihotri, N.P., Pandey, S.Y., Jain, H.K. and Srivastava, K.P. 1977. Persistence of aldrin, dieldrin, lindane, heptachlor and DDT in soil. *J. ent. Res.* **1**: 89-91.

Agnihotri, N.P. 1983. Monitoring of pesticide residues in the environment. *Pestic. Inf.* **6**: 64-79.

Agnihotri, N.P., Gajbhiye, V.T., Kulshrestha, G. and Mahapatra, S.P. 1995. Monitoring of pesticides in Ganga river water and agricultural fields on its bank at Furrukhabad, UP, India. *In*: S. Walia and B.S. Parmar (eds). *Pesticides, Crop Protection and Environment.* SPS Publication No. 6, Society of Pesticide Science, India, Oxford & IBH Publishing Co. Pvt. Ltd., New Delhi, India, pp. 477-498.

Alam, S. and Karim, A.N.M. 1977. Brown planthopper-a probable threat to rice cultivation in Bangladesh. *Second Bangladesh Annual Sci. Conf.*, Mymensing, Bangladesh, Jan. 23-26, 1977.

AICRIP (All India Co-ordinated Rice Improvement Project) 1978. *Progress Report, Kharif* Indian Council of Agricultural Research, New Delhi, India.

Anonymous 1984. Pesticide residues in food commodities in India and Nepal. *Final Report of Food Contamination Study.* GCP/PAS/078/ No. 2, Department of Entomology, Punjab Agricultural University, Ludhiana. Sponsored by FAO, United Nations, Rome.

Anonymous 1988. Pesticide residues in food items available in Bombay market. *In*: *Pesticide Residues in Food.* Consumer Education Research Centre, Ahmedabad, India, pp. 304-309.

Anonymous 1996. *World Crop Protection News*. Agrow. No. 258, 20pp.

Awasthi, M.D. 1987. *Pesticide Residues in Horticultural Crops*. Technical Bulletin No. 2, Indian Institute of Horticultural Research, Bangalore, India.

Bartlett, B and Ewart, W.H. 1951. Effect of parathion on parasites of *Coccus hesperidum*. *J. econ. Ent.* **44**: 344-347.

Bates, R. 1990. Pesticide residues-the arithmetic of risk. *Int. Pest Cont.* **32**(4): 94.

Battu, G.S. 1992. Potential of baculoviruses on lepidopteran pests in Punjab. *In*: T.N. Ananthakrishnan (ed.). *Emerging Trends in Biological Control of Phytophagus Insects*. Oxford & IBH Publishing Co. Pvt. Ltd., New Delhi, India, pp. 173-181.

Bhatia, S.K. 1986. Pesticide resistance in agricultural pests in India. *Proc. Indian natn. Sci. Acad.* **52**(1): 148-164.

Bottrell, D.G. and Rummel, D.R. 1978. Response of *Heliothis* populations to insecticides applied in an area wide reproduction diapause boll weevil suppression programme. *J. econ. Ent.* **71**: 87-92.

Burges, H.D. (ed.). 1981. *Microbial Control of Pests and Plant Diseases 1970-1980*. Academic Press, New York, USA.

Carson, R. 1962. *Silent Spring*. Hamish Hamilton, London, UK.

Chelliah, S. and Heinrichs, E.A. 1984. Factors contributing to brown planthopper resurgence. *Proc. FAO/IRRI Workshop on Judicious and Efficient Use of Insecticides on Rice*. International Rice Research Institute, Los Banos, Philippines.

Cole, H., Mackenzie, D., Smith, C.B. and Bergman, E.L. 1968. Influence of various persistent chlorinated insecticides on macro and micro element constituents of *Zea mays* and *Phaseolus vulgaris* growing in soil containing various amounts of these materials. *Bull. Envir. Contam. Toxicol.* **3**: 141-153.

DeBach, P. 1974. *Biological Control of Insects, Pests and Weeds*. Reinhold, New York, USA.

Denholm, I. and Rowland, M.W. 1992. Tactics for managing pesticide resistance in arthropods. Theory and practice. *A. Rev. Ent.* **37**: 91-112.

Dev, P.G., Hasan, S.B. and Majumdar, S.K. 1988. Toxicity and suitability of some insecticides for household use. *Int. Pest Cont.* **30**(3): 118-129.

Dhaliwal, G.S. and Singh, B. (eds). 1993. *Pesticides: Their Ecological Impact in Developing Countries*. Commonwealth Publishers, New Delhi, India.

Dittrich, V. 1966. N-(2-methyl-4-chlorophenyl)-N', N'-dimethyl–formamidine (c-8514/Schering 36268) evaluated as a new acaricide. *J. econ. Ent.* **59**: 889-893.

Dittrich, V., Streibert, P. and Bathe, P.A. 1974. An old case reopened; mite stimulation of insecticide residues. *Envir. Ent.* **3**: 534-540.

Dybas, R.A., Hilton, N.J., Babu, J.R., Preiser, F.A. and Dolce, G.I. 1989. Novel second generation avermectin insecticides and miticides for crop protection. *In*: A.L. Smith, G.A. Somkuti, J.C. Hunter and H.W. Rossmore (eds). *Novel Microbial Products for Medicine in Agriculture*. Elsevier/Society for Industrial Microbiology, New York, USA, pp. 203-212.

EPA (Environmental Protection Agency). 1974. *Strategy of the Environmental Protection Agency for Controlling the Adverse Effects of Pesticides*. Office of Pesticide Programs, Washington, DC, USA.

Faust, E.W. 1990. Staying alive in the 21st century. *In: World Environment, Energy and Economic Conference*. October 17-20, 1990, Winnipeg, Canada, pp. 1-13.

Federici, B.A. 1991. Microbial insecticides. *Pestic. Outlook* **2**(3): 22-28.

Fischer, H.P. and Bellus, D. 1983. Phytotoxicants from micro-organisms and related compounds. *Pestic. Sci.* **14**(3): 334-346.

Georghiou, G.P. and Legunes-Jejida, A. 1991. *The Occurrence of Resistance to Pesticides in Arthropods*. Food and Agriculture Organization of the United Nations, Rome.

Gerling, D. and Horowtiz, R. 1984. Yellow traps for evaluating the population levels and dispersal pattern of *Bemisia tabaci*. *Bull. ent. Soc. Am.* **77**: 753-759.

Greig-Smith, P.W., Thompson, H.W., Hardy, A.R., Bew, M.H., Findlay, E. and Stevenson, J.H. 1994. Incidents of poisoning of honeybees (*Apis mellifera*) by agricultural pesticides in Great Britain 1981-1991. *Crop. Prot.* **13**(8): 567-581.

Hall, C.V. 1984. Pesticide waste disposal in agriculture. *ACS Symp. Ser.* **259**: 27-36.

Heinrichs, E.A., Reissig, W.H., Valencia, S. and Chelliah, S. 1982. Rates and effect of resurgence-inducing insecticides on population of *Nilaparvata lugens* (Homoptera: Delphacidae) and its predators. *Envir. Ent.* **11**: 1269-1273.

Huffaker, C.D. (ed.). 1980. *New Technology of Pest Control*. John Wiley & Sons, New York, USA.

IRRI (International Rice Research Institute). 1974. *Annual Report*. Los Banos, Philippines.

Isman, M.B. 1994. Botanical pesticides. *Pestic. Outlook* **6**: 26-31.

Jayaraj, S. (ed.). 1985. *Microbial Control and Pest Management.* Tamil Nadu Agricultural University, Coimbatore, India.

Johnson, R.D. and Manske, D.D. 1977. Pesticides and other chemical residues in total diet samples (XI). *Pestic. Monti. J.* **11**: 116-131.

Johnson, R.D., Manske, D.D., New, D.H. and Podreba, D.S. 1981. Pesticides, heavy metals and other chemical residues in infant and toddler. Total diet sample - (11)-August 1975 - July, 1976. *Pestic. Monit. J.* **15**: 39-50.

Kalra, R.L. and Chawla, R.P. 1983. *Final Technical Report of the PL-480 Project.* Punjab Agricultural University, Ludhiana, Punjab, India.

Kalra, R.L. and Chawla, R.P. 1985. Pesticidal contamination of foods in the year 2000 AD. *Proc. Indian natn. Sci. Acad.* **B52**: 188-204.

Kalra, R.L. and Krishnamurthy, B.S. 1965. Studies on the insecticide susceptibility and control of bed bugs. *Bull. Indian Soc. Malariol. Commun. Dis.* **2**: 223-239.

Krishnakumari, M.K. and Venkataraman, L.V. 1995. Biopesticides - A newer perspective. *Pest Mgt.* **8**: 1-10.

Kulshrestha, S.K. 1989. *Technical Report on Monitoring and Surveillance of Residues of Organochlorines and Heavy Metals in Fishes from Selected Environmental Components of Madhya Pradesh and Rajasthan.* Ministry of Environment and Forests. Government of India, New Delhi, India.

Lal, R., Narayanarao, U.V.S. and Krishnamoorthy, V. 1989. Pesticide residues in food: An overview of studies taken in India. *In*: *Pesticide Residues in Food.* Consumer Education and Research Centre, Ahmedabad, India, pp. 99-113.

Levine, R.S. 1986. *Assessment of Mortality and Morbidity due to Unintentional Pesticide Poisoning.* WHO/VBC/86,929, World Health Organization, Geneva.

Lisansky, S. 1989. Biopesticides: The next revolution? *Chem. Ind.* **15**: 478-482.

Lisansky, S.G. and Coombs, J. 1992. Technical improvements to biopesticides. *Proc. Brighton Crop Prot. Conf. Pests and Diseases* **1**: 345-350.

Mathews, R.E.F. 1982. Classification and nomenclature of viruses. *Intervirol.* **17**(3): 33-35.

McClure, M.S. 1977. Resurgence of the scale, *Florinia externa* (Homoptera: Diaspidae) on hemlock following insecticide application. *Envir. Ent.* **6**: 480-484.

McDonald, D. 1991. Biopesticides-pesticides with a bright future. *Int. Pest Cont.* **33**(2): 33-35.

Mehrotra, K.N. 1989. Pesticide resistance in insect-pests. Indian Scenario. *Pestic. Res. J.* **1**(2): 95-103.

Mehrotra, K.N. 1984. Insecticide resistance and insect pest management. Sixth Dr C.P. Alexander Memorial Lecture, 18 January, 1994, Delhi University, Delhi, India.

Melander, A.L. 1914. Can insects become resistant to sprays? *J. econ. Ent.* **7**: 167-173.

Mordue (Luntz), A.J. and Blackwell, A. 1993. Azadirachtin: An update. *J. Insect Physiol.* **39**: 903-924.

Mukherjee, S.K. 1986. Plant protection chemicals in 2000 AD. *Proc. Indian natn. Acad.* **B52**(1): 35-48.

Nagata. T. and Moringa, S. 1974. Resistance in brown planthopper *Nilaparvata lugens* (Stal) to lindane. *J. appl. Ent.* **18**: 73-80.

NAS (National Academy of Science). 1986. *Pesticide Resistance: Strategies and Tactics for Management.* National Academy Press, Washington, DC, USA.

Oka, I.N. 1978. Paper presented at the FAO panel of experts on integrated control of pests. September 4-9, 1978, Rome, Mimeo. 28p.

Oka, I.N. and Pimentel, D. 1974. Corn susceptibility to corn leaf aphids and common corn smut after herbicide treatment. *Envir. Ent.* **3**(6): 911-915.

Ozaki, K. 1989. Effective application of mixtures-insecticides. *Japan Pestic. Inf.* **54**: 17-22.

Pawar, V.M. and Thombre, U.T. 1992. Prospects of baculoviruses in integrated pest management of pulses. *In*: T.N. Ananthakrishnan (ed.). *Emerging Trends in Biological Control of Phytophagous Insects.* Oxford & IBH Publishing Co. Pvt. Ltd., New Delhi, India, pp. 165-171.

Pillai, M.K.K. and Agarwal, H.C. 1979. *DDT Residues and its Degradation in Soil, Water and Few Aquatic Animals of Jamuna in New Delhi.* Department of Science and Technology, Government of India, New Delhi, India.

Pimentel, D. and Edwards, C.A. 1982. Pesticides and ecosystem. *BioScience* **32**: 595-600.

Pimentel, D., Shoemaker, C., LaDue, E.L., Rovinsky, R.B. and Russell, N.P. 1977. *Alternatives for Reducing Insecticides on Cotton and Corn: Economic and Environmental Impact.* Report on Grant No. R

802518-02, Office of Research and Development, Environmental Protection Agency, Washington, DC, India.

Pradhan, S., Jotwani, M.G. and Sarup, P. 1964. Save your Singhara crop. *Indian Fmg.* **14**: 33-35.

Raghavan, N.G.S., Wattal, B.J., Bhatnagar, V.N., Choudhury, D.S., Joshi, G.C. and Krishnan, K.S. 1967. Present status of susceptibility of arthropoda of public health importance to insecticides in India. *Bull. Indian Soc. Commun. Dis.* **13**:209-245.

Reddy, A.S., Rosaiah, B., Bhaskararao, T., Ramarao, B. and Venugopalrao, N. 1986. The problem of white fly on cotton. *Seminar on Problems of Whitefly on Cotton.* 14 March, 1986, Pune, India.

Regupathy, A. and Kuttalam, S. 1992. Monitoring of HCH and DDT residues in certain avian eggs in Tamilnadu. *Pestic. Res. J.* **4**: 149-154.

Ripper, W.E. 1956. Effect of pesticides on balance of arthopod population. *A. Rev. Ent.* **1**: 403-428.

Roush, R.T. and Tabashnik, B.E. 1990. *Pesticide Resistance in Arthropods.* Chapman & Hall, New York, USA.

Saxena, R.C. 1986. Neem seed oil - a potential antifeedant against insect pests of rice. *In*: R. Greenhalgh and T.R. Roberts (eds). *Pesticide Science Biotechnology.* Blackwell Scientific Publications, Oxford, UK, pp. 139-144.

Shetty, K.M., Subbiah, V.K. and Panickar, K.K. 1975. A comparative evaluation of fenitrothion, dursban and malathion for the control of bedbugs. *Indian J. Pub. Hlth.* **19**: 79-83.

Shishido. T. 1990. Future view of pesticides. *Japan Pestic. Inform.* **57**: 19-21.

Singh, B. and Dhaliwal, G.S. 1993. Pesticide residues in fatty foods in developing countries. *In*: G.S. Dhaliwal and B. Singh (eds). *Pesticides: Their Ecological Impact in Developing Countries.* Commonwealth Publishers, New Delhi, India, pp. 131-161.

Singh, S.P. 1995a. Chemical pesticides and biotic agents. *In*: S. Walia and B.S. Parmar (eds). *Pesticides, Crop Protection and Environment.* SPS Publication No. 6, Society of Pesticide Science, India, Oxford & IBH Publishing Co. Pvt. Ltd., New Delhi, India, pp. 165-190.

Singh, S.P. 1995b. Compatible natural enemies and synthetic pesticides for use in integrated pest management in India. *Pestic. Res. J.* **7**(1): 8-28.

Stapley, J.H., Jackson, Y.Y. and Golden, W.G. 1979. Observation in the Solomon Islands on the Control of brown planthopper,

Nilaparvata lugens. In: *Brown Planthopper: Threat to Rice Production in Asia*. International Rice Research Institute, Los Banos, Philippines.

Tomar, S.S. and Parmar, B.S. 1993. *Evolution of Agrochemicals in India - An Overview*, A Special Publication for Participants of CASAFA-ISSS-TWAS Workshop on Integration of Natural and Man Made Chemicals for Sustainable Agriculture in Asia. September 13-16, 1993, Indian Agricultural Rcsearch Institute, New Delhi, India.

Tsuboi, A., Hiramatsu, T., Watanabe, A. and Nagai, K. 1981. Effect of pesticide combination on the development of resistance in green rice leafhopper, *Nephotettix cincticeps* Uhler - laboratory selection by propahos-carbaryl and propaphos-BPMC, insecticide resistance in green rice leafhopper in Japan. Japan Plant Protection Association, pp. 48-54.

Walia, S. and Parmar, B.S. 1993. Decontamination and disposal of pesticides and related xenobiotics. *In*: V.P. Kudesia and M.V. Charaya (eds). *Pesticide Pollution*. Pragati Prakashan, Meerut, India, pp. 171-192.

Winterlin, W.L., Seiber, J.N., Craigmill, A., Baler, T., Woodrow, J. and Walker, G. 1989. Degradation of pesticide waste taken from a highly contaminated soil evaporation pit. *Arch. Envir. Contam. Toxicol.* **18**: 734-747.

8

BIOTECHNOLOGY AND INSECT PEST MANAGEMENT

T.N. ANANTHAKRISHNAN
Entomology Research Institute, Loyola College, Chennai - 600 034, India

8.1. INTRODUCTION

The age old practice of selectively collecting seeds from the best yielding crops that withstood the rigours of biotic and abiotic stresses, has in recent years been supplemented by using techniques to incorporate genes conferring pest resistance into cultivars. It is now possible to characterize and modify genes at the molecular level and recombinant DNA technology allows engineering and transfer of genes across formerly insurmountable biological barriers to hybridization (Goodman *et al.*, 1987). Genetic erosion as a result of extensive spread of high yielding varieties is typical in many tropical crops leading to their being vulnerable to insect pests. In order to establish durable resistance to pests, novel approaches through the involvement of genetic engineering techniques are in vogue. Unlike classical plant breeding, the possibility of transferring genes for pest resistance from among various taxa, is great. Not only is it possible to incorporate a single foreign resistance factor, but also a diverse variety of factors. An aspect of significance is that while different cultivars produce diverse allelochemics or secondary plant compounds, genetically engineered plants are capable of enhanced production of very potent chemical resistance factors which could pose a real threat to pests. As many biologically active chemicals are produced as a result of intense selection over the years, the intensity of insect - plant interactions has naturally led to increased diversity of these chemicals, to the extent that insects tend to affect the frequency of plant genes conferring resistance. However, intrinsic immunity to insect adaptation is evident in several instances wherein pest resistant crops have been grown for several years without pest adaptation. Such pest resistance genes are important natural resources (Brattsten, 1991).

Genetic engineering offers wide scope to transfer derived genes that synthesize, store and degrade secondary products to other plants without the co-transfer of other undesirable traits. This results in the acceleration of the development of new varieties, besides allowing the transfer of genes across incompatibility barriers. Further, genes can also be introduced from sources that are totally unavailable to conventional plant breeding. In short, the strategy in genetic engineering is to assemble gene packages in host plants whose products are targeted to skew off the different biochemical and physiological processes in insects. Through this approach, a multimechanistic form of resistance can be tailored to different crops, it is possible to establish multiline mixtures in an otherwise individual genetic background by creating new isogenic lines containing different resistance genes. The use of multilines forms part of the IPM strategy to improve the durability of resistant varieties and thwart the development of biotypes in insects (Chelliah, 1994).

Ever since the first transgenic tobacco plants expressing foreign proteins were obtained, progress in genetic engineering to develop insect resistant plants has been rapid. Most of the early successes have been generated through the mediation of *Agrobacterium tumefaciens*, a soil phytopathogen. Transgenic plants are generated in nature by *A. tumefaciens* through the transfer of some of its own DNA into susceptible plants. The genes are carried on a plasmid, which can be manipulated for use as a general vector for the transfer of specific segments of foreign DNA. By moving them into vulnerable plants, their increased protection against insects has been demonstrated (Schell, 1987). In subsequent years many microorganisms colonizing plant surfaces have been recognized such as *Pseudomonas, Rhizobium, Erwinia,* etc., having the potential to function as delivery systems for plant protection.

8.2. CHEMICAL FACTORS

Qualitative and quantitative enhancement of a plant's defence mechanism against insect pests in response to insect feeding stress may invoke cellular chemical changes, as well as changes in cells adjacent to damaged tissues. Generalized changes in the entire plant inclusive of phytochemicals detrimental to insects are also evident (Edward, 1992). Substances like gallic acid, vanillic and salicylic acids besides resorcinol, phloroglucinol and gossypol of cotton have been known to be inimical to pests, promoting resistance (Ananthakrishnan, 1992 a,b, 1993, 1994; Ananthakrishnan *et al.*, 1990). In other cases, defense genes are activated which may function as deterrents or antifeedants or genes may encode for enzymes such as phenylalanine ammonia lyase (PAL) which synthesizes defense-related products such as phytoalexins, caryophyllene, polygodial, warburganal, etc. Different defense genes are activated simultaneously within the same cell and this activation may involve encoding a set of enzymes that function within a specific metabolic pathway (Ayers, 1992). Genes encoding for structural and functionally diverse products are also activated. Different classes of genes relating to resistance strategy are expressed sequentially. Some of the more important defense related proteins involve hydroxyproline rich glycoproteins (HRGP), glycine rich glycoproteins

(GRP) which act directly as deterrents, and the proteins functioning as enzyme inhibitors (amylase and proteinase inhibitors). Other proteins can be toxic proteins such as lectins and thionins; hydrolases such as β, 1-3, glucanases, proteinases and antimicrobial compounds like phtytoalexins. Others Like PR proteins or pathogenesis related proteins, are of unknown function, being products of genes that accumulate in wound response (Bennet and Wallsgrove, 1994).

The sequence of events leading from defense responses such as the simultaneous synthesis of diverse proteins and the nature of their expression at various loci, is basic to further research on their significance in terms of insect-plant interactions (Bowles, 1990). Defense depends on a battery of mechanisms, some constitutive, becoming enhanced later and induced ones occurring after the event. Another gene marker exercise relates to the induction of ethylene forming enzymes. They regulate ethylene biosynthesis during a wound response, which in turn triggers the activity of phenolic enzymes and phenols. Similarly, proteinase inhibitors are known antifeedants and have been shown to exhibit toxic effects when genes encoding them have been expressed in transgenic plants. Even intracellular signalling events are recognised, linking recognition of oligosaccharides at the cell surface to changes in gene activity. Injury leads to the release of cell wall fragments and the impact of these on ion transport may lead to systemic signalling effects (Bowles, 1990, 1992).

Recent research has shown that lipoxygenases catalyse the oxidation of polyunsaturated fatty acids like linolenic acid by molecular oxygen, giving rise to the formation of hydroperoxidases, which can be further converted enzymatically into other compounds (Ayers, 1992). One of them is jasmonic acid or methyl jasmonate which has hormone-like actions eliciting proteinase gene inhibitor expression and is exhibited in interplant signalling. Methyl jasmonate, a common secondary plant substance, induces synthesis of proteinase inhibitors in the treated and nearby plants (Farmer and Ryan, 1990; Anathakrishnan, 1994).

Investigations on defense-related lectins that have carbohydrate binding properties and specificity towards foreign glycan structures, have indicated their protective potential. Lectins are the only class of plant proteins capable of recognising and binding glycoconjugates present on the surface of the intestinal tract. Lectins are known to cause deleterious effects and a study of the effect of several commercially available plant lectins on the cowpea weevil has indicated that only those from peanut, wheat germ, potato and thorn apple had a significant inhibitory effect on the development of weevil larvae. Of all lectins only ricin from castor was 100 per cent lethal to the larvae of *Diabrotica undecimpunctata* Mannerheim, when applied topically to the artificial diet. An increased insect resistance has also been reported for tobacco plants expressing the pea lectin gene. The effect of pea lectin was additive to that of the cowpea trypsin inhibitors in tobacco plants expressing the genes of both the lectin and protease inhibitors (Peumans and Van damme, 1995). More recently, mannose-binding agglutinins from garlic have shown anti-insect properties on sucking insects (Gatehouse, 1993). It is likely that toxicity of lectins is based on highly specific binding to receptor molecules that are located in the gut of insects. Lectins may act synergistically with the protective proteins like protease inhibitors, *alpha*-amylase inhibitors, ribosome inactivating proteins and thionins, establishing a highly efficient defense system against insects.

Further, plants being a source of natural resistance mechanisms which have the potential of being translated into new insect control chemical products, provide the genetic elements of resistance suitable for incorporation into desirable cultivars (Bowers, 1988).

8.3. TRANSGENIC CROPS

Recent advances in molecular genetics have shown that genes regulating basic developmental processes have been modified to control responses to biotic stresses. Genes coding for phenylalanine ammonia lyase are developmentally regulated and are normally expressed at particular phases in the development of healthy plants (Liang *et al.*, 1989). It becomes increasingly possible that genes regulating resistance could be manipulated to increase or intensify their expression, to the advantage of man (Ayers, 1992).

Transgenic plant technology has emerged as a useful tool in producing pest resistant crops by introducing entirely new genes into plant genes. Inserting genes into crop plants involved initially the soil bacterium, *A. tumefasciens*, as vector, wherein during infection transfer of foreign DNA occurs along with a portion of its own plasmid DNA with a plant cell. The discovery that the aerobic, grampositive spore-forming bacterium, *Bacillus thuringiensis* Berliner (*Bt*) which produces a toxic protein during sporulation, has introduced a new era of a genetically engineered biocontrol agent. Moving *Bt* insecticide genes from the original bacterium to plants, through the mediation of a vector, *Pseudomonas flourescens*, allows crop protection from insect attack. Another class of proteins, the proteinase inhibitors, are highly effective against certain insects, when combined with *Bt* endotoxins. The advantages of transgenic plants involve improved targeting of toxins, season- long protection in all plants without the need for expensive equipment and less toxicity exposure, advantages that are both economic and environmental. The toxin can be delivered to the pest directly because the active ingredient is expressed within the plant (Dilawari and Dhaliwal, 1996).

Transgenesis is aimed at altering or improving the product derived from crop plants. Modification of metabolic pathways by the introduction of different enzymes from other species, can result in a different product. Plants genetically modified to improve fitness advantage in some environments, such as tolerance to environmental stresses and pest resistance have been produced (Rogers and Parker, 1995). The ability to move bacterial genes encoding natural insecticidal proteins into plants enables the development of crops intrinsically resistant to insect attack, with advantage over conventional chemicals (Brunke and Meeusen, 1991).

More than 30 crop species including some major field crops-corn, cotton, soybean and rice have now been successfully transformed with insect crystal protein (ICP) genes to become insect resistant. Receptor binding accounts to a greater extent each individual type of ICP. Binding studies have shown that one insect may be susceptible to different ICPs depending upon the presence of different receptors on the midgut wall. For example, Cry1A (b) and Cry1B recognised different receptors in *Pieris brassicae* (Linnaeus). These receptor molecules appear to be glycoproteins. Upto now seven different types of Lepidoptera have been defined and for each type at least one amino acid sequence is known. *Gossypium hirsutum* has been genetically modified to incorporate recombinant DNA from *Bacillus thuringiensis kurstaki*(Btk) which codes for *delta*-endotoxin proteins. Transgenic *Btk* cotton expressing Cry I(A) (b) and Cry IA (c) insecticidal proteins have shown significant protection against *Helicoverpa zea* (Boddie) and *Heliothis virescens* (Fabricius) (Halcomb *et al.*, 1996). While insect resistant plants with ICP genes are expected to be commercialised, concern is expressed that a large scale introduction of crops containing a monogenic resistance trait could rapidly lead to the development of resistance within an insect population. This is true of field resistance of *Plutella xylostella* (Linnaeus) to *Bt* (Van Rie, 1991).

Crystal proteins and their genes have been classified based on their structure and activities spectrum, into Cry I (Lepidoptera species), Cry II (Lepidoptera and Diptera), Cry III (Coleoptera) and Cry IV (Diptera), each of these are functionally divided into several toxin types. This heterogeneity in toxin production is responsible for some of the diversity in activity spectrum among strains. There is also a high degree of heterogeneity among binding sites in some species, suggesting that some sites may bind a single toxin whereas others may bind two or more toxins (McGaughey and Whalon, 1992).

While the relatively narrow range of target insects is affected by a single protein toxin, their strong insecticidal activity and the single gene nature of protein are clear advantages for genetic transfer of Bt endotoxins to plants. However, sufficient evidence is accumulating that widespread use of *Bt*-protected crops may select for resistant insects. The diamondback moth, *P. xylostella* is notable as having definitely evolved high levels of resistance in the field. Besides, lower levels of resistance shown by the Indian meal moth, *Plodia interpunctella* (Hubner), low resistance levels in *H. virescens* and *Leptinotarsa decemlineata* (Say) have also been reported. Possible physiological mechanisms causing low levels of resistance to *Bt delta*-endotoxins include a change in gut pH or in enzymes that would affect dissolution and activation of the proteinaceous crystal. In *Plodia* and *Plutella*, resistance is affected by the binding affinity of receptors or binding sites on the brush border membrane of the insect midgut (McGaughey and Whalon, 1992).

Insect susceptibility to entomopathogens may be influenced by allelochemicals of the host plants. Sivamani *et al.* (1992) have demonstrated that both resorcinol and gallic acid increased the effect of *Bt* on *Helicoverpa armigera* (Hubner). Ludlum *et al.* (1991) has also shown that toxicity of crystal proteins of *Btk* was increased when incubated with chlorogenic acid and polyphenol oxidase, the principal resistance factors of tomato.

8.4. BACULOVIRUSES IN INSECT CONTROL

8.4.1. Genetic Implications

Of all insect pathogenic viruses, the family Baculoviridae is the commonest characterised by their circular, double stranded DNA genome and rod-shaped enveloped virion. These viruses are known to contain more than 100 different genes in the genome, which consists of unique sequences that may encode structural or non-structural polypeptides. Use of viruses as biological control agents, especially nuclear polyhedrosis viruses (NPV) and granulosis virus (GV) for several pest-crop systems has become an increasingly common technique. Genetic manipulations made possible by recombinant DNA technology permit even greater efficiency in these systems. One of the most powerful approaches is the ability to manipulate persistent viral infections in pest populations. Virus persistence is well known in wild insect populations and currently attempts are being made to initiate persistently infected cell lines. A beginning has been made towards the induction of a non-occluded baculovirus persistently infecting *H. zea*, by using *H. armigera* and *Trichoplusia ni* (Hubner) NPV. The ability to trigger these levels before pest populations reach damaging levels has significant implications (Dougherty, 1987).

It is well known that all insect virus vectors give transient gene expression and do not stably transfer either somatic or germ line insect tissue. One unique property of baculoviruses is that they serve as genetic sinks for transposon like elements. Transposable elements involve naturally occurring DNA segments that act as discrete units and are capable of changing their position in the cell. It is important to establish the number of transposable elements existing in lepidopteran tissue, their specificity for various tissues, genome integration sites and the function they serve. Whether these transposable elements can serve as genetic vectors having specificity for germ line tissues are vital aspects if future control strategies call for altering the genetic make up of pest insects (Dougherty, 1987).

Baculoviruses are, therefore, unique in that they form a useful expression vector. It has been suggested that by introducing appropriate foreign genes into the baculovirus genomes, pathogenicity and insecticidal effectiveness can be increased (Roberts and Granados, 1988). In the formation of a baculovirus insecticide, a recombinant baculovirus of increased toxicity is made by introducing appropriate foreign genes into the highly expressed polyhedrin gene site. Infection with such recombinant viruses may cause direct toxicity, alter behaviour or arrest development of insects (Maeda, 1988). Stability of the baculovirus under field conditions is dependent upon the presence of occlusion bodies made up of polyhedrin. A recombinant virus still retaining its ability to produce polyhedrin can be created by introducing the foreign gene in the polyhedrin site without removing the polyhedrin gene itself (Dougherty, 1987).

8.4.2. Vectors for Neurohormonal Genes

Recent advances in recombinant DNA technology and the genome engineering of insect baculoviruses have made possible the use of insect baculoviruses as an expression vector for neurohormone genes. Neurohormones, by virtue of their peptidic nature, are amenable to applications using recombinant DNA and genetic engineering technology. The protein nature of neurohormones makes them amenable to insect pest control approaches that involve recombinant DNA technology and genetic engineering. For example, insect neurohormone genes might be inserted into the genomes of crop plants so that transformed plants would produce the hormone and disrupt the physiological balance of phytophagous insects (Schneiderman, 1984). Alternatively, bacteria transformed with insect neurohormone genes could be applied to crops or to the soil so that ingestion of the hormone-producing bacteria by feeding insects would result in physiological imbalance.

Alternatively, insect viruses may prove useful as vectors for neurohormone genes in a genetic engineering approach to pest insect control. Baculoviruses of Lepidoptera have proven recently to be highly efficient cloning-expression vectors for foreign genes (Smith *et al.*, 1993). Baculovirus expression vectors could carry neurohormone genes into target insects, and during replication, the transformed virus may produce and release large quantities of the hormones from within infected cells of the insect's body. An endocrine imbalance combined with viral infection may enhance the efficiency of the virus. It has also become feasible to involve genes that code for 'antipeptides' against neurohormones which could result in products that bind with neuromones and block their physiolocial action. Though speculative, using neuroendocrine manipulation for insect pest control could be a profitable exercise, since the pace at which insect neuroendocrine research is advancing, leaves no doubt about their utility in insect control (Keeley and Hayes, 1987).

8.5. PROTEINASE INHIBITORS

Of the many natural defense mechanisms involved in plants against herbivorous insects, the synthesis of proteinase inhibitors is an important one (Green and Ryan, 1972). In contrast to the *Bt* endotoxins, these proteins have anti-metabolic activity against a wide range of insects. They are present in the tissues of some plants at relatively high concentrations, participating in complex defense responses with other molecules produced by the plant. The introduction of specific proteinase inhibitors into plants which do not produce these molecules, is an alternative approach for obtaining crops that are resistant to insect attack. Four classes, serine, thiol, metallo and aspartyl-proteinase inhibitors, have been identified (Brunke and Meeusen, 1991). Presence of trypsin inhibitor in the diet of insects reduces the effective concentration of trypsin available for digestion. This triggers a series of events including the inability to obtain amino acids from ingested food.

The advantages of using proteinase inhibitors as insect-control agents are that they are active against a wide range of insects, besides, their use as a second mechanism to help prevent development of insects that are resistant to *Bt* endotoxin. The need for high levels of protein for insect killing, and the potential need to regulate protein expression to specific plant organs, are major disadvantages.

The first gene of plant origin that was successfully transferred to another plant species through genetic engineering resulting in enhanced insect resistance was that isolated from cowpea, encoding a trypsin inhibitor. The trypsin inhibitor is an effective antimetabolite against a number of lepidoperan, coleopteran and orthopteran field and storage pests. The coding sequence of cowpea trypsin inhibitor (CPTI) has been successfully inserted into tobacco plants and such plants afforded protection against *H. zea*, *Spodoptera litura* (Fabricius) and *Manduca sexta* (Johannsen). Besides, the genes encoding protease inhibitors isolated from cowpea conferred resistance when expressed in tobacco. The tomato inhibitor II gene, when expressed in the tobacco has also been shown to confer insect resistance. The tomato and potato inhibitor

II gene encodes a trypsin inhibitor and expression of the gene in tobacco results in increased levels of protection against the tobacco hornworm, *M. sexta.*

Ever since the first transgenic tobacco plants expressing foreign protein virus were obtained, progress in genetic engineering to develop resistant plants has been considerable. Most of the engineered plants, have been generated through *A. tumefaciens*-mediated gene transfer, since it is a very efficient and versatile vector for the stable introduction of genes into plants. Currently, around 25 different plant species have been transformed using different systems (Gasser and Fraley, 1992).

8.6. CONCLUSIONS

The take-off stage in genetic engineering or biotechnology in insect pest management has been initiated, but increased efforts have to be made towards a better understanding of the biological regulation of plant-gene action in order to produce ecologically accepted crops. Emphasis should be on the genetics of plant-herbivore interaction in natural and agricultural systems and on the genetics of plant secondary metabolism (Gould, 1988). Genetically engineered, pest resistant cultivars, would tend to limit the use of synthetic chemicals, making pest resistance genes as an important natural resource. While the challenge of moving toxin-coding DNA sequences into several crops has been met, there is a need to move these toxin-coding groups into more crops enabling these genes to be expressed at higher levels. However, the production of crops that express resistance coding DNA sequences should be undertaken only when needed. Considering the rapid pace at which genetic engineering techniques are emerging, the development of plants that produce toxins in specific tissues during the growing season, to control insects, deserves high consideration.

Resistance management should involve tactics which are coordinated against the pest, in particular transgenic plants, where a combination with *Bt* or other allelochemical mortality factors could be most useful. The increasing attention given to biotechnology may enable more effective production of predators and parasites, and genetic manipulation may produce more competitive organisms. The incorporation of genes conferring pesticide resistance into predators and parasites would facilitate their integration with the use of pesticides.

While genetically modified plants can express a transgenic stably without affecting other agronomic characters, cases of instability and pleiotropic effects exist, resulting in a transgenic becoming silent for several generations and then becoming activated (Rogers and Parkes, 1995). The likelihood of transgenes coding for toxins through time, becoming incorporated within the DNA of wild relatives and increasing the opportunities for pest species' exposure to the toxin, has to be kept in mind. It has been suggested that by intercropping transgenic plants with unmodified plants, it should be possible to create an environment unfavourable to pest colonization (Thacker, 1993, 1994).

REFERENCES

Ananthakrishnan, T.N. 1992a. *Dimensions in Insect- Plant Interactions.* Oxford & IBH Publishing Co. Pvt. Ltd., New Delhi, India.

Ananthakrishnan, T.N. 1992b. Chemodynamics of insect-plant interactions. *Palaeobotanist* **41**: 144-148.

Ananthakrishnan, T.N. 1993. Changing dimensions in the chemical ecology of insects: Role of infochemicals on the behavioural diversity. *In*: T.N. Ananthakrishnan and A. Raman (eds). *Chemical Ecology of Phytophagous Insects.* Oxford & IBH Publishing Co. Pvt. Ltd., New Delhi, India, pp. 1-20.

Ananthakrishnan, T.N. 1994. Viewpoint in the chemical ecology of insects. *Phytophaga* **6**: 69-72.

Ananthakrishnan. T.N., Annadurai, R.S., Senrayan, R. and Murugesan, S. 1990. Antibiotic effects of resorcinol, gallic acid and phloroglucinol on *Heliothis armigera* Hubner. *Proc. Indian Acad. Sci.* **99**: 39-52.

Ayers, P.G. (ed.). 1992. *Pests and Pathogens.* Bios Scientific Publishers, Oxford, UK.

Bennett, R. and Wallsgrove, R.M. 1994. Secondary metabolites in plant defense mechanisms. *New Phytol.* **127**: 617-633.

Bowers, W.S. 1988. Allelochemicals in insect management. *In*: D.W. Roberts and R.R. Granados (eds). *Biochemistry, Biological Pesticides and Novel Plant-Pest Resistance for Insect Pest Management.* Boyce Thompson Institute for Plant Research at Cornell University, Ithaca, New York, USA, pp. 123-131.

Bowles, D. 1990. Defense related protein in higher plants. *A. Rev. Biochem.* **59**: 873-907.

Bowles, D. 1992. Signals in the wounded plant. *In*: P.G. Ayers (ed.). *Pests and Pathogens.* Bios Scientific Publishers, Oxford, UK, pp. 33-38.

Brattsten, L.B. 1991. Bioengineering of crop plants and resistant biotype evolution of insects: Counting coevolution. *Arch. Insect Biochem. Physiol.* **17**: 253-267.

Brunke, K.J. and Meeusen, R.L. 1991. Insect control in the genetically engineered crops. *Tibtech.* **9**: 197-200.

Chelliah, S., 1994. Current developments in genetic manipulation of crops for insect resistance. *Phytophaga* **6**: 89-93.

Dilawari, V.K. and Dhaliwal, G.S. 1996. Biotechnology and host plant resistance to insects: Opportunities and challenges. *In*: T.N. Ananthakrishnan (ed.). *Biotechnological Perspectives in Chemical Ecology of Insects*. Oxford & IBH Publishing Co. Pvt. Ltd., New Delhi, India, pp. 36-53.

Dougherty, E.M. 1987. Insect viral control agents. *J. Ind. Microbiol.* Suppl. **2**: 63-75.

Edward, P.J. 1992. Resistance and defence; the role of secondary plant resistance. *In*: P.G. Ayers (ed.). *Pests and Pathogens*. Bios Scientific Publishers, Oxford, UK, pp. 19-83.

Farmer, E.E. and Ryan, C.A. 1990. Interplant communications: Airborne methyl jasmonate induces synthesis of proteinase inhibitors in plant leaves. *Proc. natl. Acad. Sci.* USA **87**: 7713-7716.

Gasser, C.S. and Fraley, R.T. 1992. Trangenic crops. *Scient. Am.* **266** (6): 34-39.

Gatehouse, J.A. 1993. Approaches to insect resistance using transgenic plants. *Phil. Trans. Royal Soc.* London **B342**: 279-286.

Goodman, R.M., Hawptli, H., Crossway, A. and Knauf, V.C. 1987. Gene transfer in crop improvements. *Science* **236**: 343-353.

Gould, F. 1988. Evolution of biology and genetically engineered crops. *BioScience* **38**: 26-33.

Green, T.R. and Ryan, C.A. 1972. Wound-induced proteinase inhibitors in plant leaves: A possible defense mechanism in insects. *Science* **175**: 776-777.

Halcomb, J.L., Benedict, J.H., Cock, B. and Ring, D.R. 1996. Survival and growth of bollworm and tobacco budworm on transgene and transgenic cotton expressing a CryIA insecticidal proteins (Lepidoptera: Noctuidae). *Envir. Ent.* **25**: 251-255.

Keeley, L.L. and Hayes, T.K. 1987. Speculation biotechnology application for insect neuroendocrine research. *Insect Biotechnol.* **17**: 639-651.

Liang, X., Dron, M., Gramer, C.L., Dixin, R.A. and Lamb, C.J. 1989. Differential regulation of phenylalanine ammonia lyase genes during development and by environmental clues. *J. Biol. Chem.* **264**: 14486-14492.

Ludlum, C.T., Felton, G.W. and Duffey, S.S. 1991. Plant defenses: Chlorogenic acid and polyphenol oxidase enhance toxicity of *Bacillus thuringiensis* subsp. *kurstaki* to *Heliothis zea*. *J. Chem. Ecol.* **17**(1): 217-237.

Maeda, S., 1988. Genetic engineering of baculoviruses. *In*: D.W. Roberts and R.R. Granados (eds). *Biotechnology, Biological Pesticides and*

Novel Plant - Pest Resistance for Insect Pest management. Boyce Thompson Institute for Plant Research at Cornell University, Ithaca, New York, USA, pp. 17-21.

McGaughey, W.H. and Whalon, M.E. 1992. Managing insect resistance to *Bacillus thuringiensis. Toxins Sci.* **258**: 1451-1455.

Peumans, W.J. and Van damme, E.J.M. 1995. The role of lectins in plants defence. *Hist. J.* **2**: 253-271.

Roberts, D.W. and Granados, R.R. (eds). 1988. *Biotechnology, Biological Pesticides and Novel Plant-Pest Resistance for Insect Pest Management.* Boyce Thompson Institute for Plant Research at Cornell University, Ithaca, New York, USA.

Rogers, H.J. and Parker, H.C. 1995. Transgenic plants and the environment. *J. expt. Biol.* **46**: 467-488.

Schell, J.S. 1987. Transgenic plants as tools to study the molecular organization of plant genes. *Science* **237**: 1176-1183.

Schneiderman, H.A. 1984. What entomology has in store for biotechnology? *Bull. ent. Soc. Am.* **30**: 55-61.

Sivamani, E., Rajendran, N., Senrayan, R., Ananthakrishnan, T.N., and Jayaraman, Kunthala 1992. Influence of some plant phenolics on the activity of endotoxin of *Bacillus thuringiensis* var. *galleriae* on *Heliothis armigera. Ent. exp. Appl.* **63**: 243-248.

Smith, G.E., Summers, M.D. and Fraser, M.J. 1993. Production of human lecta interferon in insect cells infected with a baculoviruses expression vector. *Molec. Cell Biol.* **3**: 2156-2165.

Thacker, J.R.M. 1993. Transgenic crop plants and pest control. *Sci. Prog.* **776**: 207-219.

Thacker, J.R.M. 1994. Transgenic plants. *Trends Ecol. Evol.* **9**: 486.

Van Rie, J. 1991. Insect control with transgenic plants: Resistance proof? *Tibtech.* **9**: 177-179.

9

BEHAVIOUR AND DEVELOPMENT MODIFYING APPROACHES FROM PLANT DEFENSE MECHANISMS

OPENDER KOUL

Insect Biopesticide Research Centre, Jalandhar - 144 003, India

9.1. INTRODUCTION

Among naturally occurring plant-insect systems, a defensive strategy of the plants involves the synthesis and accumulation of chemicals in their tissues that prevents insect herbivory. Additionally, certain regulatory mechanisms maintain the activities of insects below the point at which plant reproduction is severely impaired. As a result of these regulatory mechanisms, mortality, reduced fecundity, delayed development and the insect's inability to find suitable food may occur, all of which profoundly affect populations. On the whole, the morphology, physiology, chemistry and ecology of plants include many traits that interfere with most aspects of insect life (Hedin, 1983). This has led to the view of population biology that strongly emphasizes regulation of lower trophic levels, i.e., the availability and suitability of host plants (Cates, 1980; Rhoades, 1983).

Various studies concerning insect-plant interactions have emphasized that the end product of these interactions is an ecologically and evolutionally stable relationship and involves behaviour and developmental changes to maintain this relationship. The purpose of this chapter is to consider the critical issues involved in using behaviour and development modifying approaches from plant defense mechanisms in insect pest management strategies.

The role of plant-insect interactions in agriculture may be divided into two general approaches (Raffa, 1986):

- Directed approach with an objective to develop practices that protect crops.
- Basic approach with an objective to develop new knowledge about insect-plant interactions.

The major strategies in the directed approach are the identification of resistant lines, extraction of natural products and using them as molecular models for insecticides, antifeedants and repellents, insect growth regulators, and cultural practices. In the basic approach, however, interactions can be described in ecological, physiological and evolutionary perspectives with an aim to identify key processes associated with insect or plant success. The strategies to achieve this are the evaluation of environmental and genetic factors affecting host location and acceptance by insects, quantification of biochemical mechanisms that could reduce insect success and the analysis of detoxification, sequestration, excretion and interference with plant defense chemicals.

Whatever may be the approach, though each have their own strengths and weaknesses, there is ample evidence that to find something efficacious the goal could be achieved through two basic ways, viz., behaviour modifications and developmental modifications. The former can be well evidenced by deterrent allelochemicals that act on the peripheral nervous system, chiefly on the chemosensory apparatus and result in a behavioural change. The effects on the chemosensory system have been interpreted as phytochemical psychomanipulation which describes the behaviour modifying effects of such chemicals acting on the central nervous system (Murdock *et al.*, 1985). In contrast, the latter could be physiological toxins disrupting cellular, biochemical or physiological processes in an insect system.

9.2. BEHAVIOUR MODIFICATIONS

Behaviour is generally equated with the movement of an organism or body parts and the cessation of such movement. To be classified as behaviour, movement requires the participation and direction of the central nervous system (CNS). Simple coordination of muscles and body parts results in behaviour ranging from the fundamental walking and grooming to the complex navigation and learning. Because behaviour depends on the proper functioning of every neuron in its expression, disruption of any kind could have severe consequences. These consequences, however, may be important without being obvious. For example, greatly increased locomotor activity might be detrimental because the insects will have an abnormally high level of acetylcholine (Ach) in the synaptic cleft. This will subsequently lead to prolonged stimulation of the post-synaptic cell and ultimately the dysfunction of the physiological process controlled by the cholinergic input.

Ethological studies of insects have provided important insight into the nature of environmental cues that elicit certain behaviour responses in these animals. The key stimulus that results in a specific behavioural response is often remarkably simple in nature. This seems to be true not only of physical stimuli, but also of chemical signals.

A number of chemicals are involved in the conveying of information in intra-and interspecific interactions between individual organisms. Chemical communication is a major means of information transfer in the insect community and is especially highly developed in locating a mate, food and an oviposition site. Such communicating chemicals, the semiochemicals, include intraspecific (pheromones) and interspecific (allelochemicals) substances. Allelochemicals are further subdivided into those that benefit the stimulus emitter (allomones), the stimulus receiver (kairomones) or both (synomones). The response induced by these may be immediate behaviour reaction or long lasting physiological changes. According to Takahashi (1993) six categories encompass most behaviour modication active compounds (Fig. 9.1).

I. LOCOMOTOR STIMULANTS
Cause kinetic reaction in an insect by increasing locomotion

II. ARRESTANTS
Cause kinetic reaction in an insect by decreasing locomotion or slowing the turning rate

III. ATTRACTANTS
Cause an insect to make movements towards them

IV. REPELLENTS
Cause an insect to make movements away from them

V. FEEDING, MATING AND OVIPOSTION STIMULTANTS
Elicit any of the reactions

VI. FEEDING, MATING AND OVIPOSITION DETERRENTS
Inhibit/suppress one of the behaviour reactions

Fig. 9.1. Six categories of behaviour modifying chemicals

The outstanding advances in chemical technology during the past two decades have made possible the discovery, identification and synthesis of many behaviour modifying chemicals (BMCs). This work was accelerated by the introduction of ecological work on the role of chemical cues, and the term “Chemical Ethology” has been coined for such studies (Takahashi, 1991). For example, this strategy can be explained by the stimulation and the suppression of wing-raising activity in the male *Nauphoeta cinerae* (Olivier) (Lobster cockroach). Male-wing raising response is fully elicited upon antennal contact with excised parts of the female body. Accordingly, washings of the antennae were analysed and found to be C_{31} to C_{53} saturated hydrocarbons. At the same time wing-raising suppressants such as nauphoetin have been identified (Fig. 9.2). (Takahashi, 1993) which at the 5.0 mg level suppresses wing raising by 100 per cent.

$$CH_3 - (CH_2)_{13} - CH = CH\text{-}(CH_2)_7 - CO_2 - (CH_2)_{17} - CH_3$$

Fig. 9.2. Chemical structure of nauphoetin

Similarly, oviposition stimulants in coccid cuticular waxes are utilized by parasitic wasps to locate and identify their host insects. Compounds like cerorubenol -I, -II and -III (Fig. 9.3) are ovipositional stimulants in the scale insect, *Ceroplastes rubens* Maskell, for the encyrtid parasitoid, *Anicetus beneficus* Ishii & Yasumatsu. A mixture of ceriferol and ceriferol-I elicit ovipositional behaviour in *A. beneficus* targeting *C. cereferus* Comstock and *C. japonicus* Green.

A braconid wasp, *Apanteles kariyai* Watanabe, is a gregarious parasitoid of the common armyworm, *Mythimna separata* (Walker). *A. kariyai* shows a sequential oviposition behaviour, i.e., recognizing a host habitat through antennal perception of chemical cues (arrestants) present in the feeding trace left by host larvae. On perceiving the arrestants, response is in an orthokinetic manner, involving stopping, slow walking and probing with antennae for the host larvae. Compounds like 2-undecyl-5-eicosyl-tetrahydrofuran, 2-tridecyl-5-octadecyl-tetrahydrofuran and 2-pentadecyl-5-hexadecyl-tetrahydrofuran, 2-tridecyl-5-octadecyl-tetrahydrofuran in a ratio of 1:2 or 8:1 show arrestant activity. Walking speed increased and the time spent and distance walked on the treated patch decreased. These results indicate that an arrestant is an important factor in the host searching behaviour of the braconid wasp, *A. kariyai*. Thus, treatment of the host habitat with such an arrestant would improve the use of parasitoids for pest management.

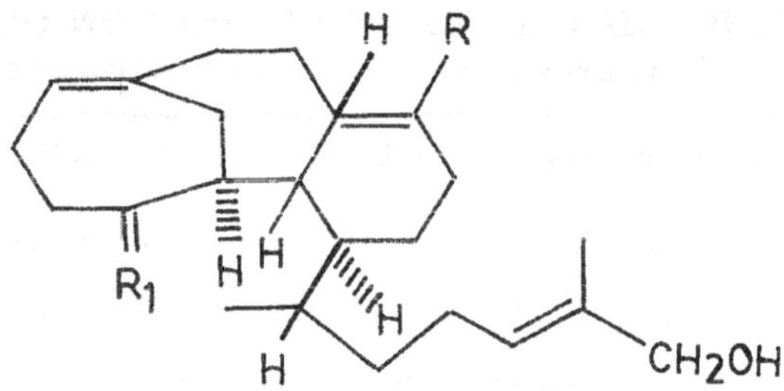

CERORUBENOL-I; R = CH_3, R_1 = CH_2

CERORUBENOL-II; R = CH_2, R_1 = CH_2

CERORUBENOL-III; R = CH_2, R_1 = H

Fig 9.3. Ovipositional stimulants cerorubenol-I, -II and -III

Similarly, the antimones are very important as BMCs that affect host searching behaviour. It can be explained by again taking the example of *A. kariyai* where the arrestant of *Acantholeucania loreyi*, a nonhost noctuid larvae, evokes an arrestant reaction - a behaviour or physiological reaction that is maladaptive to both the emitter and the receiver. These arrestants belong to the chemical groups shown in Fig. 9.4.

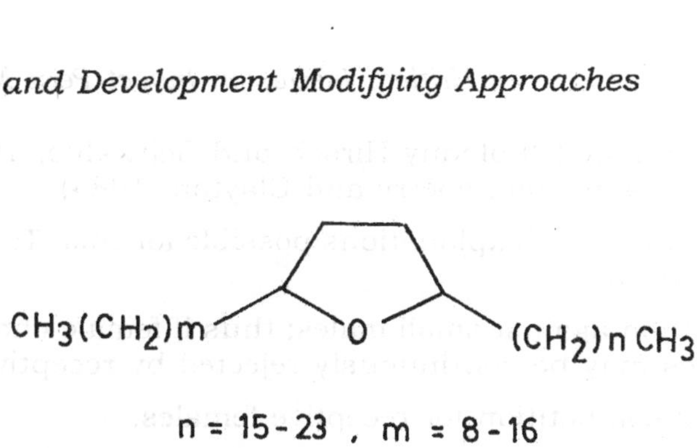

Fig 9.4. Chemical structure of arrestant compounds

9.2.1. Semiochemicals in Management Programmes

The specificity and non-toxic nature of semiochemicals makes them an appealing option within an integrated pest management programme, even though the use of BMCs as management tools has not always lived upto initial expectations (Lewis, 1981; Carde, 1990). However, they have been successfully used in management programmes for agricultural and forest pests (Jutsum and Gordon, 1988; Ridgway *et al.*, 1990). Use of semiochemicals could be considered by understanding a mate choice and male-male competition which will help in the interpretation of trap catch data and by interpreting their effect on certain biological characteristics of the target populations (McNeil, 1992). These considerations could be applied to the use of sex pheromones for lepidopteran as well as other insect orders and to the development of other BMCs.

Monitoring programmes play an important role in determining the distribution of an insect within a given geographic area (Elkinton and Carde, 1981), period of adult activity in order to predict phenology and synchronize insecticide applications (Baker *et al.*, 1980; Wall, 1988), monitoring long term trends in density to detect when populations pass from endemic to epidemic levels (Sanders, 1981) or predicting subsequent larval densities or resource losses (Allen *et al.*, 1986). In the majority of monitoring programmes, it is not necessary to know the reproductive status of the males captured. However, if the objective is to predict the density of some future developmental stage in the life cycle, such information may help explain why, in some cases (Palaniswamy *et al.*, 1990), there is a poor correlation between pheromone trap catch data and subsequent egg or larval densities.

Another factor that has implications in a monitoring system is the responsiveness to a pheromone. There is ample evidence from various studies on gypsy moth, spruce budworm and corn earworm which suggests that a significant proportion of males captured in pheromone traps may be unmated (Kolodny-Hirsch and Schwalbe, 1990; Bergh and Seakbrook, 1986; Henneberry and Clayton, 1984).

There are number of explanations possible for this. To generalise, the reasons could be:

- Discrimination against small males; thus if female choice occurs, many males may be continuously rejected by receptive females.
- Intrasexual competition for receptive females.
- Sub-lethal pathogenic infects, which may influence flight performance and thus affect the male's ability to compete with conspecifics as has been observed in the case of spruce budworm males, infected with *Nosema fumiferanae* (Wilson, 1987).

On the whole, there is very little information on the level of female choice or male-male competition in leipdoptera under field conditions, thus extent of trap catch due to pheromones remains a speculation.

The second important role of semiochemicals is mating disruption. The wide-scale application of sex pheromones may, through a number of mechanisms like peripheral reception and CNS effects, competition between natural and synthetic sources, or imbalance in sensory input, disrupt pheromone-mediated communication and successfully reduce pest densities (Carde, 1990). The repeated disruption, however, might result in the selection of resistant populations as is induced by repeated use of insecticides (Metcalf, 1980) through inter-individual variability associated with signal emission and reception or through a genetic basis or covariance between the different characters and individual fitness. This is an area of insect pheromone biology that remains to be explored. Inter-individual variation in age at which females initiate pheromone release (Kanno, 1979; Howlader and Gerber, 1986), pheromone titre (Delisle and McNeil, 1987), emission rate (Haynes *et al.*, 1984) and the proportion of different pheromone components present in the blend (Miller and Roelofs, 1980; Haynes *et al.*, 1984; Roelofs *et al.*, 1987; Witzgall and Frerot, 1989) is well documented. There is also evidence that such characters are heritable (Sreng *et al.*, 1989; Collins *et al.*, 1990; Han and Gatehouse, 1991). Similarly, male response to pheromone blend is heritable (Collins and Carde, 1990) which suggests that such traits could be subjected to selection. However, there are already questions concerning genetic relationship due to variabilities, therefore, broader data base is necessary before the possibility of resistant populations through simultaneous changes in both the emitter and the receiver are realistically recognized.

In the arctiid moths (Willis and Birch, 1982; Wunderer *et al.*, 1986) a dual pheromone-mediated mating system exists. However, very little is known about the dynamics of such dual systems. A similar dual system is observed in *Trichoplusia ni* (Hubner), where females flying upwind to males at dusk, several hours before calling females attract males (Landolt and Heath, 1990). This is an example where a female sex pheromone can be applied to the crop, as both the major components of the female sex pheromone and host plant's volatiles stimulate male sex pheromone release. Thus, this system may effectively reduce the female-mediated avenue of male location but, by selecting for the male-mediated component, have little or no impact on overall mating success (McNeil, 1992).

Thus, it is evident that the development of pheromones in pest management will not only depend upon the chemical, biochemical and neurological levels, where significant progress has been made (Tumlinson, 1988), but also depends on behavioural, ecological and genetic aspects of pheromone communication. On the whole, both approaches are complimentary and study of both approaches could provide meaningful results concerning semiochemicals for pest control.

9.2.2. Phytochemicals and Behaviour Alterations

Phytochemicals ranging from ions and free radicals to complex secondary metabolites serve as messengers to insects during plant-insect interactions. These messengers are usually volatile compounds (Norris, 1990; Liu *et al.*, 1989). In contrast, relatively non-volatile compounds serve as messengers upon the insect's contact with the source plant. These latter chemicals especially influence insect feeding and may not be important factors in the initial selection of a host plant or rejection of a nonhost plant. Whatever may be the nature of the messenger, once it arrives at a chemosensory structure on the insect, it then donates, accepts or variously shares electrons and protons with the receptor protein of that chemosensory organ (Norris, 1994).

It is now well known fact that insects use common chemical messengers in molecular communication. However, such common chemicals may not convey the same message even among biotypes of the same species (Norris, 1986). Some times, a chemical may act as a double agent, whereby it tells one insect species about its favourite host and, tells other insect species to reject the plant species. Major chemical classes of messengers include simple inorganic chemicals like sodium chloride, hydrochloric acid, mercury and sulfur and even simple sugars are feeding excitants of insects. Most classes of plant metabolites act as messengers affecting insect behaviour. These metabolites comprise of benzoquinones, naphthoquinones, alkaloids, flavonoids and terpenoids. Phytochemicals bearing one or more special functional group like alcohol, ketone, aldehyde, ester, ether or acetate may also be important messengers to insects because these groups make the molecule more volatile (Visser, 1986; Ramachandran and Norris, 1991).

According to Norris (1986) there are more than 4500 known alkaloids which may act as messengers affecting insect behaviour. These are nitrogen based compounds and highly toxic in nature. Many alkaloids are stored in plants as glycosides. This has two advantages, i.e., (i) they can be stored by plant tissues safely while otherwise they may be toxic to the source plant, and (ii) they can be hydrolysed quickly to release the toxic alkaloid when attack from a pathogen occurs. However, the presence of alkaloids does not guarantee a plant safety from insect herbivory.

Flavan-3-ol, hydrochalcones, chalcones, flavonones, leucoan-thocynidins, flavones, anthocyanidins, benzalcoumaranones, flavanonols, flavanols, etc. which possess variable oxidation states in the C_3 unit form the array of flavonoids which provide communication perceived by the insect. It has been established that plants respond to environmental stresses by making quantitative changes in their array of flavonoid messengers (Ebel, 1986; Neupane and Norris, 1990).

Terpenoids represent the most diverse class of secondary phyto-chemicals and include a range of monoterpenes, sesquiterpenes, diterpenes and a range of triterpenes. Most of these act as antifeedants (Koul, 1982). Azadirachtin (Fig. 9.5) is perhaps the most studied phytochemical messenger involved in plant-insect interactions (Koul, 1992). In fact, azadirachtin is a double messenger acting as an antifeedant as well as an insect growth regulator. Secoiridoids, also derivatives of terpenoids are highly active messengers between plants and insects. Glycosides of some steroidal alkaloids like demissine and tomatine found in Solanaceae, reduce feeding by the Colorado potato beetle, *Leptinotarsa decemlineata* (Say) (Koul, 1982).

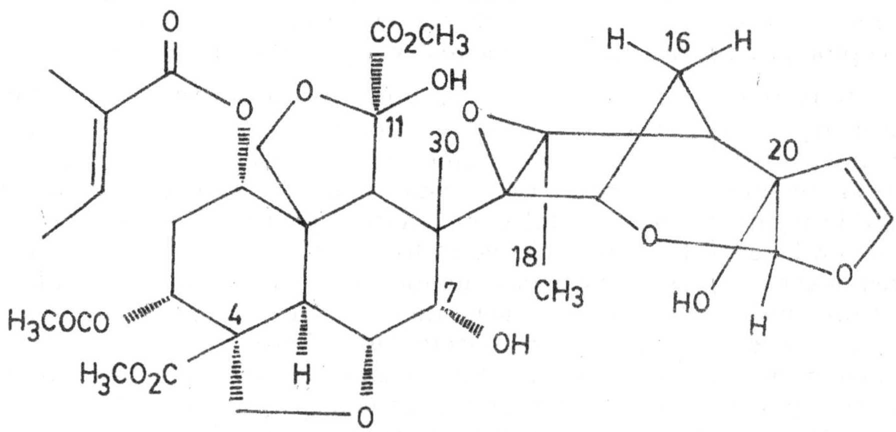

Fig 9.5. Structure of azadirachtin

Similarly, quinones, though a smaller class of phytochemicals, are important messengers, Benzoquinones are produced by approximately

20 diverse families of plants and naphthoquinones are found in at least 20 angiosperm families (Norris, 1994). These compounds are known to function through reacting via 1,4-addition and oxidative reduction chemistry with sulfhydral groups in receptor proteins of the insect's chemosensory neurons (Norris, 1988; Rozenthal and Norris, 1973; Singer *et al.*, 1975). It is now clear that most, if not all, phytochemicals serve as messengers in plant-insect communications by altering directly, or indirectly, the sulfhydryl-disulfide dependent electrochemistry in the receptor protein in the insect's chemosensory neurons (Norris, 1994).

9.2.3. Antifeedant Approach and IPM

From the above discussion it is obvious that many phytochemical messengers could be antifeedant compounds - the compounds that inhibit feeding and insects may die of starvation. The terms feeding deterrent and gustatory repellent are synonymous with antifeedant, but these terms should not be confused with olfactory repellents, which repel the insect before it starts to eat. In recent years, the use of antifeedants in pest management programmes has considerable intuitive appeal as it satisfies the need to protect specific crops while avoiding damage to non-target organisms. It has also become evident that antifeedants play an important role in the continuing battle for survival between insects and plants. Apparently all plants contain one or more substances that are unpalatable to insects and that must have evolved as a medium of defense. The details of various types of insect antifeedant compounds, their evolution or perception, have been comprehensively dealt with in many reviews (Koul, 1993; van Beek and de Groot, 1986; Schoonhoven, 1988). However, there are two somewhat different perspectives of antifeedant use in pest management. First, how three dimensional mapping of insect taste receptor sites helps in novel antifeedant development, and secondly, how these compounds have a very important role in integrated pest management.

Insects possess highly-adapted behaviours for feeding that are elicited by specific chemical signal stimuli. These signals are detected by a number of chemosensory cells that are grouped within sensilla located primarily on the mouth parts. These chemosensory cells of caterpillars in particular, that respond to taste stimuli, are among the most studied and have been selected for a biorational design of antifeedants.

The styloconica of the tobacco hornworm, like those of many catterpillars, contain cells responsive to plant sugars. Information from these cells is used by the caterpillar in the decision to feed. Although exact sensory codes are yet to be understood, some antifeedants appear to disrupt the message from sugar sensitive cells and reduce feeding (Frazier and Lam, 1990). Thus, mapping the binding requirements for the receptive sites of these sugar-sensitive cells and using this information to design/synthesize feeding inhibitors is a novel approach. This approach involves:

- Quantification of chemosensory cell activity
- Mapping the receptor site
- Synthesis of site inhibitors
- Evaluation of site inhibitors

The quantification of chemosensory cell activity can be done by application of computer aided analysis. A template matching technique is used to identify the action potentials produced by each cell. This technique separates the response of each cell (like sugar sensitive cell, salt sensitive cell, etc. on the basis of variation in spikes) type and can thus be identified and quantified. One can thus approach quantitative structure activity relationship (QSAR) studies for individual taste cells within a reasonable time frame (Frazier and Lam, 1990).

The second step will be the mapping of the taste receptor sites. The styloconic glucose-sensitive taste cells of the tobacco hornworm have served as models for receptor site mapping and for the design of candidate inhibitors. For this, various fluoro - and deoxy-glucoses have been used to determine the hydrogen bonding pattern for the medial glucose cell of the tobacco hornworm caterpillar. The rationale for determining the three dimensional aspects of the receptor hydrogen bonding relative to the plane of the stimulus, glucose combined the use of epimers at each carbon position relative to glucose. If the epimeric hydroxy at a certain position contributes to binding, the receptor site is located between the hydroxy position of the epimer and glucose. In this mode, the receptor site will be situated approximately

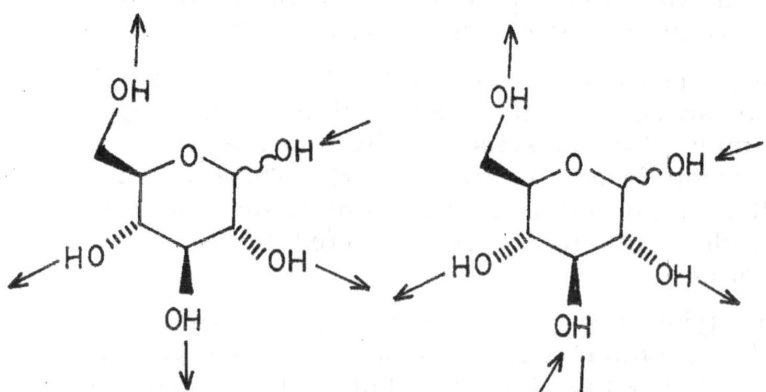

Fig. 9.6. Comparison of binding sites for the lateral and medial receptors (arrows indicate direction of bonding)

in the plane of the glucose ring. If epimeric hydroxy does not contribute to binding, the receptor site will be located on the opposite side of the epimeric position. In this mode, the receptor site will be either above or below the plane of the glucose ring.

Frazier and Lam (1990) have described three dimentional position of hydrogen bonding at each C-position for the medial glucose receptor site and have clearly shown the comparison of the binding site requirements for the medial and lateral glucose sensitive cells (Fig. 9.6).

Based on these binding site models, a series of compounds with electrophiles have been synthesized as active feeding inhibitors (Oikawa *et al.*, 1982). These compounds when tested for antifeedant activity on several species of insects like *Heliothis virescens* (Fabricius), *Spodoptera frugiperda* (J.E. Smith) and *Ostrinia nubilalis* (Hubner), reduced feeding by 15 to 77 per cent at 1000 ppm in a leaf disk test (Frazier and Lam, 1990). This shows that same compounds also exhibit low potency activities on other caterpillar species, whose receptor site geometries are not known. A conclusion which can be drawn here is that the evolutionary arena of insect taste and plant distaste is apparently even more complex than previously surmised.

According to the second perspective, antifeedants have an important role in integrated pest management. High target specificity makes feeding deterrents ideal for incorporation in these programmes. For instance, an antifeedant application to the upper leaves of the plants combined with a limited application of teflubenzuron to the lower leaves gave maximum protection to the plant foliage, and a maximum kill of insects, coupled with minimum use of insecticide (Griffiths *et al.*, 1992).

The degradability of antifeedants leads to a prospect, the persistence, which is largely unknown for plant allelochemicals. That persistence of antifeedants is affected by behavioural desensitization has been reported and a generalized model for quantifying the same using lepidopteran larvae has been developed (Raffa and Frazier, 1988).

Systemic action of antifeedants is another useful aspect of their practical application in IPM. On one hand it will exert uniform distribution within the plant and on the other it will counterbalance the phagostimulatory effects of plant surface chemicals (Chapman and Bernays, 1989). Systemic action of neem extracts and azadirachtin are well documented (Gill and Lewis, 1971; Meisner *et al.*, 1987; Osman and Port, 1990; Koul and Shankar, 1995). Similarly, extracts of *Amora ruhituka* and *A. squamosa* (Islam, 1987) are systemic antifeedants. Coumarin is transported in grass leaves and thus unpalatable to *Chorthippus parallelus* (Zetterstedt.) (Bernays and Chapman, 1975) and sinigrin is absorbed from water solution and transported to stems and leaves of various plant species.

9.2.4. Critical Issues

The problems illustrated with the behaviour modifying compounds, which play a major role in plant defense in nature are basically the critical issues for their use in IPM programmes. If such materials, derived from one plant are to be applied to another crop plant, they would have to provide continuous protection from living insects (Bernays, 1983). This implies that the compounds must either be translocated systemically to the newly growing, unprotected portion of the plant or they should induce irreversible effects in the target insect, such that it fails to recognize any part of the plant (treated or untreated) as suitable substrate. However, few chemicals, such as azadirachtin and formamidine which systemically and irreversibly prevent feeding, are available. Still this approach differs from the normal phytochemical context, in which allelochemicals are continuously synthesized, transported and sometimes actively accumulated at the site of attack (Raffa, 1986).

Similarly, the attraction of insect natural enemies by plant semiochemicals has theoretical potential but practical utilization can be difficult. Problems can be illustrated where application of kairomones derived from the host insect have been used. For instance, indiscriminate foliar application can actually hinder the ability of parasites to concentrate on the target site of preference (Chiri and Legner, 1983) and will be related to host density. Lewis *et al.* (1979) found that kairomone applications were effective at very high pest densities only. Thus, farmers/growers very rarely accept this strategy. In fact, for each crop system consideration should be that entomophagous insect populations are continuously present, and stable at high densities. Many agrosystems do not meet these criteria even under non-sprayed conditions (Stehr, 1982). Also, the range of pest species affected is usually quite narrow, and innate spacing mechanisms that improve efficiency could well be negated. This clearly implies that identification of additional active compounds is not sufficient to promote crop protection. Thus, extensive research on the integration of plant and insect-derived chemical messengers, increased releases, predator and prey distribution patterns, and available means for controlling coexisting pest species will be necessary before any practical gains can be realized (Raffa, 1986).

In recent years the synergistic activity of phytochemical defense mechanism in plants is receiving increased attention. These different compounds may function independently, thereby delaying the evolution of resistance because of the multiple changes required to confer selective advantage, or they may synergistically inhibit metabolic detoxification, again achieving the same effect. There are interesting examples of such combinations like xanthotoxin + myriticin (Berenbaum and Neal,

1985), fusaric acid + gossypol (Dowd, 1989) or *cis*-asarone + menthol (Koul *et al.*, 1990b). Phytochemical application could mimic this strategy, either by directly applying mixtures of independently active plant-derived materials or by applying the compounds that function synergistically. Although, from a physiological point of view the strategy is sound its economical competitiveness remains only a possibility. This has happened with classical insecticide synergists which show high potential value for counteracting or delaying insecticide resistance, but have not yet proven practical for widespread agricultural use (Raffa and Priester, 1985). On the whole, natural models have not proven sufficient.

Other critical issues are the lack of knowledge of the exact mode of action (particularly for antifeedant compounds), insect chemistry and physiology involved in interaction, influence on the ecological system, structure/bioactivity relationships in general and the lack of extensive field evaluations. All these parameters warn us against over-optimistic views.

9.3. DEVELOPMENTAL MODIFICATIONS

The primary adaptive role of secondary plant compounds is the repulsion of insects, either through direct repellence or indirect mechanisms like distastefulness. There are certain phytochemicals that adversely impact the developmental physiology of insects via growth inhibition or growth regulation.

9.3.1. Growth Inhibitors

Compounds that inhibit the growth of insects, irrespective of the mechanisms involved, are usually termed as growth inhibitors. There is a constitutive accumulation of toxic, low molecular weight compounds like antibiotics, saponins, alkaloids, glycosides, cyanogen generating compounds, phenolic compounds and many ribosome inactivating compounds which are implicating in conferring resistance against insects (Peumans and Vandamme, 1995; Ananthakrishnan, 1996). For example, saponins from a variety of plants have been implicated as bases of constitutive resistance to attack of insects. Saponins from alfalfa seed have been shown to inhibit growth of *Tribolium castaneum* (Herbst) larvae (Shany *et al.*, 1990) where growth reduction has been attributed to aglycone moiety, with the sapogenin medicagenic acid being most toxic. Triterpene glycosides in organ pipe cactus inhibit the maturation of *Drosophila nigrospiracula* larvae (Kircher, 1972). Glycoalkaloids from potato affect larval development of Colorado potato beetle (Sinden *et al.*, 1980) as well as nymphal survival and duration of settling of the potato leafhopper, *Empoasca fabae* (Harris) (Raman *et al.*, 1979). Tomatine, a glycoalkaloid from tomatoes inhibits nymphal growth/ larval growth, reduces pupal eclosion, adult longevity and causes developmental malformations in many insects (Harley and Thorstiensen, 1967; Campbell and Duffey, 1979; Isman and Duffey, 1982; Duffey and Felton, 1989). Similarly, flavones from corn silk affect the growth of insect larvae (Wiseman *et al.*, 1992, 1996).

These studies imply that glycoalkaloids are resistance factors in the solanaceous crops inducing physiological toxicity to the insects. In fact, two phenolics, rutin and chlorogenic acid (also developmental inhibitors) also proposed as a basis of resistance in tomatoes (Isman and Duffey, 1982), are modified in toxicity by both polyphenol oxidase activity and dietary protein. Likewise tomato proteinase inhibitors are differentially toxic to *Helicoverpa zea* (Boddie) and *S. exigua* (Hubner) and their toxicity depends on the quantity and quality of dietary protein. This indicates that before use in IPM programmes, more profound integration of plant phenology, and nutritional physiology and toxicology of the chemical involved is required. These effects can be considered antibiotic type of effects and are known for many other phytochemicals like resorcinol, gallic acid and phloroglucinol (Ananthakrishnan *et al.*, 1990) and flavonoids (Mahoney *et al.*, 1989; Elliger *et al.*, 1980).

Inhibition of growth and development by nonprotein amino/imino acids are other well known examples of developmental modifiers. The growth and development of corn earworm, *H. zea* larvae was markedly disrupted by a naturally occurring lower homologue of proline, L-azetidine-2-carboxylic acid. It caused severe retardation of larval development, larval mortality resulting from exuvial ligature, death at larval pupal ecdysis, formation of grossly deformed pupae and nonsclerotization of the anterior ventral half of the pupal integument (Adeyeye and Blum, 1989). Similarly, the oxyguanidino structural homologue of arginine, L-canavanine has been shown to disrupt growth of *Manduca sexta* (Johannsen) (Dahlman and Rosenthal, 1975) and *S. litura* (Koul, 1985) and fertility of *Dysdercus koenigii* (Fabricius) and *Periplaneta americana* (Linnaeus) (Koul, 1983 a,b).

Mimosine (Fig. 9.7) is another nonprotein heterocyclic amino acid (β-[N-(3-hydroxy-4-oxypyridyl]-α-aminopropionic acid, isolated from *Mimosa pudica* and *Leucaena leucocephala*. This amino acid inhibits growth and development of *T. castaneum* (Ishaaya *et al.*, 1991) through reduction of the trehalase enzyme.

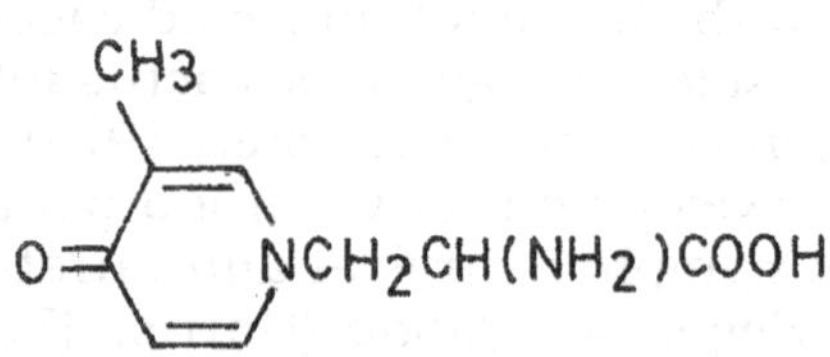

Fig. 9.7. Chemical structure of mimosine

Apparently, physiological toxicity induced by non-protein amino acids in one way will generally be manifested through their particular ability to compete with other protein amino acids in protein synthesis or in other biosynthetic pathways involving protein amino acids, and possibly through the amount that can be tolerated when incorporated into protein before malfunction becomes lethal. Incorporation of canavanine into protein can alter the conformation and disrupt the function of the native macromolecules (Rosenthal, 1991). In this context, non-protein amino acids may be viewed as intermediate between compounds like alkaloids and cyanides that simply block a particular internal physiological reaction and compounds like tannins that bind with a substrate or enzyme in the gut and prevent its use (Janzen *et al.*, 1977).

There are many other growth inhibitory compounds that show post-ingestive effects on growth and development of insects and belong to the groups like phenyl propanoids (Koul, 1995, 1996; Koul *et al.*, 1995), many nonsteroidal terpenoids like nagilactones, podolactones, glaucolides, steroidal terpenoids, gossypol, various alkaloids, lignins, tannins, benzoxazolinones and so on. It, however, will be beyond the scope of this chapter to describe all such compounds in detail, but comprehensive details are available in the literature (Reese and Holyoke, 1987).

9.3.2. Insect Growth Regulators

Toxicity, in the strictest sense, is the outcome of an interaction between a toxin and a sensitive target site in an organism, resulting in a physiological lesion. Physiological targets (usually termed as receptors-not in sensory pretext) are most often proteins found on cell/organelle membranes (Isman, 1992). Organ systems in insects which become physiological targets for plant allelochemicals are the nervous system, endocrine system, gonads, heart and epidermis. Therefore, insect growth regulators (IGRs) are referred as the compounds that interfere with the regulatory system of insects, specifically the endocrinal pathway.

Basic research into insect physiology and endocrinology has shown that plants produce secondary products that can selectively perturb important aspects of the insect endocrine system. Two important hormonal regulators in insects are the juvenile hormones and the moulting hormones (ecdysones). The juvenile hormones, by their presence or absence, regulate many aspects of insect physiology. Various compounds isolated from nature (Fig. 9.8) have been shown to mimic juvenile hormone activity. These compounds applied to developing insects, exert the same morphogenetic activity as the natural JH compounds. The resulting disruption of insect metamorphosis demonstrates the potential that these compounds could have in insect

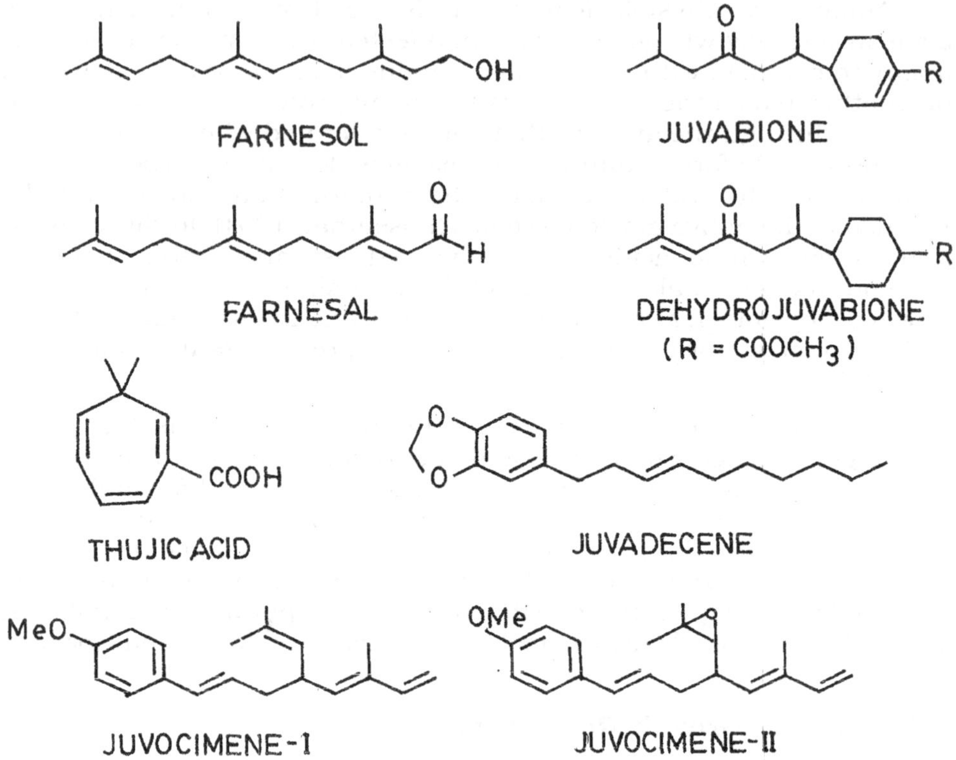

Fig. 9.8. Juvenile hormone mimics from plants

control. On the basis of these studies several juvenile hormone analogues have been developed for commercial utilization particularly for flies, fleas, and mosquitoes.

That plants have adopted anti-hormonal strategies is obvious from the discovery of anti-juvenile hormone compounds - the precocenes. The precocenes from the *Ageratum* plant (Bowers *et al.*, 1976), are simple chromenes and cause cytotoxic disruption of the corpora allata (Unnithan *et al.*, 1977) through oxidative activation into a highly reactive epoxide that destroys the corpus allatum (Bowers, 1982). This results in premature moulting to adults, entrance into diapause, or induction of sterility in insects. Even minute amounts of these plant compounds are often sufficient to disrupt the insect's physiology and development. Recently, anti-JH compounds have been isolated from the genus *Nama* (Hydrophyllaceae) that disrupt the endocrine system of the large milkweed bug, *Oncopeltus fasciatus* (Dallas) (Binder *et al.*, 1991). An interesting aspect of this study was the discovery of both JH and anti-JH compounds in three *Nama* species.

Another class of growth regulators is the phytoecdysteroids. Since the isolation of the first insect moulting hormone from plants (Nakanishi *et al.*, 1966) different phytoecdysteroids have been found in 83 different families of plants (Hikino and Takemoto, 1974). It is well known that ecdysteroids have marked physiological effects in insects. Consequently, it has been suggested that phytoecdysteroids may play an important role in the defense of the plant against attack by phytophagous insects. However, it is necessary that phytoecdysteroids should occur in sufficient amounts to contribute significantly to defense against insect herbivores. The question as to whether these compounds are able to function efficiently as defense in natural habitats still exists and evidence is required to demonstrate the occurrence of adequate amounts in the appropriate parts of the plant at the appropriate times throughout the plant's life cycle (Jones and Firn, 1978).

Similarly, a group of compounds, known generally as ecdysis inhibitors, also occur in nature. A natural plant compound that was found to inhibit ecdysis through an inhibition of chitin synthetase is the naphthoquinone, plumbagin isolated from *Plumbago capensis* which inhibits the ecdysis of a few species of lepidopteran agricultural pests (Kubo *et al.*, 1983).

Another interesting natural product with an IGR activity and which has developed a commercial potential in recent years is azadirachtin from the well known neem tree, *Azadirachta indica* A. Juss. The compound prevents apolysis and ecdysis, induces pharate morality or sometimes induces permanent larvae. These actions are dose and time dependent and relative to mode of application (Koul *et al.*, 1987). However, it has been shown unequivocally that azadirachtin does not act directly on the prothoracic glands or prothoracicotropic hormone (PTTH). The corpus cardiacum has been suggested to be the target for azadirachtin in terms of PTTH, eclosion hormone, bursicon or general tropic hormone release (Mordue and Blackwell, 1993). On the whole, azadirachtin induces variable effects at the endocrinal level which are apparently relative to the mode and time of application and physiological state at the time of azadirachtin treatment. This may also have to correspond with the timing of juvenile hormone levels relative to ecdysteroid levels and will be of critical importance in regulating the expressions of the hormonal milieu of insects due to varied inter-relationships.

9.3.3. Critical Issues

It is quite clear that from the developmental point of view there are many defense molecules in plants that inhibit or regulate the physiological processes in insects. There is also a widespread interpretation that this represents coevolutionary adaptation. But then the activity alone does not assure absolute crop protection. There are compounds that show absolute toxicity at very low concentrations, yet they do not economically protect the crop from the insect attack even if the rate of concentration is increased 100-fold. The limitation is that they act slowly, because the insect must first arrive at a susceptible stage, i.e., moult, before the material is effective. In a natural situation, from agricultural point of view, this is often an intolerable situation. So, relying on a natural model is not enough. Take the example of the highly polar nature of phytoecdysteroids which severely limits their movement through the insect cuticle, thus hindering their commercial utilization. The complex structure of ecdysones seriously limits their availability and the possibility of isolating them from plants in commercial quantities is unrealistic.

In general, there have been nagging problcms with increased susceptibility to some insects or microorganisms, loss of desirable agronomic characteristics and insect biotype evolution. The evolution of insect biotypes poses a special problem for pest management.

Therefore, what should the pest management tactics comprise? Insect control strategy must initially provide consistent economical plant protection to be adopted. This is true for both natural plant defense mechanisms as well as for grower practices. While no strategy can be called stable, if it is not successful in the short run it can induce instability (Raffa, 1986). the question is, can this be avoided in devising pest management schemes based on plant-insect interactions.

The first point to consider is system stability. A stable system is at equilibrium and will return to the original variables when disturbed (Raffa, 1986). According to Berryman (1981) negative feedback pathways in nature, occur through biotic changes caused by insect activities, such as resource depletion, natural enemy reproduction and host plant alterations that render their habitat less suitable. Our agricultural systems, however, do not generally have these properties. For example, resource depletion is an unacceptable form of insect control.

The traditionally applied control measures have successfully provided the immediate control, but these perturbations have not resulted in a return to the original stage. Each new chemical, or for that matter any other strategy has placed an intense and unidirectional selection pressure on the insects. The objective, then, is to incorporate factors into the insect's environment that reduce the fitness of genetic adaptations to the control strategy. Can resistance to the control strategy be at least partially nonadaptive? The conifer bark beetle system provides a natural model for this and has been fully described (Raffa and Berryman, 1982). This nature based example provides support for the views of Kogan and Ortman (1978) and Gould (1983, 1984) who have argued that insect aversion behaviour and plant antibiotic constituents need to be considered in an inclusive fashion. In fact, the systems involving both may not promote resistant biotype evolution, because there would be no advantage in possessing either reduced nonpreference or increased detoxification capacity in the absence of the other trait. Those insects that were only resistant to plant's or chemical's antifeedant properties would be poisoned, and those that were only metabolically resistant would never exploit the host.

Recent findings on plant-insect interactions provide some ideas to how stabilizing elements could be introduced into agricultural systems. For example, a phenomenon of aggressive chemical mimicry is well known in predaceous insects, i.e., they produce allomones which resemble pheromones of their prey (Koul and Smirle, 1993). A similar example is the production by wild potatoes of allomones that mimic aphid alarm pheromones (Gibson and Pickett, 1983).

Thus, such systems act as natural repellents and could be a potential defense strategy by placing the insect in a position of conflicting selective pressures.

9.4. CONCLUSIONS

Many plant defense mechanisms provide cues for insect control and ideas for insect pest management strategies. But agricultural tactics must be efficacious and economical in the short run to be adopted. There should be an integrated approach for plant protection tactics avoiding unidirectional selection pressures. This will lessen the genetic resistance among target species. Functional dynamics of phytophagous insects vis-a-vis the plant interactions need to be best understood under natural conditions and manipulations of key processes under control conditions exercised. Relative advantages and disadvantages of each control tactic using plant defense mechanisms, need to be studied in depth in each crop system and the most compatible control method adopted. Effective integration of basic and directed research ideas will definitely provide the means in devising optimum pest management strategies based on plant defense mechanisms

REFERENCES

Adeyeye, O.A. and Blum, M.S. 1989. Inhibition of growth and development of *Heliothis zea* (Lepidoptera: Noctuidae) by a non-protein imino-acid, L-azetidine-2-carboxylic acid. *Envir. Ent.* **18**: 608-611.

Allen, D.C., Abrahamson, L.P., Eggen, D.A., Lanier, G.N., Swier, S.R., Kelley, R.S. and Auger, M. 1986. Monitoring spruce budworm (Lepidoptera: Tortricidae) populations with pheromone-baited traps. *Envir. Ent.* **15**: 152-165.

Ananthakrishnan, T.N. 1996. Biochemical profiles of host plant resistance to insects. *In.* T.N. Ananthakrishnan (ed.). *Biochemical Basis of Host Plant Resistance to Insects.* National Academy of Agricultural Sciences, New Delhi, pp. 1-13.

Ananthakrishnan, T.N., Senrayan, R., Annadurai, R.S. and Murugesan, S. 1990. Antibiotic effects of resorcinol, gallic acid and phloroglucinol on *Heliothis armigera* Hubner (Insecta: Noctuidae). *Proc. Indian Acad. Sci. (Anim. Sci.)* **99:** 39-52.

Baker, T.C., Carde, R.T. and Croft, B.A. 1980. Relationship between pheromone trap capture and emergence of adult oriental fruit moths, *Grapholitha molesta* (Lepidoptera: Tortricidae). *Can. Ent.* **112:** 11-15.

Bergh, J.C. and Seabrook, W.D. 1986. The mating status of spruce budworm males, *Choristoneura fumiferana* (Clem.) (Lepidoptera: Tortricidae), caught in pheromone-baited traps. *J. Ent. Sci.* **21:** 254-262.

Berenbaum, M. and Neal, J.J. 1985. Synergism between myristicin and xanthotoxin, a naturally occurring plant toxicant. *J. Chem. Ecol.* **11:** 1349-1358.

Bernays, E.A. 1983. Antifeedants in crop pest management. *In:* D.L. Whitehead and W.S. Bowers (eds). *Natural Products for Innovative Pest Management.* Pergamon Press, New York, USA, pp. 259-271.

Bernays, E.A. and Chapman, R.F. 1975. The importance of chemical inhibition of feeding in host plant selection by *Chorthippus parallelus* (Zetterstedt.). *Acrida* **4:** 83-93.

Berryman, A.A. 1981. *Population Systems - A General Introduction.* Plenum Publication Corp., New York, USA.

Binder, B.F., Bowers, W.S. and Evans, P.H. 1991. Insect anti-juvenile hormone activity from plants in the genus *Nama. Experientia* **47:** 199-201.

Bowers, W.S. 1982. Endocrine strategies for insect control. *Ent. exp. Appl.* **31:** 3-14.

Bowers, W.S., Ohta, T., Cleere, J.S. and Marsella, P.A. 1976. Discovery of insect anti-juvenile hormones in plants. *Science* **193:** 542-547.

Campbell, B.C. and Duffey, S.S. 1979. Tomatine and parasitic wasps: potential incompatibility of plant antibiosis with biological control. *Science* **205:** 700-702.

Carde, R.T. 1990. Principles of mating disruption. *In:* R.L. Ridgway, R.M. Silverstein and M.N. Insoe (eds). *Behaviour-Modifying Chemicals for Insect Management: Application of Pheromones and Other Attractants.* Marcel Dekker, New York, USA, pp. 301-327.

Cates, R.G. 1980. Feeding patterns of monophagous, oligophagous and polyphagous insect herbivores: The effect of resource abundance and plant chemistry. *Oecologia* **46:** 22-31.

Chapman, R.F. and Bernays, E.A. 1989. Insect behaviour at the leaf surface and learning as aspects of host plant selection. *Experientia* **45:** 215-222.

Chiri, A.A. and Legner, E.F. 1983. Field application of host searching kairomones to enhance parasitization of the pink bollworm (Lepidoptera: Gelechiidae). *J. econ. Ent.* **76:** 254-255.

Collins, R.D. and Carde, R.T. 1990. Selection for increased pheromone response in the male pink bollworm *Pectinophora gossypiella* (Lepidoptera: Gelechiidae). *Behav. Genet.* **20**: 325-331.

Collins, R.D., Rosenblum, S.L. and Carde, R.T. 1990. Selection for increased pheromone titre in the pink bollworm moth, *Pectinophora gossypiella* (Lepidoptera: Gelechiidae). *Physiol. Ent.* **15:** 141-147.

Dahlman, D.L. and Rosenthal, G.A. 1975. Non-protein amino acid-insect interactions I. Growth effects and symptomology of L-canavanine consumption by tobacco hornworm, *Manduca sexta* (L.). *Comp. Biochem. Physiol.* **51A:** 33-36.

Delisle, J. and McNeil, J.N. 1987. Calling behaviour and pheromone titre of the true armyworm *Pseudaletia unipuncta* (Haw.) (Lepidoptera: Noctuidae) under different temperature and photoperiodic conditions. *J. Insect Physiol.* **33:** 315-324.

Dowd, P.F. 1989. Fusaric acid: A secondary fungal metabolite that synergises toxicity of co-occurring host allelochemicals to the corn earworm, *Heliothis zea* (Lepidoptera). *J. Chem. Ecol.* **15:** 249-254.

Duffey, S.S. and Felton, G.W. 1991. Enzymatic antinutritive defenses of the tomato plant against insects. *In:* P.A. Hedin (ed.). *Naturally Occurring Pest Bioregulators.* Am. Chem. Soc., Washington, DC, USA, pp. 166-197.

Ebel, J. 1986. Phytoalexin synthesis: biochemical analysis of the induction process. *A. Rev. Phytopath.* **24:** 235-264.

Elkinton, J.S. and Carde, R.T. 1981. The use of pheromone traps to monitor distribution and population trends of the gypsy moth. *In:* E.R. Mitchell (ed.). *Management of Insect Pests with Semiochemicals: Concepts and Practice.* Plenum Press, New York, USA, pp. 41-55.

Elliger, C.A., Chan, B.C. and Waiss, A.C., Jr. 1980. Flavonoids as larval growth inhibitors. *Naturwissenschaften* **67:** 358-360.

Frazier, J.L. and Lam, P.Y.S. 1990. Three dimensional mapping of insect taste receptor sites as an aid to novel antifeedant development. *In:* L. Crombie (ed.). *Recent Advances in the Chemistry of Insect Control II.* Royal Society of Chemistry, London, UK, pp. 247-255.

Gibson, R.W. and Pickett, J.A. 1983. Wild potato repels aphids by release of aphid alarm pheromone. *Nature* **302:** 608-609.

Gill, J.S. and Lewis, C.T. 1971. Systemic action of an insect feeding deterrent. *Nature* (Lond.) **232:** 402-403.

Gould, F. 1983. Genetics of plant-herbivore systems: interactions between applied and basic study. *In:* R.F. Denno and M.S. McClure (eds). *Variable Plants and Herbivores in Natural and Managed Systems.* Academic Press, New York, USA, pp. 563-599.

Gould, F. 1984. Role of behavior in the evolution of insect adaptation to insecticides and resistant host plants. *Bull. ent. Soc. Am.* **30:** 34-41.

Griffiths, D.C., Maniar, S.P., Marrit, L.A., Mudd, A., Pickett, J.A., Pye, B.J., Smart, L.E. and Wadhams, L.J. 1992. Laboratory evaluation of pest management strategies combining antifeedants with insect growth regulator insecticides. *Crop. Prot.* **11:** 145-151.

Han, E.N. and Gatehouse, A.G. 1991. Genetics of precalling period in the oriental armyworm, *Mythimna separata* (Walker) (Lepidoptera: Noctuidae) and implication for migration. *Evolution* **45:** 1502-1510.

Harley, K.L. and Thorsteinson, A.J. 1967. The influence of plant chemicals on the feeding behaviour, development and the survival of two striped grasshopper, *Melanoplus bivittatus* (Say) (Acrididae). *Can. J. Zool.* **45:** 315-319.

Haynes, K.F., Gaston, L.K., Pope, M.M. and Baker, T.C. 1984. Potential for evolution of resistance to pheromones: interindividual and interpopulational variation in chemical communication system of pink bollworm moth. *J. Chem. Ecol.* **10:** 1551-1565.

Hedin, P. 1983. *Plant Resistance to Insects.* ACS Symp. Ser. 208, American Chemical Society, Washington, DC, USA.

Henneberry, T.J. and Clayton, T.E. 1984. Time of emergence, mating, sperm movement and transfer of ejaculatory duct secretory fluid by *Heliothis virescens* (F.) (Lepidoptera: Tortricidae), caught in pheromone-baited trap. *J. Ent. Sci.* **21:** 254-262.

Hikino, H. and Takemoto, T. 1974. Ecdysones of plant origin. *In:* W.J. Burdette (ed.). *Invertebrate Endocrinology and Hormonal Heterophylly.* Springer-Verlag, New York, USA, pp. 185-203.

Howlader, M.A. and Gerber, G.H. 1986. Effects of age, egg development and mating on calling behaviour of the bertha armyworm, *Mamestra configurata* Walker (Lepidoptera: Noctuidae). *Can. Ent.* **118:** 1221-1230.

Ishaaya, I., Hirashima, A., Yablouski, S., Tawata, S. and Eto, M. 1991. Mimosine, a non-protein amino acid, inhibits growth and enzyme systems in *Tribolium castaneum*. *Pestic. Biochem. Physiol.* **39:** 35-42.

Islam, B.N. 1987. Use of some extracts from Meliaceae and Annonaceae for control of rice hispa, *Dicladispa armigera* and pulse beetle, *Callosobruchus chinensis*. *Proc. 3rd Int. Neem Conf.*, Nairobi, GTZ, Eschborn, Germany, pp. 217-242.

Isman, M.B. 1992. A physiological perspective. *In:* B.D. Roitberg and M.B. Isman (eds). *Insect Chemical Ecology: An Evolutionary Approach.* Chapman & Hall, New York, pp. 156-176.

Isman, M.B. and Duffey, S.S. 1982. Toxicity of tomato phenolic compounds to the fruitworm, *Heliothis zea*. *Ent. exp. Appl.* **31:** 370-376.

Janzen, D.H., Juster, H.B. and Bell, E.A. 1977. Toxicity of secondary compounds to the seed eating larvae of the bruchid beetle *Callosobruchus maculatus*. *Phytochemistry* **16:** 223-227.

Jones, C.G. and Firn, R.D. 1978. The role of phytoecdysteroids in bracken fern, *Pteridium aquilinum* (L.) Kuhn as a defense against phytophagous insect attack. *J. Chem. Ecol.* **4:** 117-138.

Jutsum, A.R. and Gordon, R.F.S. 1988. *Insect Pheromones in Plant Protection.* John Wiley & Sons, New York, USA.

Kanno, H. 1979. Effects of age on calling behaviour of the rice stem borer, *Chilo suppressalis* (Walker) (Lepidoptera: Pyralidae). *Bull. ent. Res.* **69:** 331-335.

Kircher, H.W. 1972. Triterpene glycosides and queretaroic acid in organ pipe cactus. *Phytochemistry* **16:** 1078-1080.

Kogan, M. and Ortman, E.F. 1978. Antixenosis, a new term proposed to define Painter's nonpreference modality of resistance. *Bull. ent. Soc. Am.* **24:** 175-176.

Kolodny-Hirsch, D.M. and Schwalbe, C.P. 1990. Use of disparlure in the management of the gypsy moth. *In:* R.L. Ridgway, R.M. Silverstein and M.N. Inscoe (eds). *Behaviour-Modifying Chemicals for Insect Management: Application of Pheromones and Other Attractants.* Marcel Dekker, New York, USA, pp. 363-385.

Koul, O. 1982. Insect feeding deterrents in plants. *Indian Rev. Life Sci.* **2:** 97-125.

Koul, O. 1983a. L-canavanine, an antigonadal substance for *Dysdercus koenigii*. *Ent. exp. Appl.* **34:** 297-300.

Koul, O. 1983b. L-canavanine from *Canavalia ensiformis* (L): effects on fertility of *Periplaneta americana* (L). *Z. Angew. Ent.* **96:** 530-532.

Koul, O. 1985. Foliage spray tests of L-canavanine for control of *Spodoptera litura F. Phytoparasitica* **13:** 167-171.

Koul, O. 1992. Neem allelochemicals and insect control. *In:* S.J.H. Rizvi and V. Rizvi (eds). *Allelopathy: Basic and Applied Aspects.* Chapman & Hall, London, UK, pp. 389-413.

Koul, O. 1993. Plant allelochemicals and insect control : An antifeedant approach. *In:* T.N. Ananthakrishnan and A. Raman (eds). *Chemical Ecology of Phytophagous Insects.* Oxford & IBH Pub. Co. Pvt. Ltd., New Delhi, India, pp. 51-80.

Koul, O. 1995. Acorus allelochemicals: Chemistry and bioefficacy against insects. *In:* B.S. Parmar and S. Walia (eds). *Pesticides, Crop Protection and Environment.* Oxford & IBH Publishing Co. Pvt. Ltd., New Delhi, India, pp. 99-113.

Koul, O. 1996. Resistance to insects due to phenylpropanoid allelochemicals in some essential oil bearing plants. *In:* T.N. An^nthakrishnan (ed.). *Biochemical Basis of Host Plant Resistance to Insects.* National Academy of Agricultural Sciences, New Delhi, India, pp. 89-100.

Koul, O. and Shankar, J.S. 1995. Systemic uptake of azadirachtin into *Ricinus communis:* effect on larvae of *Spodoptera litura. Indian J. expt. Biol.* **33:** 865-867.

Koul, O. and Smirle, M.J. 1993. Allomones in insect plant interactions. *In:* G.T. Gujar (ed.). *Recent Advances in Insect Physiology and Toxicology.* Agricole Publ. Acad., New Delhi, pp. 14-40.

Koul, O., Amanai, K. and Ohtaki, T. 1987. Effect of azadirachtin on the endocrine events of *Bombyx mori. J. Insect Physiol.* **33:** 103-108.

Koul, O., Smirle, M.J. and Isman, M.B. 1990a. Asarones from *Acorus calamus* L. oil: their effect on feeding behaviour and dietary utilization in *Peridroma saucia. J. Chem. Ecol.* **16:** 1911-1920.

Koul, O., Smirle, M.J., Isman, M.B. and Szeto, Y. 1990b. Synergism of a natural insect growth inhibitor is mediated by bioactivation. *Experientia* **46:** 1082-1084.

Koul, O., Shankar, J.S., Mehta, N., Sharma, S.N., Singh, B. and Taneja, S.C. 1995. Resistance to insects in selective ecotypes of *Cymbopogon pendulus* (Nees Ex Steud) Wats. *Pestic. Res. J.* **7:** 132-139.

Kubo, I., Uchida, M. and Klocke, A. 1983. An insect ecdysis inhibitor from the African medicinal plant, *Plumbago capensis* (Plum–baginaceae), a naturally occurring chitin synthesis inhibitor. *Agric. biol. Chem.* **47**(4): 911-913.

Landolt, P.J. and Heath, R.R. 1990. Sexual role reversal in mate finding strategies of the cabbage looper moth. *Science* **249**: 1026-1028.

Lewis, W.J. 1981. Semiochemicals : their role with changing approaches to pest control. *In*: D.A. Nordlund, R.L. Jones and W.J. Lewis (eds). *Semiochemicals: Their Role in Pest Control.* John Wiley & Sons, New York, USA, pp. 3-12.

Lewis, W.J., Beevers, M., Nordlund, D.A., Gross, H.R., Jr. and Hagen, K.S. 1979. Kairomones and their use for management of entomophagous insects II. Investigations of vaious kairomone-treatment patterns for *Trichogramma* spp. *J. Chem. Ecol.* **5**: 673-680.

Liu, S.H., Norris, D.M. and Lyne, P. 1989. Volatiles from the foliage of soybean *Glycine max* and lima bean *Phaseolus lunatus* : Their behavioural effects on the insects *Trichoplusia* and *Epilachna varivestis*. *J. agric. Fd. Chem.* **37**: 496-501.

Mahoney, N.E., Roitman, J.N. and Chan. B.C. 1989. Structure- activity relationship of flavones as growth inhibitors of the navel orangeworm. *J. Chem. Ecol.* **15**: 285-290.

McNeil, J.N. 1992. Evolutionary perspectives and insect pest control : An attractive blend for the deployment of semiochemicals in management programmes. *In*: B.D. Roitberg and M.B. Isman (eds). *Insect Chemical Ecology: An Evolutionary Approach.* Chapman & Hall, New York, USA, pp. 334-351.

Meisner, J., Melamed-Medjar, V., Yathom, S. and Ascher, K.R.S. 1987. The influence of neem on the European corn borer (*Ostrinia nubilalis*) and the serpentine leaf miner (*Liriomyza trifolii*). *Proc. 3rd Int. Neem Conf.*, Nairobi, GTZ, Eschborn, Germany, pp. 461-477.

Metcalf, R.L. 1980. Changing role of insecticides in crop protection. *A Rev. Ent.* **25**: 219-256.

Miller, J.R. and Roelofs, W.L. 1980. Individual variation in sex pheromone component ratios in two populations of the redbanded leafroller moth, *Argyrotaenia velutinana*. *Envir. Ent.* **9**: 359-363.

Mordue (Luntz), A.J. and Blackwell, A. 1993. Azadirachtin : an update. *J. Insect Physiol.* **39**: 903-924.

Murdock, L.L., Brookhart, G., Edgecomb, R.S., Long, T.F. and Sudlow, L. 1985. Do plants psychomanipulate insects. *In*: P.A. Hedin (ed.). *Bioregulators for Pest Control.* ACS Symp. Ser. 217, Am. Chem. Soc., Washington, DC, USA, pp. 337-351.

Nakanishi, K., Koreeda, M., Sasaki, S., Chang, M.L. and Hsu, H.Y. 1966. Insect hormones. The structure of ponasterone A, an insect moulting hormone from the leaves of *Podocarpus nakaii* Hay. *Chem. Commun.* **24**: 915-917.

Neupane, F.P. and Norris, D.M. 1990. Iodoacetic acid alteration of soybean resistance to the cabbage looper (Lepidoptera: Noctuidae). *Envir. Ent.* **19**: 215-221.

Norris, D.M. 1986. Antifeeding compounds. *In*: G. Haug and H. Hoffman (eds). *Chemistry of Plant Protection. Vol. I.* Springer-Verlag, Berlin, Germany, pp. 97-146.

Norris, D.M. 1988. *Periplaneta americana* perception of phytochemical naphthoquinones as allelochemicals. *J. Chem. Ecol.* **14**: 1807-1819.

Norris, D.M. 1990. Volatile versus non-volatile allelochemicals in bean plant-insect interactions. *Symp. Biol. Hung.* **39**: 145-151.

Norris, D.M. 1994. Phytochemicals as messengers altering behaviour. *In*: T.N. Ananthakrishnan (ed.). *Functional Dynamics of Phytophagous Insects.* Oxford & IBH Publishing Co. Pvt. Ltd., New Delhi, India, pp. 33-54.

Oikawa, Y., Yoshioka, T. and Yonemitsu, O. 1982. Specific removal of O-methoxy benzyl protection by DDQ oxidation. *Tetrahedron Lett.* **23**: 885-888.

Osman, M.Z. and Port, G.R. 1990. Systemic action of neem seed substances against *Pieris brassicae. Ent. exp. Appl.* **54**: 297-300.

Palaniswamy, P., Galka, B. and Timlick, B. 1990. Phenology and infestation level of the European corn borer, *Ostrinia nubilalis* (Hubner) (Lepidoptera : Pyralidae) in Southern Manitoba. *Can. Ent.* **122**: 1211-1220.

Peumans, W.J. and Van damme, E.J.M. 1995. The role of lectins in plant defence. *Histol. Chem. J.* **27**: 253-271.

Raffa, K.F. 1986. Devising pest management tactics based on plant defense mechanisms, theoretical and practical considerations. *In*: L.B. Brattasten and S. Ahmed (eds). *Molecular Aspects of Insect Plant Associations.* Plenum Press, New York, USA, pp. 301-327.

Raffa, K.F. and Berryman, A.A. 1982. Phsiological differences between lodgepole pines resistant and susceptibile to the mountain pine beetle and associated microganisms. *Envir. Ent.* **11**: 486-492.

Raffa, K.F. and Frazier, J.L. 1988. A generalised model for quantifying behavioural de-sensitization to antifeedants. *Ent. exp. Appl.* **46**: 93-100.

Raffa, K.F. and Priester, T.M. 1985. Synergists as research tools in agriculture. *J. agric. Ent.* **2**: 27-45.

Ramachandran, R. and Norris, D.M. 1991. Volatiles mediating plant-herbivore-natural enemy interactions : electroantennogram responses of soybean looper, *Pseudoplusia includens* and a parasitoid, *Microplitis demolitor,* to green leaf volatiles. *J. Chem. Ecol.* **17**: 1665-1690.

Raman, K.V., Tingey, W.M. and Gregory, P. 1979. Potato glycoalkaloids : effect on survival and feeding behaviour of the potato leaf hopper. *J. econ. Ent.* **72**: 337-341.

Reese, J.C. and Holyoke, C.W. Jr. 1987. Allelochemicals affecting insect growth and development. *In*: E.D. Morgan and N.B. Mandava (eds). *Handbook of Natural Pesticides.* CRC Press, Boca Raton, Florida, USA, pp. 21-66.

Rhoades, D.F. 1983. Herbivore population dynamics and plant chemistry. *In*: R.F. Denno and M.S. McClure (eds). *Variable Plants and Herbivores in Natural and Managed Systems.* Academic Press, New York, USA, pp. 155-220.

Ridgway, R.L., Silverstein, R.M. and Inscoe, M.N. (eds). 1990. *Behavior Modifying Chemicals for Insect Management : Applications of Pheromones and Other Attractants.* Marcel Dekker, New York, USA.

Roelofs, W.L., Du, J.W., Linn, C., Glover, T.J. and Bjostad, L.B. 1987. The potential for genetic manipulation of the redbanded leafroller moth sex pheromone blend. *In*: M.D. Huettel and H. Oberlander (eds). *Evolutionary Genetics of Invertebrate Behaviour.* Plenum Press, New York, USA, pp. 265-272.

Rosenthal, G.A. 1991. The biochemical basis for the deleterious effects of L-canavanine. *Phytochemistry* **30**: 1055-1058.

Sanders, C.J. 1981. Sex attractant traps : their role in the management of spruce budworm. *In*: E.R. Mitchell (ed.). *Management of Insect Pests with Semiochemicals.* Plenum Press, New York, USA, pp. 75-91.

Schoonhoven, L.M. 1988. Stereoselective perception of antifeedants in insects. *In*: E.J. Ariens, J.J.S. van Reusen and W. Welling (eds). *Stereoselectivity of Pesticides: Biological and Chemical Problems.* Elsevier Sci. Publ., The Netherlands, pp. 289-302.

Shany, S., Birk, Y., Gestetner, B. and Bendi, A. 1990. Preparation, characterization and some properties of saponins from lucerne tops and roots. *J. Sci. Fd. Agric.* **21**: 131-135.

Sinden, S.L., Sanford, L.L. and Osman, S.F. 1980. Glycoalkaloids and resistance to the Colorado potato beetle in *Solanum chacoense* Bitt. *Am. Potato J.* **57**: 331-343.

Singer, G., Rozental, J.M. and Norris, D.M. 1975. Sulfhydryl groups and the quinone receptor in insect olfaction and gustation. *Nature* **256**: 222-223.

Sreng, I., Glover, T. and Roelofs, W. 1989. Canalization of the redbanded leafroller moth sex pheromone blend. *Arch. Insect Biochem. Physiol.* **10**: 73-82.

Stehr, F.W. 1982. Parasitoids and predators in pest management. *In*: R.L. Metcalf and W.H. Luckmann (eds). *Introduction to Insect Pest Management*. John Wiley & Sons, New York, USA, pp. 135-174.

Takahashi, S. 1991. Ethological aspects of insect pest control. *In*: *Abstr. 22nd Ethological Conf.*, Kyoto, p. 5.

Takahashi, S. 1993. Insect behavior modifiers. *In*: S.M. Colegate and R.J. Molyeux (eds). *Bioactive Natural Products- Detection, Isolation and Structural Determination*. CRC Press, Boca Raton, Florida, USA, pp. 241-263.

Tumlinson, J. H. 1988. Contemporary frontiers in insect semiochemical research. *J. Chem. Ecol.* **14**: 2109-2130.

Unnithan, G.C., Nair, K.K. and Bowers, W.S. 1977. Precocene induced degeneration of the corpus allatum of adult females of the bug *Oncopeltus fasciatus*. *J. Insect Physiol.* **23**: 1081-1094

van, Beek, T.A. and de Groot, A.C. 1986. Terpenoid antifeedants. Part I. An overview of terpenoid antifeedants of natural origin. *Recueil des Trav. Chimiq. des Pays Bas* **105**: 513-527.

Visser, J.H. 1986. Host odour perception in phytophagous insects. *A. Rev. Ent.* **31**: 121-144.

Wall, C. 1988. The application of sex attractants for monitoring the pea moth, *Cydia nigricana* (F) (Lepidotera : Tortricidae). *J. Chem. Ecol.* **14**: 1857-1866.

Willis, M.A. and Birch, M.C. 1982. Male lek formation and female calling in a population of the arctiid moth *Estigmene acrea*. *Science* **218**: 168-170.

Wilson, G.G. 1987. Observation on the level of infection and intensity of *Nosema fumiferanae* (Microporida) in two different field populations of the spruce budworm, *Choristoneura fumiferana*. Inf. Report FPM-X-79, Canadian Forestry Service, Ontario.

Wiseman, B.R., Snook, M.E. and Widstrom, N.W. 1996. Feeding response of corn earworm larvae (Lepidoptera : Noctuidae) on corn silks of varying flavone content. *J. econ. Ent.* **89**: 1040-1044.

Wiseman, B.R., Snook, M.E., Isenhour, D.J., Mihm, J.A. and Widstrom N.W. 1992. Relationship between growth of corn earworm and fall armyworm larvae (Lepidoptera : Noctuidae) and maysin concentration in corn silks. *J. econ. Ent.* **85**: 2473-2477.

Witzgall, P. and Frerot, B. 1989. Pheromone emission by individual females of the carnation tortrix, *Cacoecimorpha pronubana*. *J. Chem. Ecol.* **15**: 707-717.

Wunderer, H., Hansen, K., Bell, T.W., Schneider, D. and Meinwald, J. 1986. Sex pheromones of two Asian moths (*Creatonotos transiens*, *C. gangis;* Lepidoptera : Arctiidae) : behaviour, morphology, chemistry and electro-physiology. *Expt. Biol.* **46**: 11-27.

10

INTEGRATED PEST MANAGEMENT : SPRINGBOARD TO SUSTAINABLE AGRICULTURE

ELSKE VAN DE FLIERT

International Potato Centre, CIP—ESEAP Region, Jl. Bali 11, Widoro Baru, Yogyakarta 55283, Indonesia

10.1. INTRODUCTION

The previous chapters describe the opportunities and frontiers of several types of pest management practices. If we put all these practices together in a package, can we call it "Integrated Pest Management"? The answer is "No". Integrated pest management (IPM) has evolved into a pest management strategy that is more than the sum of individual pest management practices. The evolution of IPM and its present features, as successful applied in wetland rice-based cropping systems is several parts of Asia, are described in this chapter.

Rather than describing and evaluating IPM technologies and their environmental impact from an academic point of view, as has already been done by many other authors (e.g., Flint and van de Bosch, 1981; Altieri, 1995), the present chapter elaborates on the requirements, possibilities and constraints of IPM implementation at the farmer level. The final aim of technology generation is that research findings are put into practice. For sustainable agricultural practices such as IPM, in particular, it is of utmost importance that farmers understand underlying ecological principles and master methodologies. IPM methodologies that are workable, effective and efficient at the farm level, especially that of marginal farmers in risk-prone areas, are usually not served by sophisticated, input-intensive techniques. What seems to count is the farmers' ability to analyse the agroecosystem, and take adequate decisions. Developing and introducing IPM technologies that fulfill farmers' needs and consider their constraints, requires a different attitude of researchers and extension agents. In this chapter, an attempt is made to highlight the implications of IPM implementation, by analysing the nature of an IPM methodology that works at the farm level, the desired ıarmer behaviour, and the concomitant extension approach needed to capacitate farmers to implement IPM.

With its ecological basis, IPM has potential to be a major contributor to a more sustainable agriculture. Section 10.4 explains that this is not only due to its technical content, but also to improved management skills of farmers built through IPM extension. Nevertheless, since it involves many observable, and, thus, easily understandable, ecological processes, IPM is an attractive and practical entry point to introduce more sustainable forms of agriculture at the farmer level.

10.2. IPM: MORE THAN INTEGRATION OF CONTROL MEASURES

10.2.1. The Old Paradigm: Observation Techniques and ETLs

Integrated pest management (IPM) was practised long before the term existed. Reports are available on efforts that applied the same principles as IPM for the control of boll weevil on cotton in the United States in 1923 (Flint and van de Bosch, 1981), and on 'harmonious control' in Canada in the 1950s (Pickett *et al.*, 1958). Stern *et al.* (1959) was the first publication clearly defining 'integrated control' and using economic threshold as an operational concept. This paper marks the starting-point of IPM. In 1968, IPM was defined by the UN's Food and Agriculture Organization as "a pest management system that, in the context of the associated environment and the population dynamics of the pest species, utilises all suitable techniques and methods in as compatible a manner as possible, and maintains the pest populations at levels below those causing economically unacceptable damage or loss" (FAO, 1968).

In line with this definition, focusing on pest occurrence and suitable interventions to suppress the pest populations, IPM developed as a strategy to control pests by relying on natural control factors and using pesticides as a last resort. Rather than exploring how natural control could be exploited, IPM research and extension in the 1970-80s gave predominant attention to the judicious application of the last resort. It focused on the development and application of observation techniques and economic threshold levels (ETLs) as tools to be used in deciding when to apply pesticides (Chapter 1 of this book). The ETL for a certain pest is the level of pest occurrence estimated to cause an economic loss of yield equal to the assessed cost for chemical control. To be able to apply the ETL, one has to monitor a representative crop sample and calculate the average pest occurrence, usually expressed in percentage incidence or individuals per plant, per square meter, or per hectare. When the observation data reveal a pest population level that exceeds the ETL, a (chemical) control measure should be taken.

The application of the ETL concept at the farm level has long been questioned (e.g., Zadoks, 1985). Flaws identified invalidating the concept include the following:

- Prices for harvested product and pesticides tend to fluctuate over the season and regionally.

- Natural enemy populations are not considered in the ETLs.
- Many other factors that farmers usually consider in pest control decision making are not included in the ETL concept, for instance:
    - weather conditions;
    - vigour of crop;
    - availability of irrigation water;
    - availability of cash or credit facilities;
    - other investment opportunities; and
    - indigenous practices.

Despite these flaws, most IPM programmes continued to teach ETLs to farmers. It is, however, not likely that farmers will apply the ETL methodology as such in their daily farm management, for several reasons:

- Observation techniques required to obtain a representative sample are often complex and laborious.
- Pest population assessment methods apply units that farmers may not be familiar with, such as percentage, square metre, hectare, etc.
- An ETL applies only for one pest or disease, whereas farmers face a complex of pests and diseases in their crops; having to consider more than one ETL at a time is too complicated.

IPM extension through conventional extension services involved conveyance of research findings as a standard package for farmers to follow rigidly. Training of farmers took place in the classroom, often using colour slides, picture books or flipcharts (Matteson *et al.*, 1992). Enlarged pictures of insects, however, do not necessarily lead to their recognition by farmers, and their behaviour and function in the field is not demonstrated. Implementation of IPM in rice cultivation in Indonesia during the mid 1980s taught a lesson that using ETLs for pesticide-induced pests, such as the brown planthopper, brings disastrous consequences (Box 1). The same programme showed that farmers, who had received a two-day training in teaching observation techniques, ETLs, identification of pest insects and natural enemies, were not able to make independent pest control decisions based on their observations (van de Fliert, 1993). Whenever they observed alarming numbers of pest insects, they consulted the extension service.

Box 1

How IPM did not control BPH

In 1985-86, vast areas of rice fields on the island of Java, Indonesia, were destroyed by the brown planthopper (BPH), *Nilaparvata lugens* (Stal). Populations of BPH, a rapidly reproducing homopteran that sucks at the stem base of rice plants, had built up during the previous years because naturally occurring predators and parasites had been killed by increased use of pesticides in rice cultivation. In IPM demonstration areas, trained officers monitored sample fields and reported the counts to their superiors. This resulted in abundant pesticide application over whole areas at once with government aid, whenever the ETL for the brown planthopper was exceeded. Many fields received more pesticides per cropping season than under the prevailing farmers' practice. As we know now, these unlucky farmers experienced elimination of natural enemies, the most important mortality factor of BPH, followed by recurrence of the hoppers.

10.2.2. The New Paradigm: Ecological Principles and Decision Making Skills

Having learned from ten years of experience in rice IPM development and implementation in over ten Asian countries, the FAO's Inter-country IPM Programme[1] piloted a new approach in Indonesia in 1989. This approach was unconventional with regard to both its technical contents and its extension methodology. Rather than teaching sophisticated observation techniques, ETLs, and a set of possible measures to be applied for control, four principles were introduced to rice farmers:

- Grow a healthy crop
- Conserve natural enemies
- Observe fields regularly
- Farmers become IPM experts

The first two principles emphasize the importance and maintenance of naturally occurring ecological processes and defense mechanisms favouring crop production, that should be taken advantage of by the farmer. Healthy crops can resist pest attack and compensate for incidental damage, whereas a healthy ecosystem with plenty of natural enemies helps to defend against pest attack. A high diversity of genes

[1] *In full:* Inter-country Programme for the Development and Application of Integrated Pest Control in Rice-growing in South and Southeast Asia, United Nation's Food and Agriculture Organization.

and species in the crop ecosystem should be encouraged. To do so, farmers have to become ecosystem managers and independent experts in their own fields. Regular observation is the main tool to provide the information needed to take adequate decisions. With these IPM principles, farmers, rather than technologies, become the focal point in IPM programmes, with regard to both research and development of methodologies, and final field implementation. This approach to IPM was readily adopted in many other governmental and nongovernmental organizations' (NGOs) IPM programmes in the South and Southeast Asian region. A similar approach to IPM training was developed and experimented independently in Honduras through activities of the Pan American Agricultural School (Zamorano) (Bentley and Andrews, 1996).

The developments in rice IPM in Asia induced a paradigm shift with regard to IPM development and implementation (Kenmore *et al.*, 1995). Rather than focusing on how to integrate control methods available and finding out when to apply pesticide in a relatively judicious manner, IPM evolved into a problem-solving strategy, in which farmers are perceived as capable ecosystem managers. Among other changes implied by the strongly ecological IPM approach, the concept of economic threshold level (ETL), and particularly its application at the farmer level, were thoroughly reviewed and adapted to farmers' needs and abilities. Instead of using an economic threshold level that applies only for one pest, farmers should develop their personal 'experience thresholds', which are applicable under specific farm conditions. For farmers, the development of such experience thresholds is often not much more than rationalising what they have already been doing throughout their lives as farmers, adding one important decision making factor, i.e., natural enemies. Those finding it more difficult to let the ETL concept go, are seemingly the officers involved in IPM programmes, especially in countries with a well established top-down extension system in which research is seen as the major source of new technology (e.g., Bangladesh, India and Indonesia). As a matter of fact, this observation illustrates the often slow reaction of conventional research and extension programmes to paradigm changes stemming from field experiences.

A second important aspect of the IPM paradigm shift is the changed perception towards the role of biological control in IPM (Kenmore *et al.*, 1995). Whereas previously biological control was considered unreliable or unmanageable without sophisticated rearing facilities for augmentation of natural enemies, in the new paradigm it is the core of IPM. Biological control is already naturally existent in the field, as long as we do not disturb but rather conserve and enhance the ecological balance. This statement is particularly true for wet tropical agroecosystems such as rice, sweetpotato and mixed cropping systems favouring (relative) diversity of the ecosystem. The latest experiences in tropical cotton

IPM are also promising (Eveleens *et al.*, 1996). A more difficult situation with regard to ecological balance is usually found in introduced crops with a more artificial ecosystem, such as vegetables and fruits, where a message as simple as that for rice IPM may not work. Nevertheless, the four IPM principles as presented above still seem to apply, with even higher emphasis on the expertise of farmers to take adequate decision in these vulnerable agroecosystems.

The key element of IPM is now diversity, instead of single pest populations. Increased genetic and species diversity favour the resistance and resiliency of the ecosystem to pest attack, and should, therefore, be enhanced. A genetically diverse crop contains more mechanisms complementing one another to resist pest damage (Worede and Mekbib, 1993). A more diverse composition of primary (crops and weeds) prevents specialised pests from becoming abundant and provides better conditions for the survival of natural enemies (Altieri, 1995).

10.3. IPM IMPLEMENTATION

10.3.1. Characteristics of IPM

As with most agricultural technologies, implementation of IPM by farmers requires some kind of prior training. Based on the assumption that, in effective training, the nature of the training process should be consistent with the nature of the training contents (Röling and van de Fliert, 1994), we need to analyse first what the typical characteristics of IPM are, requiring an extension approach capacitating farmers to perform the desired behaviour.

Compared to many other agricultural technologies promoted to farmers, IPM is complex. It considers many factors in the agroecosystem, including abiotic and (at least three trophic levels of) biotic elements. Additionally, it considers many aspects of ecosystem management, such as crop, soil and water management, in order to promote a healthy crop. This makes IPM knowledge-and skill-intensive, requiring intensive training. Moreover, IPM is location-specific because it necessitates analysis of the prevailing conditions in each individual field, and, hence, location-specific decision-making. Therefore, it is not served by a standard set of recommendations. In spite of that, its location-specificity makes IPM compatible to most farming systems, since adapted action based on the specific analysis is required.

As stated above, effective IPM implementation is very much favoured by a high diversity of the (larger) agroecosystem, encouraging relative stability of (low) pest populations. In most intensive cropping systems in the tropics, which are predominantly located in developing countries, fields cultivated by individual farmers are very small, often smaller than 0.5 ha and sometimes not larger than 500 m^2. Pest and natural enemy populations in one field are strongly influenced by those in the neighbouring fields. Individual implementation of IPM, in this context, may be less effective than when implemented collectively over large areas, requiring some sort of community organization.

The impact of IPM implementation is usually not dramatically visible, no miraculous yield increases occur, as happened when high-yielding varieties and chemical fertilisers were introduced, and hardly tangible in farm economics (profit increase is likely due to reduced inputs, but is often proportionally low). According to the conventional theories of adoption of innovations (Rogers, 1983), low observability is deemed to lead to low adoptability. It should be realised here, however, that these theories were developed based on observations of transfer-of-technology models of extension, implying a linear, top-down flow of information from researches via extension officers to farmers. With technologies emphasising sustainability and farmers' expertise, such as IPM, these theories do not necessarily apply. In Indonesia, farmers expressed that 'knowledge increase' was the main benefit they obtained from IPM training, as an indicator for having gained more confidence in themselves as independent and capable farm managers (van de Fliert, 1993).

Characterised as being knowledge-and skill-intensive, location-specific, aiming at sustainability, requiring group action, and having low observability, IPM demands dedication from farmers in training and implementation. What are the characteristics that a good IPM farmer should have?

10.3.2. Characteristics of an IPM Farmer

In conventional top-down extension programmes introducing new agricultural technologies, such as the World Bank's Training-and-Visit (T&V) system, evaluation was focused on farm level effects such as yield, production and farm income, and tried to show causal relationships between these effects and the extension services (Murphy and Marchant, 1988). The main variable for evaluation of the extension process was the number of visits by the extension worker to the farmers, whilst no attention was paid to more crucial aspects such as farmer practice and decision-making leading to the farm level effects measured. Procedures recently applied in T&V evaluation measure extension effects by determining the adoption rate of the innovation as a key factor. Judging whether a farmer adopted an innovation or not (or partly), however, is very tricky, since adoption of a technology, as such, is not necessarily the best thing a farmer could do under prevailing conditions of his or her farm. Additionally, reasons for non-adoption are not understood but may be very crucial, e.g., no access to factors (such as credit, markets, etc.) other than those provided by the intervention only (Röling,

1992). Such evaluation procedures *de facto* are in correspondence with the main objectives of technological programmes, i.e., increase of production.

In determining evaluation criteria for IPM implementation, as a basis for planning an intervention, we should go back to the basic principles of IPM, as stated in Section 10.2.2, describing a focus on farmer competence and behaviour. Based on these principles, and the characteristics described for IPM, we could characterise a good IPM farmer as someone who:

- has practical knowledge of the characteristics of a healthy crop, can identify major pests, diseases, infestation symptoms and natural enemies, understands life cycles and interactions of pests and natural enemies, and is familiar with practices that prevent damage and control pests and diseases;
- does regular field observation;
- can analyse the situation in the field and take adequate decisions on pest management measures needed;
- applies management practices timely and adequately;
- can do a simple farm economic analysis;
- can design and conduct experiments to understand and solve occurring problems, as well as to test and adapt technologies;
- seeks information actively;
- seeks contact and collaboration with other farmers.

It seems obvious that such competence and behaviour can not be achieved through brief, instructional, classroom training, as is in most places still the predominant training mode of agricultural extension.

10.3.3. IPM Farmer Field School: Learning by Doing

A training model that provides opportunities for farmers to become competent IPM implementors, was developed by the aforementioned FAO Inter-country Programme for rice IPM in Asia (see footnote 1), and piloted in Indonesia in 1989 (Kenmore, 1991). The model was called the "Farmer Field School". The philosophy behind this name is that farmers go back to school, a place reputed for learning, which is located in the field, being the farmers' daily work arena. The training strategy, having its foundation in nonformal education principles, emphasizes learning by doing, and empowering farmers to actively identify and solve their own problems. Participation, self-confidence, and collective action and decision making are fostered during the experiential learning process. This approach seems highly consistent with the requirements of active ecosystem management by farmers implied by IPM.

An IPM field school lasts for a whole growing season (10-12 weeks in the case of rice, 13-17 sessions over 23-25 weeks in the case of sweetpotato), involving a group of at most twenty-five farmers in weekly sessions of, on average, four hours. The trainer is a facilitator of the experiential learning process, not an instructor. Each training session contains the following activities:

- Field monitoring of observation plots in small groups, considering environmental factors, crop development, pest and natural enemy occurrence and interaction, and damage symptoms on the plants.
- Agroecosystem analysis, in which drawings of observation data are made, and conclusions about crop status and possible measures are drawn together.
- Presentation of the agroecosystem analyses, and discussion to come to a collective agreement on what crop management measures should be taken during the coming week.
- Special topics dealing with locally occurring field problems, or providing opportunities to discover processes, causes and effects of phenomena occurring in the field; the 'insect zoo' (an enclosure) is a tool often used by farmers in the field school to study plant-herbivore-predator relations.
- Group dynamics exercises to enliven the school, strengthen the coherence of the group, and make the members better aware of the importance and dynamics of group processes.

The field school model is designed to allow farmers to learn to take informed pest management decisions based on their own observations and analysis. The ability often gives them more confidence in their own farm management skills, which, in turn, is reflected in an improved, overall farm management (see also Box 2). A second important purpose of the field school is that farmer groups are stimulated to take collective action, which in certain cases may be indispensable for effective IPM implementation, particularly for pests such as rodents. A successful IPM field school often results in follow-up activities, spontaneously organised, and funded by the field school graduates, such as IPM clubs in Vietnam (Eveleens *et al.*, 1996). During these follow-up activities, the farmers may study newly occurring cultivation problems, organise collective control measures, and even get into wider aspects of community development, such as rice-fish culture and collective marketing of produce (van de Fliert and Wiyanto, 1996).

10.4. BEYOND IPM

10.4.1. More Than Bugs That Bother Farmers

The major difference between farmers and agricultural scientists is that farmers are generalists, whereas researchers are (mostly) specialists. Technologies generated by the specialists may seem 'narrow', and not always compatible with other farm management practices, for the generalists who have to implement them. Interest in learning about a new technology covering a 'narrow' topic that is not perceived a main issue or problem by a farmer, can hardly be expected. This was, for instance, the case when Indonesian farmers were invited to participate in a pilot sweetpotato IPM farmer field school development project[2] (van de Fliert *et al.*, 1996). Even though several pests, such as the sweetpotato weevil, can cause substantial loss, farmers either do not consider the problem severe enough to tackle, or they think that their current control practices are adequate. Additionally, the farmers' perception of sweetpotato as a relatively labour- and input- extensive crop, with return very much depending on strongly fluctuating prices, very much influences their attitude towards investing time in training.

Sweetpotato farmers, however, did become interested in a field school programme, when it was agreed upon to tackle a wider range of crop management practices, causing the title of the project to change from IPM to ICM (Integrated Crop Management). Basically, the first principle of the IPM farmer field school, *Grow a healthy crop,* already implies a broader scope than pest problems only, although in practice the focus is often predominantly on (insect) pest management. The greatest need identified together with sweetpotato growers was improved plant nutrition, showing considerable potential for higher yields at a lower level of chemical inputs. Immediate adoption of fertiliser recommendations from the rice intensification technologies to sweetpotato, without any knowledge on composition of fertilisers and their mode of action in plants, resulted in a highly inefficient use of phosphate (see Box 2). Only with better knowledge of what is needed by different crops and what is provided by the different fertilisers, rather than just conveying the recipe for good fertilisation in a certain soil type, can farmers be expected to test for themselves what will be the most efficient application.

---

[2] The 'Sweetpotato Integrated Crop Management (ICM) and ICM Training Development Project in East and Central Java, Indonesia' is a collaborative effort of the International Potato Centre (CIP, Bogor) - ESEAP Region (Bogor), Mitra Tani Foundation (Yogyakarta), Dutta Wacana Christian University (Yogyakarta), and Research Institute for Legumes and Tuber Crops (Malang), funded by UPWARD and CIP.

Box 2

TSP for phosphate or calcium?

An analysis of cultivation practices of sweetpotato farmers in East and Central Java showed, that very high application rates (an average of 60 kg/ha and up to 225 kg/ha) of triple super phosphate fertiliser (TSP) were applied to fields with clay soils. According to several agronomic references, sweetpotato plants can efficiently extract phosphate from the soil and generally do not need a phosphate application when grown in rotation with rice, which is given phosphate. Sweetpotato is a potassium-demanding crop. Farmers, however, never apply potassium fertiliser (KCI) to sweetpotato, and hardly ever to rice. They have experienced that the soil becomes sticky when they apply potassium, and, thus, requires more labour to cultivate.

Farmers usually understand that urea is some kind of 'food' for the crops, because there is an immediate response after application (crop becomes greener and lusher). On the contrary, TSP and KCI are often seen as some kind of optional 'medicine' for the crop, and perceived as interchangeable. Effects of both TSP and KCI are measured by changes in soil structure rather than by their impact on crop development, in that TSP makes the soil loose whereas KCI makes it sticky. Therefore, farmers prefer to apply TSP, and the more they can afford the better. Apparently, what makes the soil sticky as a result of KCI application is a replacement of calcium with potassium ions at the clay minerals. A higher buffer of calcium in the soil could prevent this process and providing calcium is exactly what TSP does, since it consists of some 50 per cent of calcium. Calcium, however, could be provided much more cheaply by applying gypsum or lime. Additionally, after 25 years of phosphate fertiliser use in intensive rice growing areas in Indonesia, most soils seem to have achieved a level of saturation.

The danger of environmental pollution because of unnecessary TSP applications was recognised by the government, after a study was presented showing that one TSP application per four seasons at the current recommendation level for rice (100 kg/ha) would be sufficient to replenish uptake by the crops. Rather than providing information to farmers about phosphate behaviour in soils and plants, and concurrent fertilisation requirements, it was decided to formulate, and make exclusively available in the market, a new fertiliser with a reduced phosphate concentration (36% instead of 46%). The two reasons for this decision included (i) anticipated difficulties with instructing farmers to reduce TSP applications (showing recognition of the inefficiency of the conventional extension system), and (ii) non-acceptance by the chemical industry of reduced fertiliser use (illustrating the political power of the industry). Finally, farmers are now paying the same per kg price for 10 per cent less nutrients in their fertiliser, and some complain that "it is not as good as it used to be". The time is ripe for farmers to know better.

Considering the identified needs of farmers, the central focus determined for sweetpotato ICM is 'health', with specific attention to health of soil, seed, crops and ecosystem, and automatically including

IPM. The farmers' interest in sweetpotato IPM grew even more when, as a reaction to the steep drop of sweetpotato prices, marketing and processing issues were included in the curriculum (Braun *et al.*, 1995).

Because most of the commercially grown sweetpotato in Java is sold as a standing crop to traders, farmers have few incentives to increase their yields. They do not estimate yield before agreeing on a price with a trader. On the contrary, traders are better skilled in yield assessment but tend to offer payment based on the past history of the field. In this context, farmers, first, need to obtain the same, or better, yield assessment skills as traders, and, second, organise themselves to counteract the power of the trader networks.

When farmers have learned to learn from their experiences in IPM field schools, it seems a logical reaction that they want to apply the newly obtained analysis and decision-making skills to practices other than pest management. Successful IPM field school groups in Indonesia and Vietnam were observed to continue to meet, and study together, for instance, the effects of planting distance, fertiliser application rates, and weed control practices (Eveleens *et al.*, 1996). Particularly in NGO IPM programmes, where long-lasting relationships with communities usually exist, training topics tend to diverge to much wider range of problems relating to agricultural (if not community) development. This phenomenon of spontaneously emerging follow-up activities definitely indicates that a need for more and better applied information with regard to a wider range of topics exists among farmers. Additionally, the field school model, more than any other information source, seems to provide applicable answers, and serves as a platform for farmer-to-farmer communication.

10.4.2. From IPM to Sustainable Agriculture

In order to assist farmers in solving cultivation problems other than those caused by pests and diseases, however, in a sustainable manner, we have to expand our focus to a wider range of ecological processes than pest-natural enemy interactions, including nutrient and water cycling, competition, and succession. Maintenance and improvement of soil fertility appear to be major issues of concern to both farmers and scientists, since soils in many parts of the world, especially in the developing countries, are rapidly deteriorating as a result of the intensification of agriculture.

Sustainable practices for nutrient, water and weed[3] management share some of the same features as IPM: they are knowledge-intensive,

---

[3] Weed management is, strictly speaking, part of IPM. In practice, however, very little attention is given to weeds in IPM programmes, not unlikely because IPM has primarily been entomologists' business. Special attention to weeds, considering both competitive and contributive aspects, in sustainable agricultural activities is justifiable, especially in intensive, monoculture cropping systems.

location-specific, and may demand collective action under certain circumstances. Therefore, farmers implementing sustainable nutrient, water and weed management practices can be expected to share many characteristics with farmers who practise IPM (Box 3). They should be competent ecosystem managers, able to independently analyse farm conditions, make informed decisions, and collaborate with fellow farmers to improve the quality of the greater ecosystem. Capacitating farmers in this regard, as is the case for IPM, requires an extension approach that focuses on learning through discovery, skill attainment, and group processes.

The great test difference between general crop management practices and IPM, however, is that underlying ecological processes are much less observable for nutrient, water and weed management than the pest-natural enemy relations supporting IPM. This lower observability has implications when introducing the concepts to farmers. How can farmers discover nutrient cycles in the ecosystem, and understand what fertilisation practices favour sustained soil fertility? How can they prove for themselves the relation between a diverse landscape structure and soil humidity? Most of these processes have long-term effects, which cannot be shown within the limited amount of time generally available for training activities. Consequently, observable indicators and concrete experiments need be developed for farmers to become aware of more processes in nature that can be taken advantage of. For example, a simple nutrient omission trial with an appropriate test plant (e.g., maize) (Asher and Grundon, 1993) can readily illustrate the general symptoms that occur on the vegetative parts when crops are deficient of a certain nutrient. Having observed the development of the symptoms over several weeks, they can easily recognise similar symptoms on crops in their own field.

Farmer experimentation plays a crucial role in the development of sustainable agricultural practices, due to location-specificity. Different types of experiments serve different purposes, such as (i) illustrating principles and processes, (ii) solving problems, and (iii) testing new methods and fine-tuning guidelines. In IPM field schools, experiments are primarily set up to observe processes, for instance, predator-prey relations in insect zoos, or effects of pesticides on natural enemies, and, hence, serve a demonstration purpose. In order to illustrate the ecological processes underlying the broader issues of crop management, and the effects induced by management practices, longer term experiments have to be designed and conducted. Crop performance under certain treatments needs to be followed and analysed throughout the season, up to and including the harvest.

Farmers usually do experiments, but not always in a systematic way. They may compare treatments over seasons or in different fields,

Box 3

Ecological and economic sustainability

Rice IPM farmer field schools, organised by a local NGO in the district of Magelang, Central Java, Indonesia, stimulated several unanticipated changes in the farming system of participating farmers (van de Fliert and Wiyanto, 1996). IPM training was initiated as a reaction to increasing rat damage in rice. A small group of farmers participated in the first field school making them aware that organisation of the whole community was required to eradicate rats. Therefore, more intensive IPM training was needed to get more people actively involved. This required more competent trainers who were not available within the NGO. It was decided to organise a training-of-farmer-trainers for a group of selected farmers of the initial IPM group. The training lasted for five months, the growth duration of the traditional rice variety chosen by the group of the field school plot, during which the farmers met for half a day per week.

Although the training-of-trainers was called an IPM field school, the scope of training was broader than IPM, and aimed at providing opportunities for the farmer-trainers to know more about better farm management, in general. They learned to keep daily records of all farming practices and expenses. These records showed that growing local varieties, despite the long growth duration, was more profitable under the prevailing conditions than the high-yielding varieties, due to lower cultivation cost and higher market price. In addition, these varieties were more resistant to pests, and tastier. Farmers learned how to design, implement and evaluate experiments. By conducting a fertiliser experiment during the training, they proved that current fertiliser doses were far too high, implying a waste of money and a risk for environmental pollution.

One trainee reported, that by applying the knowledge and skills he had learned in the field school, he managed to reduce chemical fertiliser doses on high-yielding rice varieties by 25 per cent. He stopped using pesticides, intensified fish cultivation in the rice field, and began to purposely conserve natural enemies in his field. The farm records he had taken over several seasons, made him aware, that he could never achieve an economically sustainable farming system under the current sharecropping agreement. All the income from the one-third share of the harvest was spent on cultivation costs. Therefore, he decided to face the landlord and frankly refuse to continue working under this agreement. He suggested 50-50 share of the harvest, with all costs being borne by himself. The landlord agreed.

The trained farmer-trainers began facilitating IPM field schools for other farmers in surrounding villages. Soon, many farmers followed the examples of practices leading to a more ecological and economical rice farming, such as rice-fish cultivation, growing of traditional varieties, more efficient use of inorganic fertillisers, and, to the dismay of the landlords, demand for a more just sharecropping agreement.

in which many variables besides the one studied cannot be kept constant. This does not mean that farmers should conduct sophisticated split-plot trials to be able to reach conclusions that are sufficiently reliable and applicable within the context of their own farm management. A basic understanding about a few principles of experimental design proved to be enough for farmers to obtain consistent results from their own experiments (van de Fliert *et al.*, 1996). These principles, or 'rules of the game', include:

- select one variable to test;
- select 3-4 sensible treatments, considering current practices, new guidelines and overall farm conditions;
- have (at least) two replications per treatment;
- make small plots (minimum of 10 m^2) in the same or adjacent fields; and
- keep all other factors constant.

Besides demonstrating processes and effects, such experiments provide a tool for testing new ideas. Conducting the experiments in a collaborative manner, for instance in a field school, enhances the learning process and a possible risk is distributed among the participants. By experiencing the process of designing, conducting, observing, analysing and evaluating their own experiments, farmers become able to replicate this process in their own fields, under their own location-specific conditions, to fine-tune guidelines they perceive worth trying (Box 3). Competence of farmers to adequately *adapt* methodologies (rather than *adopt*) should be a key parameter used to measure the success of programmes that aim at achieving sustainable agricultural systems.

10.5. CONCLUSIONS

Despite the additional requirements for introducing a wider range of sustainable agricultural practices to farmers, and capacitating them to become ecosystem managers in the broader sense, IPM remains an appropriate technology. Due to the concrete and well observable phenomena involved, farmers become excited and eager to learn more, and, as a result, they easily grasp the logic of the ecological processes implied by IPM. After all, they have spent most of their time in fields and observed similar things, but never rationalised them. This process of rationalisation is likely to provide a solid basis for farmers to understand more complex and abstract features of ecosystem management. Having understood basic principles, and gained basic analytic and decision-making skills, farmers can be expected to adequately adapt to changing conditions and opportunities to make their farming systems increasingly and sustainably profitable. Only with farmers as central actors in technology development and implementation can agriculture become truly sustainable.

ACKNOWLEDGEMENTS

The author is grateful to all the farmers who shared their experience with her, and to Dr Ann R. Braun and Dr Patricia C. Matleson for their valuable and constructive comments on drafts of this paper.

REFERENCES

Altieri, M.A. 1995. *Agroecology. The Science of Sustainable Agriculture.* Westview Press, Boulder, USA, 433 pp.

Asher, C.J. and Grundon, N.J. 1993. *Diagnosis of Nutritional Limitations to Plant Growth by Nutrient Omission Pot Trials.* Unpublished Manual. Department of Agriculture, The University of Queensland, Brisbane, Australia.

Bentley, J.W. and Andrews, K. 1996. *Through the Roadlocks : IPM and Central American Smallholders.* The Gatekeeper Series No. SA56, International Institute of Environment and Development, London, UK, 20 pp.

Braun, A., van de Fliert, E., Wheatley, C., Prain, G. and Widodo, Y. 1995. Improving profits from sweetpotato by linking IPM with diversification of markets. CIP Circular (21) **3**: 8-15.

Eveleens, K.G., Chisholm, R., van de Fliert, E., Kato, M., Nhat, P.T. and Schmidt, P. 1996. *Mid-term Review of Phase III.* Report FAO Inter-country Programme for the Development and Application of Integrated Pest Control in Rice in South-east Asia. P.O. Box 3700 MCPO, 1277 Makati, Metro Manila, Philippines.

FAO (Food and Agriculture Organization). 1968. *Report of the First Session of the FAO Panel of Experts on Integrated Pest Control.* FAO Meeting Report No. PL/1967/M/7, Rome.

Flint, M.L. and van de Bosch, R. 1981. *Introduction to Integrated Pest Management.* Plenum Press, New York, USA.

Haverkort, B., van der Kamp, J. and Waters-Bayer, A. (eds). 1991. *Joining Farmers' Experiments. Experiences in Participatory Technology Development.* IT Publications, London, USA, 269 pp.

Kenmore, P.E. 1991. How rice farmers clean up the environment, conserve biodiversity, raise more food, make higher profits. *Indonesia's Integrated Pest Management - A Model for Asia.* FAO Inter-country Programme for the Development and Application of IPM in Rice-based Cropping Systems in South and Southeast Asia, Manila, Philippines.

Kenmore, P.E., Gallagher, K.D. and Ooi, P.A.C. 1995. Empowering farmers: experiences with integrated pest management. *Entwicklung Landlicher Raum.* **1/95**: 27-28.

Matteson, P.C., Gallagher, K.D. and Kenmore, P.E. 1992. Extension of integrated pest management for planthoppers in Asian irrigated rice: Empowering the users. *In*: R.F. Denno and T.J. Perfect (eds). *Ecology and Management of Planthoppers.* Chapman & Hall, London, UK, pp. 656-685.

Murphy, J. and Marchant, T.J. 1988. *Monitoring and Evaluation in Extension Agencies.* World Bank Technical Paper No. 79, Monitoring and Evaluation Series. World Bank, Washington, DC, USA.

Pickett, A.D., Putman, W.L. and Le Roux, E.J. 1958. Progress in harmonizing biological and chemical control of orchard pests in eastern Canada. *Proc. Tenth Int. Congr. Ent.* **3**: 169-174.

Rogers, E.M. 1983. *Diffusion of Innovations.* 2nd ed., Free Press, New York, USA.

Röling, N. 1992. The effects of applied agriculture research and extension: Issues for knowledge management. *J. Ext. Syst.* **8**(1/2): 166-184.

Röling, N. and van de Fliert, E. 1994. Transforming extension for sustainable agriculture: The case of integrated pest management in Indonesia. *Agric. Human Values* **11**(2/3): 96-108.

Stern, V.M., Smith, R.F., van de Bosch, R. and Hagen, K.S. 1959. The integrated control concept. *Hilgardia* **29**(2): 81-101.

van de Fliert, E. 1993. *Integrated Pest Management: Farmer Field Schools Generate Sustainable Practices.* A Case Study in Central Java Evaluating IPM Training, Doctoral Dissertation, Wageningen Agricultural University Papers, PUDOC, Wageningen, The Netherlands.

van de Fliert, E. and Wiyanto 1996. A road to sustainability. *ILEIA Newsl.* **12**(2): 6-8.

van de Fliert, E., Asmunati, R., Wiyanto, Widodo, Y. and Braun, A.R. 1996. From basic approach to tailored curriculum: Participatory development of a farmer field school model for sweetpotato. *In*: UPWARD *Proceedings of Fourth Review and Planning Conference: From Diagnosis to Action Research.* October 2-5, 1995, Yogyakarta, Indonesia. UPWARD, Los Banos, Philippines (In press).

Worde, M. and Mekbib, H. 1993. Linking genetic resource conservation to farmers in Ethiopia. *In*: W. de Boef, K. Amanor, K. Wellard and A. Bebbington (eds). *Cultivating Knowledge. Genetic Diversity, Farmer Experimentation and Crop Research.* IT Publications, London, pp. 78-84.

Zadoks, J.C. 1985. On the conceptual basis of crop loss assessment: the threshold theory. *A. Rev. Phytopathol.* **23**: 455-473.

11

IPM IN THE 21ST CENTURY : CHALLENGES AND OPPORTUNITIES

E.A. HEINRICHS
Department of Entomology
University of Nebraska, Lincoln, NE 68585 - 0816, USA

11.1. NEED FOR AN INTEGRATED PEST MANAGEMENT (IPM) APPROACH

Pests have been estimated to cause losses of about 30 per cent of potential world food, fibre and feed production, and continue to be major constraints to crop production (NRI, 1992). Changing pest populations attest to the need for dynamic crop production programmes including new and novel approaches to the development and implementation of IPM (Heinrichs, 1994). Pest/host interactions are influenced by a myriad of factors, many of which involve human activity (Bosque-Perez and Buddenhagen, 1992). Immigration of new pests, introduction of new crops, and cropping intensification have led to significant changes in pest populations in crops and thus the need for an IPM approach in crop production systems.

11.1.1. Immigration of New Pests

The rice water weevil, *Lissorhoptrus oryzophilus* Kuschel, is an example of the immigration of a new pest. This insect, indigenous to America was first discovered in Aichi Prefecture in central Japan in 1976 (Tsuzuki and Isogawa, 1976) and spread rapidly throughout the rice growing regions of Japan. The application of insecticides to control this pest has resulted in a distinct increase in production costs (Kiritani and Baba, 1994) and the widespread use of insecticides is of environmental concern. The rice water weevil later invaded Korea (Lee, 1992) where it has caused major damage.

11.1.2. Introduction of New Crops

Many pest problems are new-encounter situations where the crop has been introduced into an area and pests that have not evolved with the crop, attack the introduced crop (Buddenhagen and de Ponti, 1983). For example, the rice delphacid, *Tagosodes orizicolus* (Muir), a native of South America, has become a severe pest after the introduction of rice into South and Central America (Bosque-Perez and Buddenhagen, 1992). In addition to causing direct damage, through the removal of plant sap, this planthopper is also a vector of hoja blanca, the most serious rice virus disease in the Western Hemisphere.

Re-encounter situations occur where a pest follows a crop host to a new location. Such situations may occur after historical separations such as that of the mealybug, *Phenacoccus manihoti* Matile-Ferero, which was accidentally introduced from South America to Africa in the 1970s after being separated from cassava host, of American origin, for about 400 years (Herren and Neuenschwander, 1991). The absence of natural enemies, resulted in the spread of the mealybug across Africa where major outbreaks and devastating losses caused food shortages (Bosque-Perez and Buddenhagen, 1992).

11.1.3. Crop Intensification

Intensification involves (i) an increase in the number of crops grown per year, (ii) the planting of large, contiguous fields instead of small, isolated fields, (iii) a change from polycropping to monocropping, (iv) an increase in the use of agricultural chemicals, and (v) the planting of modern high yielding cultivars responsive to fertilizers (Loevinsohn, 1984). The goals of agricultural intensification are being undermined by pest problems which are the inadvertent result of that very process (Thomas and Waage, 1994). As Altieri and Nicholls have stated in Chapter 3, modification of plant communities to meet the food and fibre needs of humans often results in heavy pest damage. There is ample evidence that cropping intensification has been responsible for changes in insect pest populations. Pest responses to environmental changes are dynamic and a change that is detrimental to one species may offer new opportunities to another, leading to changes in species composition and abundance.

As discussed in Chapter 5, cropping intensification may result in insect resistant crop cultivars becoming susceptible to pests through the development of resistance-breaking biotypes or strains. This factor has greatly complicated the breeding of resistant cultivars for the management of certain pests and has shown the need for an IPM approach in the deployment of pest resistant cultivars (Heinrichs, 1988).

Pesticides are causing harm, especially in developing countries where there is unacceptable human morbidity and mortality, and where pest populations are becoming resistant and natural control agents are being destroyed (see Chapter 7). Natural enemy destruction has led to the resurgence of pest populations and outbreaks of secondary pests (Heinrichs and Mochida, 1984).

11.2. ROLE OF IPM IN CROP PRODUCTION

IPM is acclaimed as a means for the development of ecologically sound and sustainable agricultural practices to mitigate crop production losses and for strengthening farming communities' participation in that development. The United Nations Conference on Environment and

Development (UNCED) assigned a pivotal role to IPM in the agricultural programmes and policies envisaged as part of its Agenda 21 which called for network of farmers, extensionists, and researchers to implement participatory IPM as the best, most sustainable crop protection option for the future (Kenmore, 1996b). IPM is a key component in the battle for increased food production and has the potential to play a much greater role in the future.

In spite of significant progress, the need for increased production of food, feed and fibre continues unabated. The situation is especially acute in Africa where population growth will continue to outstrip growth in food production for many years unless more is done to increase agricultural growth. If current trends continue, by 2025 Africa could have a food gap of 185 million tons, which will be 2.5 times production (Conway *et al.*, 1994). The poorest segment of the population is increasing the fastest. Most of the poor and malnourished live in rural areas and are commonly women and children.

The desire for food security has adversely affected the environment. Globally about 2 billion ha have become degraded since 1945. Of additional concern is the evidence of a yield growth rate stagnation in those countries where the 'Green Revolution' had its greatest impact. World growth in rice yields fell from 2.9 to 1.9 per cent per year during the late 1980s and wheat from 4.4 to 2.2 per cent (Tribe, 1994).

The Green Revolution created a dilemma in which it was believed that mass starvation could only be avoided by the widespread adoption of technology that was in some cases harmful to the natural environment. Higher yields were achieved but increased land productivity was accompanied by increased use of chemical inputs without sufficient regard to the ecological impact and the health of farmers. It is now recognised that the real costs of pesticides associated with packaged technologies are high and increasing (Kenmore, 1996a). What is required is another Green Revolution that is green in terms of conserving natural resources and the environment, or a "Super-Green" revolution as Conway *et al.* (1994) have called it. According to Conway *et al.*, "the Super-Green Revolution will require a repeat of the successes of the Green Revolution on a global scale, in many diverse localities and be equitable, sustainable and environmentally friendly."

Lessons from the Green Revolution have led to a demand for a more sustainable approach to agricultural development; an approach that balances the needs of food security and environmental conservation or regeneration. There have been two responses to this demand which can be labelled 'doing it better', and 'doing it differently'. The first response, adopted by organisations that gained status under the Green Revolution, includes the development of better cultivars, chemicals and extension services. The second strategy has involved a paradigm shift, that seeks answers in new places, reducing reliance on technological packages and those who produce and market them (Kenmore, 1996a).

From an ecological perspective, this paradigm shift involves support for naturally occurring processes, such as biological control of pests rather than trying to replace these processes with high levels of external inputs. From a sociological perspective, the paradigm shift involves support for local expertise, meaning decision-making by farmers as individuals and groups, rather than replacing this with messages from outside experts, to whom farmers often do not have access. IPM based on ecological analysis and decision-making by farmers does not involve a rejection of inputs (e.g., modern cultivars, pesticides and fertiliser) but it instead places them in a new perspective; they are something to be understood, tested and freely chosen by farmers. This is in contrast to the way in which technology has often been promoted in the past, where production targets have been based on results from research stations or commercial interests. External inputs will in most cases, still be used by IPM farmers, but the most important inputs which they use will be their own understanding of the agroecosystem and their ability to make location-specific decisions (Kenmore, 1996a).

IPM adds the dimensions of *Ecological sustainability,* as it relies on environmentally benign processes such as resis–tant cultivars, natural enemies and cultural control;

***Social stability,* because it is institutionalized at the level of the farming community and local government;**

***Economical sustainability,* as it reduces farmers' dependence on procured inputs.**

11.3. CONSTRAINTS TO IPM IMPLEMENTATION

Successful implementation of IPM programme is dependent on a number of factors. First of all, the IPM technology must be appropriate for farmers and thus should be developed in close cooperation with them. Even with the availability of appropriate technology, there are numerous constraints to the implementation of IPM. These include institutional constraints (informational, sociological and economic issues), and political constraints (see Chapter 1).

IPM implementation should be under the guidance of well trained extension personnel who can deliver IPM strategies in an unconventional manner to small-scale farmers who can neither read nor write (Kumar, 1990). Although extension services are considered essential for the introduction of IPM technology, they are, in many countries, ineffective (Zethner, 1995). They may be limited in number and are unable to serve even a small proportion of the farmers. Their salaries and transport are poor and most have not received any IPM training. There are too few women extension agents which poses a communication problem with many of the farmers who are women.

The training of extension services who in turn train the farmers is a major challenge to the adoption and spread of IPM practices among farmers in the third world. Such a programme is best adapted to farmers' needs if it is part of an interdisciplinary crop strategy and if research and extension cooperate closely. To promote IPM, governments and donors must put more emphasis on IPM component and implementation in agricultural projects. IPM should be considered as the best ecological approach and a competitive economic approach considering the increasing threat from pesticide use. In subsistence crops, where there is little or no use of pesticides, IPM implementation should be one component of a crop production system and thus a strategy to improve farmers' income. Otherwise it is difficult to quantify the benefits of IPM and thus convince farmers, governments and donors of the benefits (Zethner, 1995).

11.4. GLOBAL IPM IMPLEMENTATION STRATEGIES

Although a number of effective IPM programmes are now underway in Asia, the realization of IPM programmes at the community level has, in most other regions of the developing world, proceeded too slowly relative to the number of implementation opportunities available. Implementation of IPM programmes has proven to be highly complex, requiring specialised technical expertise, dedicated and skilled field trainers, detailed analysis of policy, social and economic issues at the national and local levels, political commitment, and careful attention to details in project design, operationalization, and monitoring (Persley *et al.*, 1996).

Global IPM field implementation gained momentum when the FAO Intercountry Programme for IPM in Asia (ICP) organised a Global IPM Meeting in 1993. Participants from various regions of the world familiarised themselves with the achievements and approach of the collaborating national IPM programmes in Asia. Pilot activities that were subsequently initiated in Africa yielded promising results and stimulated requests for additional assistance (FAO, 1997).

In 1994 an Inter-Agency Task Force was established by FAO, World Bank, UNDP and UNEP with assistance from the ICP (FAO, 1994). From its discussions a consensus emerged around the four key elements that are required for a concerted effort to expand and intensify IPM implementation.

- *Farmer participation* in IPM development and on-farm research and training in farmers' fields, with the involvement of local nongovernment organisations (NGOs) and extension agents;

- *Supportive national policies* that provide an enabling environment for programme implementation, including removal of subsidies for pesticide use, and availability of resources for IPM research and extension activities;
- *Financing for pilot projects* and other activities where substantial benefits generated by IPM implementation can be demonstrated; and
- *Planning and development of national IPM programmes,* supported by domestic and, if required, external resources, which serve to co-ordinate IPM activities and support new local initiatives as they arise.

Successful IPM programmes have demonstrated the advantages of adopting each of these four elements as part of a comprehensive approach to IPM development. However, lack of co-ordination among international agencies and national programmes have hampered attempts to apply this approach in some regions and in certain cropping systems. As a result, there has been unnecessary fragmentation and duplication of effort, failure to capitalise on past successes and to learn from mistakes, and lost opportunities arising from a failure to mobilise resources in a timely manner.

A model for IPM programme development as described by Kenmore (1996a) distinguishes the four empirical stages in the development of IPM field implementation:

Awareness. Activities in this stage use IPM entry points to generate human and financial commitments to testing IPM by farming communities. Awareness building activities have included workshops, field visits, and farmers' presentations. They can be local, national, or international, but usually emphasise one country's situation.

Development. Once awareness has built up and a commitment is made to field test IPM implementation, then pilot projects comprise the main activities of this stage. These pilot projects take from 18 to 30 months.

Capacity building. In the few countries that have gone through this stage, it has concentrated on preparing a national cadre of full time IPM trainers. They are mostly employed by national and local governments.

Consolidation. The highly qualified and experienced full time IPM trainers can be supported for a number of years by different kinds of projects. In consolidation stage projects, the scale of operation generates large internal rates of return, so these tend to seek or be part of projects with investment financing. Prerequisites include supportive policies, highly trained field staff, and a means for monitoring and evaluation that ensures the maintenance of technical and training quality while expanding. These will usually require a component of grant-financed technical assistance to guarantee the quality of training, assess impact, and adjust implementation through external changes in the field during implementation.

11.5. THE ASIAN PROTOTYPE FOR IPM IMPLEMENTATION

> **Building on the success of rice IPM in Asia**
>
> - **Farmers as experts**
> - **Farmer Field Schools**
> - **Participatory research and training**
> - **Elimination of pesticide subsidies**
> - **Banning of hazardous chemicals**
> - **IPM as a national policy**

The Asian prototype has been described in detail by Kenmore (1966b). IPM programmes now being implemented on a large scale throughout Asia are not limited to reducing toxic pesticides through improved pest management practices but seek to improve overall food production on a sustainable and local basis. These IPM programmes use a new model of farmer interaction in which local expertise among farmers and extension staff is improved through community-based educational programmes including such activities as Farmer Field Schools (FFS) (see Chapters 1 and 10). The underlying assumption of this community approach is that site specific education and action research are essential for improving crop protection and increasing food production. This is in contrast to conventional extension models in which centrally researched and packaged technologies are transferred to farmers by an extension system that primarily depends on non-expert field level extension staff. Examples that illustrate the need for site-specific decision making include (Kenmore, 1996a):

- *Disease management* based on local fertiliser inputs, varietal diversification and weather;
- *Insect management* based on cultivar, natural enemy communities and crop health;
- *Soil and plant nutrient management*based on site-specific soil type, cropping pattern and available inputs.

Using this approach, technology validation, development and implementation are locally controlled but inputs are obtained from a variety of organisations, other farmers and international and national research institutions.

The Asian prototype has successfully utilised the FFS approach to IPM implementation. According to Kenmore (1996b) the FFS model presumes that:

- Farmers are rational, skeptical and curious and do experiments;
- Farmers make important field decisions;
- Local ecology is significant in those decisions;
- Farmers who have concepts, theories and skills will make scientific cases by public debate on their local observations;
- These cases will consider alternative explanations, analyses, predictions and evaluations and come to decisions upon which to act.

The FFS model, first developed and implemented in Asia, has been adopted by a number of countries in Asia and is now being tested in other regions of the world. Studies from individual countries show yield increases from 4 to 13 per cent, with concurrent reductions in insecticide application frequency, and cost reductions of 30-40 per cent after the first FFS training. In all cases, farm-level profits increase, both from cost reduction and from yield increases (FAO,1993). The Asian documentation of these programmes is extensive and goes beyond many current data sets that often focus on pesticide reduction, or number of farmers trained (Kenmore, 1996b).

11.6. CONCLUSIONS

As mentioned in the opening chapter, and discussed in more detail in this chapter, pest problems are expected to become more serious as agriculture is further intensified. This will be especially evident in areas where farmers are converting from a subsistence type of agriculture to a 'modern' approach. However, IPM continues to develop as a science and will play a key role in sustainable agricultural production systems in the 21st century. This role has been fortified by the establishment of the Global IPM Facility, cosponsored by the FAO, the World Bank, the UNDP and the UNEP (Persley *et al.*, 1996).

There has been significant progress in both, the development of IPM tactics, and in the development of more effective implementation strategies. The development of botanicals, behaviour modifying chemicals and bioregulators to replace environmentally harmful pesticides will continue to be a major research thrust into the next century. The integration of genetically resistant cultivars with natural biological control agents and cultural controls will continue to be a sought after strategy. Although significant progress has been achieved in the development and commercial utilization of insect resistant rice cultivars they have the potential to play an even much greater role in rice production systems in the future. Biotechnology techniques such as wide hybridization, transformation, DNA markers, and DNA fingerprinting, are being used to overcome some of the constraints to breeding of insect resistant cultivars. The cultivars developed will contribute to an economically more efficient and an environmentally benign crop protection technology.

However, we must be actually aware of the fact that IPM technology without an effective means of transfer to the farmer, is of little practical value. Thus it is exciting to see the progress made in transferring IPM technology to the farmer level through the FFS as discussed in this chapter and in Chapters 1 and 10. Because of the dynamic nature of the FFS approach, the 21st century will see an expansion of IPM implementation to cover more crops and more regions of the world. IPM will continue to serve as an entry point for the training of farmers in a holistic approach to sustainable crop production.

REFERENCES

Bosque-Perez, N.A. and Buddenhagen, I.W. 1992. The development of host-plant resistance to insect pests: outlook for the tropics. *In:* S.B.J. Menken, J.H. Visser and P.J. Harrewijn (eds). *8th International Symposium on Insect-Plant Relationships.* Kluwer Academic, Dordrecht, pp. 235-249.

Buddenhagen, I.W. and de Ponti, O.M.B. 1983. Crop improvement to minimize future losses to diseases and pests in the tropics. *FAO Pl. Prot. Bull.* **31:** 11-20.

Conway, G., Lele, U., Peacock, J. and Pineiro, J. 1994. *Sustainable Agriculture for a Food Secure World, a Vision for the Consultative Group on International Agricultural Research.* Consultative Group on International Agricultural Research, Washington, DC.

FAO (Food and Agriculture Organization). 1993. *Global IPM: A Report on the Global IPM Field Study Tour and Meeting.* August-September, 1993, FAO, Manila, The Philippines.

FAO (Food and Agriculture Organization). 1994. *IPM Task Force Report.* FAO, Rome.

FAO (Food and Agriculture Organization). 1997. *Global IPM Facility Programme Document.* Global IPM Facility/AGP, FAO, Rome.

Heinrichs, E.A. 1994. Development of multiple pest resistant crop cultivars. *J. agric. Ent.* **11:** 225-253.

Heinrichs, E.A. 1988. Variable resistance to homopterans is rice cultivars. *ISI Atlas* Sci. *Anim. Pl. Sci.* **1:** 213-220.

Heinrichs, E.A. and Mochida, O. 1984. From secondary to major pest status: the case of insecticide-induced rice brown planthopper, *Nilaparvata lugens* resurgence. *Prot. Ecol.* **7**: 201-218.

Herren, H.R. and Neuenschwander, P. 1991. Biological control of cassava pests in Africa. *A. Rev. Ent.* **36**: 257-283.

Kenmore, P.K. 1996a. The Global IPM Facility *In: Global Integrated Pest Management Facility Partners Meeting,* May 7, 1996. FAO, Rome.

Kenmore, P.K. 1996b. Integrated pest management in rice. *In*: G.J. Persley (ed.). *Biotechnology and Integrated Pest Management.* CAB International, Wallingford, UK, pp. 76-97.

Kiritani, K. and Naba, K. 1994. Development and implementation of rice IPM in Japan. *In*: E.A. Heinrichs (ed.). *Biology and Management of Rice Insects.* Wiley Eastern, New Delhi, India, pp. 713-731.

Kumar, R. 1990. Principles and philosophy of integrated pest management. *In*: S.K. Hahn and F.E. Caveness (eds). *Integrated Pest Management for Tropical Root and Tuber Crops.* International Institute of Tropical Agriculture, Ibadan, Amarin Printing Group Co., Bangkok, pp. 14-17.

Lee, S.C. 1992. Towards integrated pest management of rice in Korea. *Korean J. appl. Ent.* **31:** 205-240.

Loevinsohn, M.E. 1984. The ecology and control of rice pests in relation to the intensity and synchrony of cultivation. Ph.D. Dissertation, Imperial College, London, UK.

NRI (Natural Resources Institute). 1992. *A Synopsis of Integrated Pest Management in Development Countries in the Tropics.* Chatham, UK.

Persley, G.J., Toenniessen, G. and Dart, P. 1996. Summary and future directions. *In*: G.J. Persley (ed.). *Biotechnology and Integrated Pest Management.* CAB International, Wallingford, UK, pp. 387-394.

Thomas, M. and Waage, J. 1994. *Integration of Biological Control and Host Plant Resistance Breeding.* International Institute of Biological Control, Silwood Park, UK.

Tribe, D. 1994. *Feeding the Greening World. The Role of International Agricultural Research.* CAB International, Wallingford, UK.

Tsuzuki, H. and Isogawa, K. 1976. The initial occurrence of a new rice insect pest, the rice water weevil, in Aichi Prefecture in Japan. *Shokubutsu - Boeki Pl. Prot.* **30**: 341 (in Japanese).

Zethner, O. 1995. Practice of integrated pest management in tropical and sub-tropical Africa: an overview of two decades (1970-1990). *In*: A.N. Mengech, K.H. Saxena and H.N.B. Gopalan (eds). *Integrated Pest Management in the Tropics.* John Wiley and Sons, New York, pp. 1-67.

INDEX

A

Abies balsamea (L.) Mill. 161
Abiotic stresses 120, 131, 209
Abscisic acid 194
Acephate 201
Acetogenins 129
Acetylcholine 222
Acorus calamus L. 161, 195
Actellic 169
Acyrthosiphon pisum (Harris) 76, 123, 133, 137
Aedes aegypti (Linnaeus) 99, 102
Aglycone 233
Aflatoxin 162, 184
Ageratum houstonianum Mill. 161
Agglutinins 88, 212
Agroecosystems 15, 40, 49, 67, 69, 74, 78, 82, 83, 188, 232, 250, 255, 258, 270
Aldrin 187
Aleurocanthus woglumi Ashby 73
Alfalfa caterpillar 72, 81
Alkaloids 122, 129, 156, 159, 227, 228, 233, 235
Allelochemicals 8, 124, 128, 129, 156, 209, 214, 222, 223, 231, 232, 235
Allethrin 3, 157, 181, 190
Alley cropping 40
Allomones 124, 129, 223, 239
Allopatric 136
Allylisothiocyanate 184
Alpha-amylase inhibitors 212
Alpha-terthienyl 160
Alternative host plants 28, 31, 32, 35
American bollworm 19
Amitraz 192
Amoorastaitin 163
Amrasca biguttula (Ishida) 33, 44, 123
Amylase inhibitors 211
Annona reticulata L. 160
Annona squamosa L. 195
Anopheles gambiae Giles 92, 95, 160
Anopheles pseudopunctipenis franciscanus McCracken 97
Anthonomus grandis Boheman 19, 33, 35, 80, 128, 134
Antibiosis 121
Antifeedants 9, 131, 180, 181, 195, 210, 211, 221, 229, 231, 233
Anti-GABA agents 192
Antijuvenile hormones 161, 180, 193, 236
Antimones 224
Antixenosis 121
Aonidiella aurantii (Maskell) 73
Aphelinus mali (Haldeman) 81
Aphis craccivora Koch 31, 43, 126, 134, 137
Aphis gossypii (Glover) 19, 137
Aphytis melinus DeBach 76
Aphytis proclia (Walker) 79
Apolysis 237
Applied biological control 68
Aproaerema modicella (Deventer) 31, 41
Aquatic organisms 188
Arginine 234
Arrestants 223, 224, 225
β-asarone 161
Asporogenous mutants 97
Atherigona naqvii Steyskal 50
Atherigona soccata Rodani 29, 38, 41, 43, 50, 134
Atherigona varia (Meigen) 123
Attractants 123, 128, 129, 223
Augmentation 75, 83, 254
Augmentative biological control 68, 74
Auxin 194
Avermectins 185, 198, 200
Azadirachta indica A. Juss. 161, 170, 181, 195, 237
Azadirachtin 163, 164, 170, 171, 228, 231, 232, 237
Azadirachtoids 163
Azadirone 162
Azugarhin 194

B

Bacillus lentimorbus Dutky 3, 87

Bacillus popilliae Dutky 3, 78, 87
Bacillus sphaericus Neide 87, 88, 95, 96, 98, 100, 103, 104
Bacillus thuringiensis Berliner 3, 4, 75, 78, 87, 88, 89, 97, 99, 100, 102, 103, 139, 181, 197, 212
Bacterial insecticides 87
Bacterial pathogens 195
Baculoviruses 196, 214, 215
Balsam fir 161
Beauveria bassiana (Balsamo) Vuillemin 74, 75, 78, 198
Bedding plant 161
Behaviour modifications 222
Behaviour modifying chemicals 8, 180, 232, 274
Behaviour modifying approaches 221
Bemisia tabaci (Gennadius) 33, 42, 186
Benzoquinones 129, 227, 228
Benzoxazolinones 235
Bialophos 200
Bioaccumulation 187
Bioallethrin 157
Biochemical factors 128
Biocides 192
Biodiversity 67, 82, 83
Biological control 1, 3, 4, 52, 67, 68, 118, 155, 181, 254, 274
Biological recycling 96
Biopesticides 195, 196, 201
Biophysical factors 125
Bioregulators 192, 274
Biotechnological approaches 8
Biotechnology 209, 217, 274
Biotic stresses 209, 212
Biotypes 8, 133, 135, 137, 210, 227, 238, 239, 268
Bombyx mori (Linnaeus) 88, 93
Bordeaux mixture 181
Botanical pest control 155
Botanical pesticides 2, 3, 4, 164, 180, 189, 194
Bracon kirkpatricki (Wilkinson) 76
Bracon mellitor Say 76
Brevicoryne brassicae (Linnaeus) 80, 127, 137
Brown planthopper 16, 127, 133, 134, 138, 186, 252, 253
Bursicon 237
Busseola fusca (Fuller) 27, 28, 45

C

Cabbage aphid 127, 137
Cabbage looper 69, 131
California red scale 73, 76
Calocoris angustatus Lethiery 29
Campoletis sonorensis (Cameron) 76
Canavanine 235
Carbamates 3, 157, 180, 181, 185, 194
Carbosulfan 201
Carrot fly 127, 128
Cartap hydrochloride 167
Caryophyllene 210
Cassava green mite 39
Cassava mealybug 39
Causes of resistance 125
Cephus cinctus Norton 43, 135
Cephus pygmeus (Linnaeus) 70
Cereal leaf beetle 134
Ceriferol 224
Ceriferol-I 224
Cerorubenol-I 224
Cerorubenol-II 224
Cerorubenol-III 224
Chaparrinone 130
Chemical ethology 223
Chemical factors 210
Chemosterilization 162
Chilo infuscatellus Snellen 28, 38, 45, 48, 199
Chilo partellus (Swinhoe) 27, 28, 34, 37, 41, 45, 124, 134, 165, 166
Chilo suppressalis (Walker) 28, 35, 71, 129, 134
Chinaberry 170
Chitin inhibitors 185
Chlordane 187
Chlordimeform 192
Chlorinated hydrocarbons 185
Chlorogenic acid 130, 214, 234
Choristoneura fumiferana (Clemens) 130, 168

Chromosomal translocations 136
Chrysanthemum cinerariaefolium Vis. 156, 195
Chrysnthemum roseum Adam 157
Chrysoperla carnea (Stephens) 70, 75, 77, 78
Cimex lectularius Linnaeus 185
Cinerin I 157
Cinerin II 157
Cinnamonitrile 130
Cis-asarone 233
Citrus blackfly 73
Classical biological control 68, 72, 73
Classification of resistance 120
Clavigralla tomentosicollis Stal 31
Clerodin 130
Cnaphalocrocis medinalis (Guenee) 164
Colorado potato beetle 181, 199, 228, 233
Components of IPM 6
Coniesta ignefusalis (Hampson) 27, 29, 38, 39, 43, 47
Conservation 78, 83
Contarinia sorghicola (Coquillett) 29, 41
Contingent valuation 10
Conventional tillage 82
Corcyra cephalonica (Stainton) 169
Corn earworm 80, 134, 225, 234
Corn leaf aphid 134, 138, 189
Corpus allatum 161, 193, 236
Cotton bollweevil 3, 5, 19, 46, 80, 128, 134, 187, 251
Cotton jassid 44
Cottony cushion scale 2, 72
Coumarins 129, 231
Cover crops 82, 83
Cowpea trypsin inhibitor 212, 216
Crop husbandry 26, 29, 33
Crop intensification 49, 268
Crop residues 27, 28, 33, 35
Crop rotation 28, 29, 30, 32, 33, 39, 82
Cropping intensity 268
Cropping pattern 273
Cropping systems 15, 40, 49, 71, 80, 83, 250, 255, 261, 272
Cross resistance 171
Cucurbitacin A 130
Cucurbitacin E 130
Cucurbitacins 129, 131
Culex pipiens Linnaeus 88, 89
Culex quinquefasciatus Say 99, 101, 103
Cultural control 1, 2, 26, 52, 155, 180, 274
Cultural practices 82, 221
Custard apple 160, 165
Cuticle 127
Cyclodiene insecticides 181
Cydia molesta (Busck) 77, 80
Cydia pomonella (Linnaeus) 77, 138, 199
Cylas formicarius (Fabricius) 32, 36
Cypermethrin 157, 160, 185, 190, 192
Cytokinin 194

D

DDT 2, 3, 78, 156, 181, 184, 185, 187, 191, 192
Decamethrin 190, 192
Decision making skills 17, 253, 261, 264
Decision making systems 6, 9
Dectes texanus Le Conte 46
Deltamethrin 171, 191
Demmisine 129, 228
Derris elliptica (Roxb.) 158
Development modifying approaches 221, 222, 233
Diabrotica undecimpunctata Mannerheim 130, 211
Diabrotica undecimpunctata howardi Barber 129
Diamondback moth 78, 102, 127, 167, 214
Diatraea grandiosella Dyar 43
Diatraea saccharalis (Fabricius) 29, 47
Dibrachys cavus (Walker) 21
2,4-dichlorophenoxy acetic acid 181, 189
Dieldrin 3, 78
2,5-dihydroxymethyl 130
3,4-dihydroxypyrroliodone 130
DIMBOA 129
Dipel 102, 103, 104, 167
Disease management 273
Dispalure 181

Diterpenes 228
Diversity 7, 40, 44, 70, 253, 254, 255
DNA finger printing 274
DNA markers 274
Doom 3
Dual discrimination theory 124
Durable resistance 139, 209
Dysdercus koenigii (Fabricius) 234

E

Earias vittella (Fabricius) 42, 123
Ecdysone 161, 193, 235, 238
Ecdysteroids 237
Ecological principles 253
Ecological sustainability 263, 270
Economic constraints 12
Economic damage 5
Economic injury level 4, 119
Economic sustainability 263, 270
Economic threshold levels 4, 7, 18, 41, 251, 252, 254
Ecosystems 1, 225, 260, 262
  cereal-based 27, 35, 36, 37, 40, 43, 44, 47, 48, 50
  cotton-based 20, 34, 36, 37, 40, 42, 44, 46, 48, 49, 50
  legume-based 34, 36, 37, 38, 41, 43, 45, 47, 49, 50
  natural 67
  root crop 34, 36, 37, 39, 40, 42, 44, 46, 48, 49, 50
Ectoparasites 70
Egyptian cotton leafworm 37
Elcar 4
Eldana saccharina Walker 27, 28, 45, 165, 166
Empoasca fabae (Harris) 126, 132, 233
Empoasca kerri Pruthi 31
Encarsia perniciosi (Tower) 73
Endoparasites 70
Endosulfan 37
Entomophagous insects 67, 68
Environment 4, 10, 12, 118, 120, 128, 131, 155, 166, 183, 198, 200, 201, 218, 238, 269, 272
Environmental economic injury level 4, 10
Environmental impact quotient 10
Ephestia kuehniella Zeller 88
Epilachna varivestis Mulsant 50
Era of IPM 2, 5
Era of pesticides 2
Eriosoma lanigerum (Hausmann) 119, 137
Erythrosin B 160
Escape 120
Essential oils 161
Ethylene 194, 211
Etiella zinckenella (Treitschke) 31
European corn borer 134, 135, 138, 158, 199
Experience threshold 254
Expression of resistance 120, 131, 132
External inputs 270

F

FAO Intercountry Programme 253, 271
Farm management practices 259
Farmer participatory research 12, 13
Farmers' Field Schools 13, 21, 257, 273
Farmers' participation 271
Farming systems 255, 263
Feeding attractants 129, 131
Feeding deterrents 131, 161, 162, 223, 229, 231
Feeding stimulants 129, 223
Fenvalerate 184, 190, 191, 192
Fertilizers 1, 28, 32, 46, 49, 82, 155, 256, 259, 261, 263, 273
First generation pesticides 180
Flavones 228, 234
Flavonoids 122, 129, 227, 228, 234
Flucythrinate 190, 191
Fluorescein isothiocyanate 101
Food chain 188
Forecasting system 6
Formamidine 232
Formulations 8, 98, 185, 198, 201, 212
Fourth generation pesticides 180
Frankliniella schultzei (Trybom) 32, 42
Functional plant loss index 123, 124
Fungal pathogens 198
Furanocoumarins 127
Fusaric acid 233

G

Gall midge 16, 134, 137, 164
Galerucella birmanica (Jacoby) 185
Galleria mellonella (Linnaeus) 104
Gallic acid 210, 214, 234
Gamma-aminobutyric acid 192
Gedunin 162
Genetic engineering 139, 209, 215, 217
Genetic recombinations 136
Genetically resistant cultivars 274
Genetics of resistance 133
Gibberelin 194
Glandular trichomes 126, 132
Glaucolides 130, 235
Global IPM Facility 4, 14, 274
Global IPM implementation 271
Glucosinolates 184
Glutamic acid 192
Glycoalkaloids 233, 234
Glycoprotein receptor 94
Glycoproteins 94, 210, 213
Gossypol 210, 233, 235
Granulosis virus 199, 214
Grape leafhopper 16, 129, 186
Grape phylloxera 3, 119, 138
Green leafhopper 81, 134, 135, 138, 164
Green peach aphid 138
Green revolution 4, 118, 155, 162, 269
Green stink bug 73
Groundnut leaf miner 41, 43
Groundnut rosette virus 36, 38
Gypsy moth 225

H

Habitat diversification 79, 82
Habitat management 68, 78, 155
Hand picking 33, 36, 37
Harrisonin 130
HCH 3, 181, 184, 185, 187
Helicoverpa armigera (Hubner) 21, 30, 33, 39, 41, 43, 50, 71, 167, 185, 196, 199, 214
Helicoverpa zea (Boddie) 19, 80, 134, 213, 214, 216, 234
Heliocheilus albipunctella (de Joannis) 29, 38, 43, 44, 47
Heliothis virescens (Fabricius) 19, 71, 72, 76, 93, 95, 103, 130, 134, 213, 231
Hellebore 159
Hessian fly 3, 70, 119, 128, 135, 137
High potency chemicals 190
High yielding varieties 1, 4, 118, 209, 256, 268
Hoja blanca 267
Holotrichia serrata (Fabricius) 30
Honeybees 159, 188
Horizontal resistance 139
Host evasion 120
Host plant resistance 52, 118, 135, 136
Household insecticides 157
Hydrellia wirthi Korytkowsky 48

I

Icerya purchasi Maskell 2, 72
Immigration 43, 267
Immunity 119
Importation 83
Indian calamus 161
Induced resistance 120
Informational constraints 11, 21, 270
Inherent toxicity 183
Insect behaviour 8, 161, 193, 227, 228
Insect biotypes 8, 135, 137
Insect growth disruptors 180
Insect growth inhibitors 221
Insect growth regulators 193, 235
Insect-plant interactions 122, 209, 211, 222, 227, 228, 239
Insect physiology 8, 161, 193, 235
Insect resistance 119, 171, 185, 201
Insecticidal proteins 88, 89
Insecticide-induced resurgence 186
Insecticide resistance 118, 185, 201, 233
Institutional constraints 11, 21
Integrated control 5, 251
Integrated crop management 259
Integrated pest management 1, 5, 21, 140, 155, 172, 231, 250, 267
Integrated tabular approach 10
Intercropping 28, 29, 30, 31, 32, 33, 40, 41, 42, 80

Introduction 267
IPM farmers 16, 17, 256
IPM implementation 6, 11, 16, 21, 250, 253, 255, 270, 271
Irrigation 32, 49, 252
Isoboldine 130
Isolan 3
Isopimpinellin 130

J

Japanese beetle 3
Jasmolin I 157
Jasmolin II 157
Jasmonic acid 211
Jugarin I 130
Juglone 130
Juvenile hormone analogs 161, 185, 236
Juvenile hormone mimics 180, 194, 236
Juvenile hormones 160, 180, 195, 235, 237

K

Kairomones 124, 129, 223, 232
Kalmitoxin I 130

L

Landmarks in IPM 3
L-canavanine 234
Lead arsenate 3, 180
Lectins 211, 212
Legislation 34
Legislative measures 14, 21
Leptinotarsa decemlineata (Say) 33, 40, 104, 214, 228
Light 131
Lignins 235
Limonin 171
Limonoids 162, 171
Linolenic acid 211
Lipaphis erysimi (Kaltenbach) 167
Lipoxygenases 211
Lissorhoptrus oryzophilus Kuschel 48, 123, 267
Locomotor stimulants 223
Locusta migratoria migratorioides (Reiche & Fairmaire) 127
Lycorine 130
Lygus lineolaris (Palisot de Beauvois) 19
Lymantria dispar (Linnaeus) 76, 93, 130

M

Major gene resistance 135
Major genes 135, 139
Maize weevil 132, 169
Mammalian safety ratio 190, 192
Management tactics 8
Manduca sexta (Johannsen) 4, 78, 95, 131, 216, 217, 234
Marangin 170
Marigold 195
Marrango 170
Maruca testulalis (Geyer) 31, 134
Mass trapping 37
Mating deterrents 223
Mating disruption 162, 226
Mating stimulants 223
Mayetiola destructor (Say) 70, 119, 131, 135, 137
Mechanical control 1, 2, 180
Medicagenic acid 233
Mediterranean flour moth 88
Melanagromyza obtusa (Malloch) 32, 39
Melia azedarach L. 170, 171
Melianone 171
Meliantriol 171
Menthol 233
Meromyza americana Fitch 38
Metamorphosis 129, 160
Metarhizium anisopliae (Metschnikoff) Sorokin 74, 75, 198, 199
Methacrifos 168
Methoprene 4, 191
Methyl jasmonate 211
Mevalonate kinase 193
Mexican bean beetle 50, 158
Microbial antibiotics 198
Microbial control 87
Microbial insecticides 20, 104, 195
Migratory locust 127
Milkweed bug 236
Mimosine 234
Minor genes 135, 139
Mixed cropping 40, 79

Monoamine oxidase 192
Monoculture 67, 71, 82
Monogenic resistance 133, 134, 135, 136, 139, 213
Monophagous insects 122
Monoterpenes 228
Moulting hormones 160, 161, 180, 195, 235
Mulching 28, 31, 32, 33, 44, 45
Multiline varieties 139
Multivoltine 70
Mylabris phalerata Pallas 39
Myriticin 232
Mythimna separata (Walker) 30, 45, 224
Mythimna unipuncta (Haworth) 45
Myzus persicae (Sulzer) 32, 48, 71, 138

N

Nagilactones 235
Naphthoquinones 227, 229, 237
National IPM programmes 16, 271
Natural biological control 78
Natural enemies 6, 9, 31, 40, 45, 49, 51, 82, 121, 140, 186, 193, 232, 238, 252, 253, 254, 262, 263, 273
Natural pest control 189, 251
Nauphoetin 223
Neem 2, 45, 164, 167, 172, 181, 194, 195, 231, 237
Neemazal-T/S 167
Neem-based insecticides 163
Neem-based pest management
  blood-sucking pests 169
  fibre crop 167
  flowers 167
  forest trees 168
  legumes 166
  maize 165
  millet 165
  ornamentals 167
  rice 164
  sorghum 165
  stored products 168
  vegetables 166
Neemix 167
Nematodes 195
Nephotettix virescens (Distant) 129, 134, 138, 164
Neurohormonal genes 215
Neurohormones 215
New chemicals 190
New crops 267
New pests 267
Nezara viridula (Linnaeus) 73
Nicotiana tabacum L. 158
Nicotine 129, 158, 172, 194
Nicotine sulfate 159, 180
Nilaparvata lugens (Stal) 123, 133, 134, 138, 253
Nimbin 162
N-methyl flindersine 130
Nomilin 171
Non-IPM farmers 17
Non-occlusion virus 199
Nonpreference 121, 239
Non-target organisms 97, 183, 188, 189
Nuclear polyhedrosis virus 78, 199, 214
Nutrient management 262, 273
Nutritional factors 124

O

Obacunone 171
Observation techniques 251, 252
β-Ocimene 130
Ocimum basiliocum L. 161
Oligogenic resistance 133
Oligophagous insects 122
Oncopeltus fasciatus (Dallas) 236
Onion thrips 138
Operational research projects 18, 21
Ophiomyia phaseoli (Tryon) 47
Organic cotton 20
Organic soil management 82
Organochlorines 157, 180, 187, 189, 192
Organophosphates 3, 157, 180, 185, 192
Orseolia oryzae (Wood-Mason) 134, 137, 165
Oryctes rhinoceros (Linnaeus) 73, 199
Oryzaephilus surinamensis (Linnaeus) 169
Ostrinia furnacalis (Guenee) 29, 41, 80
Ostrinia nubilalis (Hubner) 43, 77, 104, 134, 135, 138, 231

Oviposition deterrents 9, 129, 162, 223
Ovipositional stimulants 223, 224

P

Parapatric 136
Parasitoids 8, 18, 68, 70, 71, 162, 224
Parathion 3, 184, 191, 192
Paris green 181
Participatory research 273
Participatory technology development 13
Pea aphid 134, 137
Pectinophora gossypiella (Saunders) 33, 44, 76
Periplaneta americana (Linnaeus) 234
Permethrin 4, 157, 192
Pest forecasting system 6, 21
Pest management 5, 118, 250, 251, 261
Pest management ratings 10
Pest outbreaks 9, 16, 185, 186, 268
Pest resistance 183, 209, 213, 217
Pest resurgence 9, 155, 183, 185, 186
Pesticidal plants 155, 156
Pesticidal wastes 201
Pesticide consumption 182
Pesticide development 181
Pesticide management 180
  issues 183
  strategies 189
Pesticide market 182
Pesticide mixtures 187, 201
Pesticide poisoning 188
Pesticide residues 187, 201
Pesticide rotation 200
Pesticide subsidies 12, 14, 16, 18, 273
Pesticide suitability rating 190
Pesticide treadmill 5, 155
Pesticides 1, 4, 18, 82, 119, 155, 191, 251
Pestiferous species 82, 156
Phagorepellents 194
Phenacoccus manihoti Matile-Ferero 33, 39
Phenylalanine ammonia lyase 210, 212
Phenylhepta trinyne 130
Phenyl propanoids 235
Pheromones 180, 192, 223, 225, 226, 227, 239
Phlorizin 130
Phloroglucinol 129, 210, 234
Phostoxin 169
Photosensitizers 160
Phototoxin metabolites 160
Phthormaea operculella (Zeller) 32, 49
Physical control 1, 36
Phytoalexins 210
Phytochemicals 210, 222, 227, 228, 229
Phytoecdysteroids 237
Phytophagous insects 1, 6, 9, 128, 135, 186, 215, 237, 239
Phytosanitation 33, 34
Pieris brassicae (Linnaeus) 130, 213
Pieris rapae (Linnaeus) 76, 78, 128
Pink bollworm 3, 19, 21, 44
Piperonyl butoxide 157
Plant defense mechanisms 221, 239
Plant density 30, 32, 43
Plant diversity 79
Plant resistance 118
Plant spacing 31, 33, 43, 261
Planting depth 43
Planting time 37
Plodia interpunctella (Hubner) 102, 103, 196, 213, 214
Ploughing 2, 28, 29, 33, 46
Plumbagin 237
Plutella xylostella (Linnaeus) 78, 81, 103, 127, 167, 171, 199, 213, 214
Podocarpus nakaii Hay 161
Podolactones 235
Political constraints 12, 21, 270
Polycropping 268
Polyculture 79, 82
Polyphagous insects 122
Polygenic resistance 134, 135
Polygodial 210
Polyhedrin 215
Polyphenol oxidase 214, 234
Ponasterone A 161
Population regulation 155
Potato leafhopper 126, 132, 233
Precocene I 181
Precocene II 181
Precocenes 181, 193, 236

Predators 8, 18, 30, 44, 45, 68, 80, 162, 185, 217, 232
Propesticides 201
Protective management 5
Protein inhibitors 129, 212
Proteinase inhibitors 211, 212, 216, 234
Protelean parasites 70
Prothoracic glands 161
Prothoracicotropic hormone 237
Psallus seriatus (Reuter) 19
Pseudoresistance 44, 120
Psilia rosae (Fabricius) 127, 128
Pubescence 125
Pyrethrin I 157
Pyrethrin II 157
Pyrethroids 157, 185, 196
Pyrethrum 156, 157, 172, 180
Pyrilla perpusilla Walker 34
Pyrethrins 157, 190, 201
Pyrrhocoris apterus Linnaeus 161

Q

Quadraspidiotus perniciosus (Comstock) 73, 79, 185
Quassia amara L. 159
Quinine 130
Quinones 122, 228

R

Raspberry aphid 137
RD-9 Repelin 168
Recilia dorsalis (Motschulsky) 134
Recombinant DNA 9, 209, 213, 214, 215
Relative humidity 132
Repellents 32, 128, 161, 181, 221, 223, 229, 233, 239
Resorcinol 210, 214, 234
Resurgence 183, 186, 268
Rethrins 157, 164, 195
Rhinoceros beetle 73
Rhopalosiphum maidis (Fitch) 134, 138
Rhyzopertha dominica (Fabricius) 169
Rice-fish culture 218, 258, 263
Rice leaf folder 164, 186
Rice leaf miner 48
Rice stem maggot 137
Rice tungro virus 165
Rice water weevil 48, 267
Ricin 211
Rodolia cardinalis (Mulsant) 2, 72
Rotenoids 157, 164, 181
Rotenone 2, 157, 164, 172, 181, 194
Russian wheat aphid 131
Rutin 234
Ryania speciosa Vahl. 159, 195
Ryanodine 159, 164

S

Sabadilla 159, 164, 172
Saissetia oleae (Olivier) 138
Salicylic acid 210
Sampling techniques 6
San Jose scale 73, 185
Saponins 233
Sarcoptes scabiei (Linnaeus) 169
Schistocerca gregaria (Forskall) 130
Schizaphis graminum (Rondani) 71, 123, 124, 130, 134, 135, 138
Schoenobius incertellus Walker 38
Schoenocaulon officinale Gray 159
Schradan 3
Scirpophaga incertulas (Walker) 134, 165
Secoirdidoids 228
Second generation chemicals 180
Secondary pests 9, 185, 268
Selective chemicals 190
Semiochemicals 223, 225, 226, 227, 232
Senkirkine 130
Sesame oil 157
Sesamia botanephaga Tams & Bowden 27, 28
Sesamia inferens (Walker) 28, 48
Sesquiterpenes 228
Shiromodiol diacetate 130
Silent Spring 4, 155, 183
Silica 28, 47, 126
Singhara beetle 185
Sinigrin 231
Sitophilus oryzae (Linnaeus) 132, 137, 169
Sitophilus zeamais Motschulsky 132, 169
Social stability 270
Sociological constraints 11, 270

...ertility 120, 132, 261, 262
Soil management 273
Soil salinity 132
Sogatella furcifera (Horvath) 132, 134
Soil fauna 189
Soil flora 189
Southwestern corn borer 43
Sowing dates 28, 29, 38, 39
Soybean stem borer 46
Species diversity 82
Species richness 71
Specionon 130
Spodoptera exigua (Hubner) 71, 234
Spodoptera frugiperda (J.E. Smith) 30, 50, 123, 135, 231
Spodoptera littoralis (Boisduval) 69, 130
Spodoptera litura (Fabricius) 31, 37, 69, 130, 196, 199, 216, 234
Spotted alfalfa aphid 73, 132, 134
Spotted bollworm 19, 21
Spotted stem borer 165
Spurce budworm 168, 225, 226
Stress-insect interactions 8
Strip cropping 40
Structural activity relationships 190, 194, 233
Strychnine 130
Subsistence agriculture 52, 250, 274
Subsistence crops 271
Sugarcane leafhopper 34
Super-green revolution 269
Supportive national policies 272
Surface waxes 127
Surveillance 5, 6, 18
Sustainable agriculture 12, 155, 250, 251, 261, 264, 268, 274
Sustainable approach 1
Sustainable crop production 5, 22, 275
Sweet basil 161
Sympatric 136
Synchronous sowing 33, 39
Synergists 99, 157, 201, 232
Synomones 223
Synthetic pyrethroids 157, 168, 186, 194
Synthetic rotenoids 158, 180
Systems analysis 15, 20

T

Tagosodes orizicolus (Muir) 267
Tannins 129, 235
Teflubenzuron 231
Temperature 131
Terpenoids 131, 156, 162, 227, 228, 235
Tetranactin 200
Tetranychus urticae Koch 73, 76, 80, 131
Therapeutic tactics 8
Therioaphis trifolii (Monell) 73, 133, 138
Thickness of tissues 126
Thionins 211, 212
Thrips tabaci Lindeman 138
Thuja plicata Don. 161
Thujic acid 161, 236
Tillage 32, 44, 45
Tillage practices 82
Tissue culture 99, 139
Tobacco 2, 3, 158, 210, 211, 216
Tobacco budworm 4, 134, 187, 197
Tobacco hornworm 131, 217, 229, 230
Tolerance 121, 123, 213
Tomatine 129, 130, 228, 233
Toosendanin 171
Toughness of tissues 126
Toxaphene 3
Toxic residues 201
Toxin structure 89, 98
Traditional approach 1, 2
Transfer of technology 7, 52, 256
Transgenesis 213
Transgenic plants 20
Trap crops 31, 32, 33, 49
Trewiastine 130
Tribolium castaneum (Herbst) 233, 234
Trichogramma evanescens Westwood 79
Trichomes 79
Trichoplusia ni (Hubner) 71, 76, 77, 131, 214, 227
Triple super phosphate 47, 260
Triterpenes 228, 233
Trypsin inhibitor 212, 216
Tulirionol 130
Two spotted spider mite 73, 131, 132

U

Underhill 3, 119
Univoltine 70

V

Vanillic acid 210
Varietal resistance 5
Vectors 212, 214, 215, 217
Vedalia beetle 2, 3, 72, 181
Vepinin 162
Veratrum album L. 159
Vilasinin 163
Viral agents 196
Viteus (Phylloxera) vitifoliae (Fitch) 2, 119, 138
Volkensin 171
Volunteer host plants 28, 35

W

Walnut aphid 73
Warburganal 130, 210
WARF-antiresistant 201
Water management 32, 48, 49, 255, 261, 262
Weed destruction 30, 31
Weed managment 35, 261, 262
Weed removal 33, 82
Western corn rootworm 134
Western red cedar tree 161
Whale oil 181
Wheat stem maggot 38
Wheat-stem sawfly 43, 70, 135
White pine 161
Whitebacked planthopper 16, 132, 134, 186
Wide hybridization 139, 274
Wildlife 188
Winter Majetin 119
Withanolide E 130
Woolly apple aphid 119, 137

X

Xanthotoxin 232
Xylomonin 194

Y

Yellow clover aphid 138
Yellow fever 87, 88
Yellow stem borer 134, 164
Yponomeuta pedellus (Linnaeus) 138

Z

Zero tillage 45
Zigzag leafhopper 134